Conflict in the 21st Century

Conflict in the 21st Century

The Impact of Cyber Warfare, Social Media, and Technology

Nicholas Michael Sambaluk, Editor

An Imprint of ABC-CLIO, LLC
Santa Barbara, California • Denver, Colorado

Library of Congress Cataloging-in-Publication Data

Names: Sambaluk, Nicholas Michael, editor.
Title: Conflict in the 21st century : the impact of cyber warfare, social media, and technology / Nicholas Michael Sambaluk, editor.
Description: Santa Barbara, California : ABC-CLIO, an imprint of ABC-CLIO, LLC, [2019] | Includes bibliographical references and index.
Identifiers: LCCN 2019005231 (print) | LCCN 2019012362 (ebook) | ISBN 9781440860010 (ebook) | ISBN 9781440860003 (hardback : alk. paper)
Subjects: LCSH: Military art and science—Technological innovations—Encyclopedias. | Cyberspace operations (Military science)—Encyclopedias. | Cyberterrorism—Encyclopedias. | Social media—Political aspects—Encyclopedias.
Classification: LCC U42.5 (ebook) | LCC U42.5 .C66 2019 (print) | DDC 355.0203—dc23
LC record available at https://lccn.loc.gov/2019005231

ISBN: 978-1-4408-6000-3 (print)
978-1-4408-6001-0 (ebook)

23 22 21 20 19 1 2 3 4 5

This book is also available as an eBook.

ABC-CLIO
An Imprint of ABC-CLIO, LLC

ABC-CLIO, LLC
147 Castilian Drive
Santa Barbara, California 93117
www.abc-clio.com

This book is printed on acid-free paper ∞

Manufactured in the United States of America

About the Editor

Nicholas Michael Sambaluk, PhD, is an associate professor at Air University's eSchool of Graduate Professional Military Education and an author on military history and military technology and innovation. His first book, *The Other Space Race: Eisenhower and the Quest for Aerospace Security*, was named "Best Air Power History Book of 2016" by the Air Force Historical Foundation. He is editor of *Paths of Innovation in Warfare: From the Twelfth Century to the Present* and has authored several articles for the journals *Cyber Defense Review*, the Praeger Security International database, and *Cold War History*. He also serves as a contributing editor to the defense journal *Strategic Studies Quarterly*.

Contents

List of Entries by Section

SOCIAL MEDIA

TECHNOLOGY

Preface

The famous observation, often attributed to Plato, that only the dead have seen the end of war remains as true two decades into the 21st century as it would have been twenty-four centuries earlier. Furthermore, the nature of war endures, marked by interactivity and violence in pursuit of political goals.

That said, history shows that technology continually impacts the character of particular conflicts. In the past, the realm of the possible was radically reshaped by the introduction of a panoply of technologies, which themselves provoked or were created in response to tactical and operational contexts shaped by other forces, including other technologies. Whether the pike or the harquebus or even the archer's bow are thought to represent instruments of a "revolution" in warfare, their use in combat prompted changes in tactics. Technologies like rail mobility and steam power transformed logistics and the ability to smoothly move and sustain armies—at least up to a country's own frontiers. Breakthroughs in propellant and explosive chemistry were factors that helped foreclose the breakthroughs of armies on the battlefields of World War I. Medical advances that saved lives paradoxically also helped combatants sustain war. Myriad technologies, from the jet engine to the nuclear bomb to precision guidance and stealth and satellite communications and tracking brought rapid change to the character—but not the nature—of military affairs throughout the second half of the 20th century.

This book considers aspects of technology in the 21st century. It is organized into three overarching sections, encompassing cyber warfare, social media in security affairs, and other technologies. The introduction of new domains of fighting happens only very rarely. Combat in the sea domain began around 1200 BCE, and more than three millennia transpired before heavier-than-air vehicles allowed realistic utilization of the air domain. But barely a half-century after the first warplanes took to the sky, the United States and the Soviet Union were seeking to explore and exploit the space domain to advance national security goals, through intelligence data collection, for communications, and for guidance.

Although different voices disagree about the beginning "date" of the cyber domain, it can be thought of as coming of age in the 21st century. The use of software to exert effects on targets, and the efforts of targeted entities to avoid being victims, inspired the establishment of the cyber warfare section in this book. The notable exploitation of the cyber domain for sharing and shaping information prompted the making of a second section engaging with social media topics. Yet, just as the introduction of aerial combat in the 20th century did not in fact invalidate ground and sea fighting, exploitation of the cyber domain does not banish warfare from the other domains. The third section of this book considers various physical technology topics for this reason.

Each of these sections begins with an introductory essay, and all of these topics constitute the presentation of openly available and unclassified materials. In the case of all of the materials, the words are those of the authors and do not represent an official position by any government, instead reflecting the open data identified as being of scholarly interest. In this way, the author and the contributors hope to compile information useful for the understanding of how modern technology is impacting the conduct of war and the context of security affairs.

Section I

Cyber Warfare

INTRODUCTION

Although the nature of war continues to be defined by the application of force and the pursuit of policy objectives, the character of conflicts in the 21st century is impacted by new forms of technology, and new technologies in the digital realm are especially often noted. Events in the first two decades of the 21st century suggest the mounting and dynamic relevance of technologies and methods that might be considered part of "cyber warfare" or "cyberwar." Different institutions and various scholars present competing definitions of the term; furthermore, some scholars such as political scientist Thomas Rid assert that cyberwar will not take place, and other renowned figures in the field including Martin Libicki urge audiences to avoid holding fanciful expectations of the capabilities of cyberweapons (Libicki, 2016). The topics in the Cyber Warfare section of this reference book have been chosen to reflect a broad interpretation of the concept of cyber warfare, in part because of this debate. Broadly, cyberwar can be thought of as using digital tools and methods to exert some kind of injury on a target in connection with a policy objective, although some go more broadly still to suggest that kinetic strikes against digital systems fit under the umbrella of cyber warfare (Chen, 2017, 96–98).

A broad approach is a useful choice for another, more technical, reason. P. W. Singer and Allan Friedman have told readers that "the good news is that there are only three things [an aggressor] can do to a computer [through the cyber domain]: steal its data, misuse credentials, and hijack resources" (Singer and Friedman, 2014, 39). Although occasions of physical destruction by cyber means are so far rare, examples do already exist, including the 2007 Aurora tests conducted by the United States and the subsequent Stuxnet cyber action against Iranian nuclear centrifuges first confirmed in 2010. These represent the kind of destruction that defies neatly fitting into the three categories Singer and Friedman outline, and examples of physically destructive effects (when undertaken in connection with a political purpose) more closely fits a conventional definition of warfare.

Furthermore, from the victim's perspective, the different forms of cyber aggression can be difficult to distinguish from each other. Intrusions are often not identified until after the fact. Even when one is discovered while in process, identifying its purpose (whether to manipulate or delete data or achieve still other effects) can be very difficult. Enough similarities and ambiguities among the tools in the cyber operator's kit exist to create uncertainty as events unfold. A nondigital equivalent might be if a spy's safecracking equipment and methods were to resemble a soldier's rifle and deployment in preparation for battle. In the cyber domain, the two are frequently far harder to distinguish than their counterparts are outside it. Since espionage can be espionage—but it can alternatively be reconnaissance before an attack—some topics related to espionage nonetheless make sense to address and discuss among a list of terms dealing with cyberwar. Additionally, the difficulty of accurately predicting the results of cyberactivity means that neither the adverse impact recognized by an injured party nor the true extent of the injury necessarily provides a window on an aggressor's intent.

There are several other ways in which cyberweapons are often deemed significantly different from earlier forms of weapons. Among these is the speed with which cyberattacks occur. Indeed, examples exist of activities in the cyber domain that deliver identifiable and nearly instantaneous effects. The wave of distributed denial-of-service (DDoS) attacks against Estonia in April 2007, in which experts identified 128 separate DDoS attacks in which the volume of virtual bombardment reached 4 billion bytes per minute (Mazanec and Thayer, 2015, 18), quickly threatened connectivity and normal Internet commerce in a small but highly connected country. In August of the following year, another small country found its far more modest connectivity virtually eradicated by similar DDoS attacks, although in the case of Georgia the cyber blow fell simultaneously with a literal one, as Russian forces intervened in an existing dispute between the Georgian government and separatists sympathetic to Russia (Kaplan, 2016, 65). In February 2008, almost exactly between the events in Estonia and Georgia, Pakistan Telecom reputedly rerouted Internet traffic away from YouTube in order to prevent domestic viewing of videos that the government found contrary to Islam; when this move inadvertently misdirected the majority of *global* YouTube bound traffic, the result was a sort of accidental self-DDoS that temporarily brought down Pakistan's own network (Singer and Friedman, 2014, 21).

Such examples notwithstanding, the assumption that hostile activities in the digital domain are necessarily instantaneous is flawed. DDoS actions can be mounted quickly, and their impact can be felt abruptly. However, DDoS attacks constitute an interruption in connectedness rather than destruction. For that reason, some scholars have deprecated the practical value of DDoS launched by hives, such as the amorphous "group" Anonymous and other entities whose memberships may be ambiguous and whose level of expertise is likely to vary from member to member (Rid, 2013, 131).

Such attacks can nonetheless impede a target's ability to interact with the world and can therefore hinder a target's ability to share its message with the world or to communicate and plan an effective response to an attack. In terms of damaging an opponent, experts suggest that a DDoS is a poor substitute for an activity that causes actual damage. Although it is conceivable that actions creating physical destruction could unfold abruptly, it seems unlikely that they could be brought from concept to command instantaneously, due to the planning and preparation required. The few so-far existing examples of actual destruction via cyber means also indicate a need for a substantial investment in both time and money resources to develop an effective cyberweapon that will exert reasonably predictable forms of damage against an intended target.

To date, the quintessential case is the Stuxnet malware that was discovered in 2010. Forensics showed that the malware was designed to distinguish between various forms of supervisory control and data acquisition (SCADA) and programmable logic controllers (PLC) in order to find the type associated with Iran's suspected weapons-grade nuclear enrichment program. False instructions reportedly accelerated breakages among enrichment centrifuges and caused wastage of nuclear fuel while feeding data to controllers that simulated normal operating processes. Experts believe that development of Stuxnet malware dates to 2005, simultaneous with a breakdown in international negotiations aiming to corral Iran's nuclear weapons capability. If correct, this would indicate that this sophisticated malware, which analysis suggests also required the preliminary development of the related Duqu and Flame malware for reconnaissance purposes, was the product of about three years of development believed to require patience and resourcing that can be provided by a nation-state. The intent appears to have been an untraceable erosion in the efficiency of Iranian nuclear enrichment technologies, and by its nature such an attack was most effective when it struck gradually. Author Kim Zetter suggests that the lack of clarity about the malware's real impact might have helped prompt decisions to release an updated and more aggressive form of the malware, which spread well beyond the targeted facilities of Natanz, Iran, and brought Stuxnet to the attention of cyber forensics experts (Zetter, 2016).

Stuxnet's development and employment was patently not instantaneous, and its operational impact remains debatable almost a decade after its discovery and its strategic impact was overtaken by events, first with the controversial Joint Comprehensive Plan of Action of July 2015 and its equally controversial collapse in May 2018. Analysts have observed that despite the malware's "significant effects . . . production actually increased in the fall of 2009 and early 2010" as Iranian scientists pushed more resources to counteract the then-mysterious decline in the average efficiency of the centrifuges (Yannakogeorgos and Lowther, 2014, 135–136). It is unclear whether Stuxnet will come to set the pattern in future cyberweapons, but as the pioneer cyberweapon used to cause physical destruction, Stuxnet does not support the myth of cyberweapons being necessarily instantaneous.

A second purported truth about cyberweapons is that the global reach of cyberweapons means that distance is irrelevant in cyberwar. However, this assumption feeds overly simplistic conclusions. Although electronic communications (and therefore things like malware as well) have a global reach and can potentially be executed (but not developed) quickly, the notion that distance is irrelevant potentially implies that terrain is also an outmoded concept. Some scholars have worked to argue that terrain continues to be relevant in the cyber context, although this then demands definitions of what "terrain" constitutes. One of the few and comparatively effective ways to protect devices from cyberattack is to build "air gaps," in which a protected system is not connected to devices and systems that can harm it.

However, the 2008 events called Buckshot Yankee uncovered that malware on questionably sourced flash drives, once inserted in machines of an air-gapped network, compromised U.S. military networks (Libicki, 2016, 8). Research has also found that radio signals between proximate machines, one infected and another uninfected, can be used by specially designed malware to jump across an air gap. An appropriate conclusion might be that distance continues to matter in the cyber domain, but the ways and reasons that it matters are different from those in other fighting domains. Nor is this dynamic entirely unprecedented. Early advocates of airpower such as Giulio Douhet optimistically predicted that heavier-than-air craft would abolish the tyranny of distance (at least of friendly aircraft striking outward) and transform warfare. Aircraft changed the character of war and extended the battlefield, but they did not alter the nature of war being the leverage of violence to pursue policy goals. "Distant" targets may still be harder to strike in the cyber domain if distance is to be measured less in terms of the mileage between the origin of malware and its target and more about distances such as the targeted computer and the nearest infected device existing on the other side of the targeted entity's air gap.

Proliferation of cyberweapons advances the argument that cyberweapons are cheap and widely available. In some sense, the creation of massive botnets validates this conclusion. These utilize third-party devices without their owners' knowledge and enlist them in actions like DDoS campaigns against targets. The Russian Business Network (RBN) became infamous for selling malware packages and for creating and renting out botnets for as little as US$200 (Rosenzweig, 2013, 88), and the first decades of the 21st century showed that even following RBN's purported closure the price for renting a botnet fell. By 2010, a botnet could reportedly be rented for as little as $9 per hour, with a per-day rental price shaving the cost-per-unit-of-time further by two-thirds (Danchev, 2010). Employing botnets for simple DDoS attacks is simple and also conceivably obscures attribution. For example, although a Russian nationalist group claimed responsibility for the 2007 DDoS against Estonia, 25 percent of the computers participating in the attack were bots in the United States, whose owners probably were not even aware that their machines were involved in the attack (Singer and Friedman, 2014, 73). It is less clear whether more sophisticated malware will be readily converted and proliferated to become cheap and available for copycat attacks. Media reports on the WannaCry ransomware that appeared in the first half of 2017 had been developed through the leverage of sophisticated cyber tools developed but then leaked to other parties. Analysts associated North Korea with the launching of the ransomware.

In contrast, many experts braced in 2010 for the disclosure of the Stuxnet virus to lead to a wave of attacks by third parties piggybacking on the years of development that went into that virus in order to modify their own attacks for various criminal or political purposes. This concern has not so far been met with acknowledged copycat malware, despite the passage of nearly 10 years. Zero days, the obscure vulnerabilities that exist in software but are unknown by software vendors and the users who constitute potential victims, carry highly lucrative prices among cybercriminals intent on developing new malware and also on the "gray market" in which buyers are intelligence agencies that intend to use zero-day vulnerabilities to develop exploits from which there will initially be no protection (Zetter, 2016, 101–104). Perhaps partly as a result of the remaining time and monetary resources that are required to transform a pricey zero-day vulnerability into an actionable and profitable piece of malware, the first part of the 21st century has reportedly witnessed the conversion of sophisticated nation-state-developed cyberweapons into derivative weapons, although the process appears so far to be far less than endemic. Most malware does not employ a single zero-day exploit, let alone the four that were identified as part of the Stuxnet malware.

The ability to successfully convert a digital tool developed by a nation's intelligence community points to a fading myth about cyber warfare. The notion that cyberweapons could be used only a single time before being countered with effective software patches holds less cache than it did earlier in the century. Time and again, various examples of malware have been used for which software patches and other defenses have already been developed. The WannaCry ransomware was one example. Another was the malware used to target Ukraine's electrical grid in December 2015. The vulnerability targeted in that action had been identified by computer scientists fifteen years earlier, but it had never been patched (Giannetti, 2017, 97). SCADA supports power grids and other infrastructure using control systems that run on legacy versions of operating systems that vendors no longer support, and the result is that while patches are sometimes unavailable their application is typically a potentially risky process since "buggy patches . . . might crash . . . critical systems" on which people rely for necessities such as electrical power or clean water (Zetter, 2016, 147).

In the absence of Stuxnet and similar demonstrations such as the Aurora test at the Idaho National Laboratory in 2007 (Kaplan, 2016, 167–170), cyberweapons have been directed at information itself and the information-processing ability of other computers. This point is what has led figures like Thomas Rid to famously author an article and a book each bearing the title *Cyber War Will Not Take Place.* War's relationship with violence and death, and the absence of lethality in historic aggressions in cyberspace, lead Rid to argue that calling cyber actions "war" "has more metaphorical than descriptive value" (Rid, 2013, 9). Stuxnet constitutes a singular but potent exception to the rule that things have not been broken, as former NSA director Michael Hayden observed that "someone had crossed the Rubicon" (Hayden, 2016, 152). The increasing ubiquity of connected devices and the anticipated emergence of the Internet of Things raise the prospect of a fundamental broadening in the attackable surface, as both the number of connected devices is expected to expand exponentially and the kinds of targets can be expected to diversify enormously as well. This can be expected to exert an important impact on the character of wars in the rest of the century.

Actions in the cyber domain have also been used in concert with other methods, notably in Operation Orchard. In September 2007, Israeli malware was directed against Syrian air defense radar and was executed so that four Israeli Air Force attack planes were able to enter Syrian airspace undetected. They proceeded to the facilities of the Kibar nuclear facility under construction in Syria with North Korean assistance. This action has been described as an example of what "operational cyberwar" might look like later in the century (Libicki, 2016, 18–19).

These developments, and particularly the aspects that lend prominence to the offensive capabilities of actions in the cyber domain, have encouraged a mystique around the offensive in the cyber domain and assertions that the cyber domain is immune from traditional conceptions of war such as Carl von Clausewitz's dictum that the defense is the stronger form of warfare even though the offense is the only form that can achieve an actual outcome (Clausewitz, 1984, 84, 392, 600). Writings such as by Chris Demchack do indicate scholarly interest in the development of resilience as a means of addressing the presumed supremacy of the offensive, and other literature argues in favor of formal or informal norms of behavior that might curtail aggression in the digital realm (Demchack, 2011; Manzanec, 2015). P. W. Singer and Allan Freidman are among those who have argued that the enthusiasm for (and, alternatively, the fear of) the presumed supremacy of offensive cyber action constitutes a 21st-century equivalent to the "cult of the offensive" that led strategists and tacticians on the eve of World War I to develop bloodily unrealistic plans and conceptions of future war. They argue that "the best defense actually is a good defense," although the arms race in cyber is complicated and heightened by the presence of more participants than simply political entities (Singer and Friedman, 2014, 153–157). The first two decades of the 21st century have demonstrated the capacity of hostile acts in cyber to be a potent part of modern warfare and to impact the course and character of contemporary conflicts. These events provide evidence for consideration about precisely how this impact might be manifested in the century's remaining decades.

FURTHER READING

Chen, Dean. *Cyber Dragon: Inside China's Information Warfare and Cyber Operations.* Santa Barbara: Praeger, 2017.

Clausewitz, Carl von. *On War*, ed. and trans. Michael Howard and Peter Paret. Princeton, NJ: Princeton University Press, 1984.

Danchev, Dancho. "Study Finds the Average Price for Renting a Botnet." *ZDNet*, May 26, 2010, https://www.zdnet.com/article/study-finds-the-average-price-for-renting-a-botnet.

Demchak, Chris. *Wars of Disruption and Resilience.* Athens: University of Georgia Press, 2011.

Giannetti, William. "A Duty to Warn: How to Help America Fight Back against Russian Disinformation." *ASPJ* (Fall 2017), 95–104.

Hayden, Michael. *Playing to the Edge: American Intelligence in the Age of Terror.* New York: Penguin Press, 2016.

Kaplan, Fred. *Dark Territory: The Secret History of Cyber War.* New York: Simon & Schuster, 2016.

Libicki, Martin. *Cyberspace in Peace and War.* Annapolis, MD: Naval Institute, 2016.

Mazanec, Brian M. *The Evolution of Cyber War: International Norms for Emerging-Technology Weapons.* Lincoln: University of Nebraska Press, 2015.

Mazanec, Brian M., and Bradley A. Thayer. *Deterring Cyber Warfare: Bolstering Strategic Stability in Cyberspace.* New York: Palgrave Macmillan, 2015.

Rid, Thomas. *Cyber War Will Not Take Place.* Oxford: Oxford University Press, 2013.

Rosenzweig, Paul. *Cyber Warfare: How Conflicts in Cyberspace Are Challenging America and Changing the World.* Santa Barbara, CA: Praeger, 2013.

Singer, P. W., and Allan Friedman. *Cybersecurity and Cyberwar: What Everyone Needs to Know.* Oxford: Oxford University Press, 2014.

Yannakogeorgos, Panayotis A., and Adam B. Lowther, eds. *Conflict and Cooperation in Cyberspace: The Challenge to National Security.* Boca Raton, FL: Taylor & Francis, 2014.

Zetter, Kim. *Countdown to Zero Day: Stuxnet and the Launch of the World's First Digital Weapon.* New York: Crown, 2016.

Advanced Persistent Threat

A malicious entity characterized by technological sophistication and resources that support extensive surveillance and exfiltration of data. Although advanced persistent threat (APT) can refer to espionage threats in the physical realm, the term is most commonly associated with the digital realm. The People's Republic of China (PRC) is frequently cited as an example of an APT, and the country and its Shanghai-based digital espionage Unit 61398 were famously identified in the APT1 Report released by Mandiant in 2013, but experts identify other nation-state efforts as constituting APTs as well. APTs are considered especially virulent adversaries for a variety of reasons.

One obvious source of an APT's virulence entails its sophistication. An APT's toolkit includes complex malware that, once a targeted system has been compromised, facilitates the APT's ability to examine data and explore systems without its activities being identified. Developing these tools requires significant financial resources on a scale that is out of reach of most individuals, so an APT's activities tend to point to support from a group such as the government of a nation-state.

Forensic analysts report that although Stuxnet included code that tracked the efficiency and activities of Iranian nuclear enrichment technologies, Stuxnet aimed fundamentally to exert physical destruction to degrade nuclear enrichment capabilities, and that the surveillance aspects of the code existed to ensure that only the intended target systems were damaged and to degrade systems in a manner that would be difficult for the target to identify and trace as the result of hostile cyber action. Some cyber analysts have pointed to Stuxnet as fitting the definition of an APT's activities, despite the apparent subordination of its surveillance aspects.

Persistence is a characteristic of APTs that strongly correlates with governmental or other organizational backing. Individuals, frequently aiming to compromise and explore a targeted system or motivated by profit or thrills, rarely practice the patience that is demonstrated by an APT's activities. Nation-states deliberately choose their targets and invest the time and financial resources necessary to establish deep access to a targeted system. Gaining this access often requires a variety of tools and methods, and deploying these requires time. Although a nation-state or wealthy group can conduct actions in line with an APT, it is likely to select targets carefully so that the advantages that are derived exceed the inherent effort of unleashing APT actions.

After selecting a target, an APT conducts its initial compromise. Typically, this will entail social engineering work to increase the chances that an authorized individual working in or with the target will be susceptible to efforts such as spear phishing or will trust and follow a link to a website that secretly loads malware onto the user's computer. Alternatively, an APT may rely more heavily on the technology itself and deploy sophisticated malware that may even use zero-day exploits to evade a target's antivirus systems.

Having established its first access into a target, an APT next solidifies this access. This frequently includes the establishment of unobtrusive backdoors so that the APT can maintain access to the target even in the unlikely event that the target manages to identify and respond to the initial access event. The APT also expands its access by raising the administrator privileges of those accounts that the APT has compromised. In these two actions, the APT ensures that it will continue to have access into the target and that more and more of the target will be susceptible to its surveillance. This process is advanced even further as the APT explores and maps the digital infrastructure of the target and gains control of additional workstations and accounts.

Armed with this invasive presence, the APT becomes able to discretely observe data throughout the target's workstations and across its network. The depth of its access and privileges means that the target is unlikely to yet suspect the existence, let alone the scale, of the breach that has occurred. There is therefore little reason for an APT to conduct its efforts in haste, since it is not yet racing against detection.

The extraction phase is often the point at which a victim becomes aware of a breach. APTs frequently conduct operations as part of strategically relevant military or corporate espionage. Often the APT wants to find valuable data, extract, and exfiltrate it. Conceivably, an alerted victim might derive information about the origin of an APT by watching the outgoing data traffic, although APTs reportedly tend to be sophisticated enough to exfiltrate data indirectly so that the origin of an APT action is less straightforward to discern. The very size of the exfiltrated data sets, however, can be extremely difficult to conceal, and for this reason, the large-scale exfiltration stage is the point at which an APT's actions are most likely to be identified by a target.

APTs occupy an ambiguous space between cyberespionage and cyberwar. They tend to require a scale of resources most often available only to nation-states. They most often target entities whose information is of economic or military interest to a nation-state. They usually, but not universally, focus on the discovery and exfiltration of strategically important information, rather than on data deletion or physical destruction.

Nicholas Michael Sambaluk

See also: APT1 Report; GhostNet; Malware; Mandiant; Operation Aurora; Operation Shady RAT; People's Liberation Army Unit 61398; Stuxnet; Zero-Day Vulnerability

Further Reading

Allsopp, Wil. *Advanced Persistent Threat Modeling: Defending against APTs*. Sevastopol, CA: O'Reilly, 2013.

Chandra, J. Vijaaya, et al. "Intelligence Based Defense System to Protect from Advanced Persistent Threat by Means of Social Engineering on Social Cloud Platform." *Indian Journal of Science and Technology* 8, no. 28 (October 2015), 1–9.

John, Jeslin Thomas. *State of the Art Analysis of Defense Techniques against Advanced Persistent Threats*. Seminar Future Internet, 2017.

Singer, P. W., and Allan Freidman. *Cybersecurity and Cyberwar: What Everyone Needs to Know*. Oxford: Oxford University Press, 2014.

Symantec. *Advanced Persistent Threats: A Symatec Perspective, Preparing the Right Defense for the New Threat Landscape*. Mountain View, CA: Symantec, 2011.

Wrightson, Tyler. *Advanced Persistent Threat Hacking: The Art and Science of Hacking Any Organization*. New York: McGraw-Hill, 2015.

APT1 Report

Released in March 2013 by the Mandiant Corporation, a massive report on a previously unknown "advanced persistent threat." The report, whose full title was "APT-1: Exposing One of China's Cyber Espionage Units," identified Unit 61398 of the People's Liberation Army (PLA) of the People's Republic of China (PRC). It detailed a massive campaign of economically motivated cyberespionage (EMCE) that broadly targeted major corporations, primarily in Western countries, with the intention of stealing intellectual property, technological specifications, and corporate data used for international agreements.

Mandiant Corporation was founded by Kevin Mandia in 2004. The company focused on cybersecurity needs for businesses. Unlike many companies in the cybersecurity sector, which rely on passive defenses such as firewalls, Mandiant Corporation tended to rely on aggressive detection of intrusion attempts followed by an active response against specific threats. The company quickly built a reputation for thorough forensic analysis of cyberattacks, particularly for companies that had been targeted by large-scale cyber campaigns. In December 2013, FireEye purchased Mandiant Corporation for $1.05 billion.

In 2012, the *New York Times* published a series of reports detailing the enormous wealth accumulated by family members of Wen Jiabao during his stint as the premier of the PRC. Shortly after the articles began, the newspaper began experiencing a protracted cyber campaign that appeared determined to shut down the company's computer networks. The *Times* hired Mandiant Corporation to eliminate the threat. Specialists from Mandiant first assessed the size and scope of the campaign, and then commenced aggressive operations to backtrack the attacks to their source. Over a period of weeks, the Mandiant operatives managed to uncover a massive PRC cyber campaign, and to determine the location of the PLA unit conducting the attacks. Mandiant found that the cyber attackers had grown so overconfident in their own abilities that they took only the most basic precautions to hide their own identities. As a result, the Mandiant specialists could actually penetrate the attackers' networks and observe the cyberattacks being launched in real time.

When the Mandiant APT-1 report was released, it caused significant embarrassment for the PRC. Although the Chinese government denied any culpability for the attacks, the level of detail from the reports proved difficult to deny. The report certainly makes the case that the PRC ordered the attacks. When the *Times* published the results of the report, the group identified in it essentially disappeared from the cyber domain for a year. In 2014, the United States indicted five members of Unit 61398 for their role in EMCE. The public denunciation of cyberattacks may have contributed to the agreement between U.S. president Barack Obama and Chinese leader Xi Jinping that neither nation would engage in EMCE against the other. It remains to be seen if that agreement will have the desired effect.

Paul J. Springer

See also: Mandiant; Operation Byzantine Hades; Operation Titan Rain; People's Liberation Army Unit 61398

Further Reading

Lindsay, Jon R., Tain Ming Cheung, and Derek S. Reveron, eds. *China and Cybersecurity: Espionage, Strategy, and Politics in the Digital Domain.* New York: Oxford University Press, 2015.

Mandiant Corporation. "APT-1: Exposing One of China's Cyber Espionage Units." March 2013. http://intelreport.mandiant.com/Mandiant_APT1_Report.pdf.

Army Cyber Institute (ACI)

Organization headquartered at West Point to provide strategic insight and advice on cyber-related issues affecting U.S. national security and promote the exchange of information across the military, government, academia, public, and private sectors. Nascent activities anticipating a cyber think tank at West Point date to about 2000, but planning officially began in 2012 under the direction of the Army Chief of Staff and the organization formally opened in October 2014. The ACI employs more than fifty multidisciplinary civilian and military cyber experts to facilitate cyber-focused educational enrichment for cadets and reports to the U.S. Military Academy (USMA) Superintendent.

Education and activities expose cadets to computer science, engineering, and cyber history, policy, ethics, and law to enhance preparedness in the cyber warfighting domain. Research assists cyber, signal, and electronic warfare professionals in understanding the challenges and complexities of the electromagnetic spectrum. The Internet of Things (IoT) lab analyzes the impact of Internet-connected devices and systems in future operating environments. To facilitate cooperation on research across government and industry, the ACI hosts the CyCon U.S. conference in partnership with the NATO Cooperative Cyber Defence Center. CyCon U.S. is intended to complement the annual NATO CyCon held in Tallinn, Estonia. A partnership with Citigroup established in 2013 promotes the Army Cyber Institute mission through joint working groups and infrastructure protection exercises. Threatcasting efforts in conjunction with Arizona State University's School for the Innovation of Society seek to anticipate threats and develop strategies to mitigate effects. The ACI also publishes *The Cyber Defense Review*, a scholarly journal to promote dialogue on strategic, operational, and tactical aspects of the cyber realm.

Melia T. Pfannenstiel

See also: USCYBERCOM

Further Reading

Conti, Gregory, and David Raymond. *On Cyber: Towards an Operational Art for Cyber Conflict*. New York: Kopidion Press, 2017.

Healy, Jason. "A Brief History of US Cyber Conflict." In *A Fierce Domain: Conflict in Cyberspace, 1986–2012,* ed. Jason Healy. Arlington, VA: CCSA/Atlantic Council, 2013, 14–87.

Sambaluk, Nicholas Michael. "Perspectives on Training Cyber Warriors." In *Cyber War: A Reference Handbook,* ed. P. J. Springer. Santa Barbara, CA: ABC-CLIO, 2015, 116–120.

Assange, Julian

Born Julian Paul Hawkins, Townsville, Queensland, Australia, July 3, 1971. Australian computer programmer and founder and editor of WikiLeaks, an international nonprofit organization that obtains and publishes classified information. Assange created WikiLeaks in 2006, but the organization first garnered international attention in 2010, when it published hundreds of thousands of classified documents relating to American military operations. The documents were leaked by U.S. Army Private Chelsea Manning, who was court-martialed and imprisoned in 2013 for espionage. The leaked materials related primarily to U.S. operations in Iraq and Afghanistan after September 2001.

The U.S. government condemned the leaks and sought to prosecute Assange, but many activists and international political figures praised his actions as information transparency and as resistance to global U.S. military activity. In November 2010 Swedish authorities issued a warrant for Assange's arrest on sexual assault charges. Assange avoided extradition from the United Kingdom to Sweden when he was granted asylum by the government of Ecuador in August 2012. He has since lived at the Ecuadorian embassy in London. WikiLeaks has continued its efforts, and in the years since 2010 has frequently leaked material deemed beneficial to the Russian government and embarrassing to the Obama administration and the Hillary Clinton presidential campaign.

Daniel Campbell

See also: Hacktivism; Manning, Chelsea/Manning, Bradley; Snowden, Edward; WikiLeaks, Impact of

Further Reading

Brevini, Benedetta, Arne Hirtz, and Patrick McCurdy, eds. *Beyond Wikileaks: Implications for the Future of Communications, Journalism, and Society.* London: Palgrave Macmillan, 2013.

Greenberg, Andy. *This Machine Kills Secrets: How Wikileakers, Cypherpunks, and Hacktivists Aim to Free the World's Information.* New York: Penguin, 2012.

Leigh, David, and Luke Harding. *WikiLeaks: Inside Julian Assange's War on Secrecy.* London: Guardian Books, 2011.

Botnet

An amalgamation of the two elements required to bring one into existence: bots and a network. Before the network can be created, bots have to be gathered. A "bot" is an electronic device, typically a smartphone or a computer, that through some malicious software is now controlled by a third party. The third-party person typically gains control of the device because of a security breach in the device's protocols. Malware, malicious software, is usually used by the third party to gain possession of the device, most of the time without the owner's knowledge or consent.

Once the third-party person has an indeterminant number of bots, a network is formed so that the bots can operate in unison. The person, or entity, creating the network of bots is sometimes referred to as a "bot herder." Historically, these networks were created using standard protocols. Two of the standard protocols are Hypertext Transfer Protocol (HTTP) and Internet Relay Chat (IRC). In this configuration, a central server exists that issues commands to the client bots. IRC remains the favored means of controlling the botnet because of the low bandwidth and the fact that simple messages can be sent to all the client bots through the central server.

Recently, the network structure of botnets changed. The presence of a central server to control the network made it easier to discover the botnet. It also makes the network more vulnerable since taking out the central server can cripple the botnet. In place of this type of network architecture, bot herders have started using peer-to-peer protocols to control their botnets. The lack of a central server makes the botnet more difficult to detect. By making botnet detection harder, it also means that targeting the botnet becomes complicated as well.

After gathering bots and establishing a network, the bot herder typically rents out the use of the botnet for criminal purposes. When hired, botnets usually perform distributed denial-of-service (DDoS) attacks on targeted networks. During a DDoS, the computers on the botnet issue as many requests or inquiries as possible to the targeted server, which overloads the server and prevents it from taking in legitimate requests. An example of such an attack occurred in 2007 when a suspected Russian botnet targeted the country of Estonia. During these attacks, rented botnets targeted servers that hosted websites for the parliament, banks, ministries, and public information.

Sometimes bots can be referred to as "zombie computers." The idea is that the owner is unaware the computers or smart device is now acting like a zombie. When linked in a network, zombie computers can perform functions similar to those of a botnet. Keeping with the zombie theme, when used to conduct DDoS attacks, the zombie network attack is commonly referred to as a "zombie horde attack."

Another trend seen is botnets and zombie computers being used for hacktivism. Although previously these networks were used for financial gain of the bot herder, the networks are now being used to conduct attacks for social or political causes.

Melvin G. Deaile

See also: Distributed Denial-of-Service (DDoS) Attack; Estonia, 2007 Cyber Assault on

Further Reading

Karuppayah, Shankar. *Advanced Monitoring in P2P Botnets: A Dual Perspective.* New York: Springer, 2018.

Libicki, Martin C. *Cyberspace in Peace and War.* Annapolis, MD: Naval Institute Press, 2016.

Singer, P. W., and Allan Friedman. *Cybersecurity and Cyberwar: What Everyone Needs to Know.* Oxford: Oxford University Press, 2014.

Code Red Worm

The first major malware event of the 21st century, a malware discovered in mid-2001 that wrought an estimated $2.6 billion in economic damage worldwide. The Code Red Worm, along with its subsequent strains (CRv1, CRv2, and CRII), engaged in website defacement, distributed denial-of-service (DDoS) attacks, and created "backdoors" to computers allowing access for future attacks. Primarily targeting small business and home computer users, it also impacted governmental and corporate websites.

As with other worms, Code Red was a self-replicating malware type that did not rely on infected files spreading via user "cooperation," such as opening an e-mail attachment or launching a program, as is the case with computer viruses. The Code Red Worms initially exploited a known software vulnerability and then scanned random IP addresses in order to spread the infection creating a "botnet," or network of infected computers (bots). Botnets can be controlled by a single computer that can initiate an attack on a chosen victim or victims using DDoS attack. In this type of attack, the bots bombard the victim(s) with traffic in order to quickly overwhelm the bandwidth/memory capacity of servers, shutting them down.

CRv1 was able to infect and deface the home pages of 10,000 servers before discovery. At the height of its efficiency, CRv2 infected new machines at a rate of 2,000 per minute, reaching 359,000 computers within 14 hours. Fewer than half of these computers were estimated to be in the United States. CRv2 had significant international impact due to the large volume of infected computers and exponential propagation creating Internet traffic that overwhelmed bandwidths, causing slow service and widespread outages. CRv2 crashed or rebooted devices with Internet interfaces, such as printers, routers, and DSL modems, and coordinated a focused attack on one of the servers that hosted the U.S. White House website. CRII installed a mechanism that allowed remote, administrator-level access to the infected computer. CRII installed "backdoors" in infected machines that would allow any code to be executed for future attacks.

The Code Red Worm strains provided lessons regarding the dangers of failing to react quickly to public warning of vulnerabilities, the international scope such attacks can have, the speed at which widespread software vulnerabilities can be exploited and incapacitated, the degree small business and home users are integral to the global Internet, and the importance of developing Internet worm malware mitigation techniques beyond host patching.

Diana J. Ringquist

See also: Botnet; Distributed Denial-of-Service (DDoS) Attack; Malware

Further Reading

Dolak, John C. *The Code Red Worm.* Bethesda, MD: SANS Institute, 2001.

Rhodes, Keith A. *Information Security: Code Red, Code Red II, and SirCam Attacks Highlight Need for Proactive Measures.* Washington, D.C.: Government Accountability Office, 2001.

Schauer, Renee C. *The Mechanisms and Effects of the Code Red Worm. SANS Institute InfoSec Reading Room.* Bethesda, MD: SANS Institute, 2001.

Zou, Cliff C., et al. "Monitoring and Early Detection for Internet Worms." *University of Massachusetts Amherst Computer Science Department Faculty Series* (2004), 1–16.

Computer Emergency Response Team (CERT)

An association of cybersecurity experts organized to respond to cybersecurity emergency events. Although the overall mission of a CERT involves prevention and mitigation of further attacks, this may involve activities such as running simulations and other tests. National-level response entities, compassing experts contracted by the government or employees of pertinent corporations, exist for some fifty-five countries.

The first CERT was established in the wake of the Morris Worm, ostensibly a graduate research project launched in November 1988 whose rapid spread transformed into a denial-of-service attack across the Internet and infected an estimated 10 percent of the world's then total of 60,000 connected machines. The worm's designer became the first person convicted under the Computer Fraud and Abuse Act of 1984. Efforts to coordinate a response included the establishment of the phage mailing list by a young Purdue University professor Gene Spafford; the mailing list ran for six months after the release of the Morris Worm and facilitated the early sharing of information. To pave the way for a more permanent entity capable of leading response efforts at the time and in the future, the Defense Advanced Research Projects Agency (DARPA) funded the creation of the CERT Coordination Center (CERT/CC), to be located at Carnegie Mellon University.

As demonstrated by CERT/CC, an important element of a CERT's activities can involve the coordinating of private sector and government entities to respond to identified software vulnerabilities. Part of the coordination process advanced by CERT/CC is known as "responsible coordinated disclosure," which arranges for vulnerabilities to be discretely addressed before public notification leads to wider awareness that might invite exploitation. CERT/CC also makes testing tools freely available to security researchers, and the organization conducts some training sessions targeting individuals and organizations interested in establishing product incident response teams able to help address particular incidents.

In the 21st century, CERT/CC is sometimes conflated with the country's national counterpart, US-CERT. The latter was created in 2003 and is part of the Department of Homeland Security (DHS) within its National Cybersecurity and Communications Integration Center (NCCIC), and consequently focuses on threats directly impinging on the United States. As is the case with other counterpart entities, US-CERT helps respond to cyberattacks while also participating in and facilitating collaboration and the sharing of information among relevant entities and sectors. Its efforts include research into an array of computer network defense (CND) elements, the preparation of analysis and reports about cyber threats and response resources, and other work that helps arrange for the use of countermeasures for responding to threats. Both the US-CERT and the Industrial Control Systems CERT (ICS-CERT), which focus on topics such as threats to supervisory control and data acquisition (SCADA) systems, also interface with international counterparts abroad.

In addition to their work in addressing specific vulnerability incidents, CERTs are reported to engage in other efforts connected to combating illicit online activities. During 2014, CERT/CC took part in efforts to reduce the anonymization of The Onion Router (TOR), which although created in the early 2000s for the purpose of protecting free speech and democracy advocates in autocratic polities has since also been used by criminal entities making use of its anonymization functionality. Additionally, CERT/CC worked against SilkRoad 2.0, a hidden service utilizing Tor in order to facilitate the illicit drug trade.

The proliferation of malware and of an ever-growing number of cyber incidents and a continually broadening range of cyber threats have prompted the establishment of increasing numbers of CERTs. Nation-level CERTs may frequently work with other CERTs belonging to nongovernmental entities. CERT-EU for the European Union dates to June 2011, and within months the total number of governmental and other CERTs across European Union member states had been found to be 140.

The 2007 cyberattacks on the eastern European country of Estonia swiftly brought high-level attention to cyber threats from nation-states, and by May 2008 the North Atlantic Treaty Organization (NATO) had not only opened a cyber defense center but had selected the Estonian capital city of Tallinn as its headquarters. Estonian officials had recommended the creation of such an entity and that their capital host it since as early as 2003, in the run-up to that country's joining NATO. Estonia's own CERT played an important role in the country's resisting cyberattacks attributed to Russia in 2007, and the following year Estonia's CERT provided assistance to the Caucasian country of Georgia when that country also suffered cyberattacks coinciding with a brief kinetic conflict against Russia.

As a result, the term "CERT" and the related title of Computer Security Incident Response Team (CSIRT) have come into more common usage. Sometimes an entity's CERT develops into a center specializing in monitoring and defense of the information systems and devices across that organization's enterprise.

Nicholas Michael Sambaluk

See also: Computer Network Defense (CND); Estonia, 2007 Cyber Assault on; Georgia, 2008 Cyber Assault on; Malware; National Cybersecurity Center; Supervisory Control and Data Acquisition (SCADA); The Onion Router (TOR)

Further Reading

Ashmore, William C. *Impact of Alleged Russian Cyber Attacks.* Ft. Leavenworth, KS: School of Advanced Military Studies, 2009.

Bada, Maria, et al. "Computer Security Incident Response Teams (CSIRTs) An Overview." *Global Cyber Security Capacity Centre* (2014), 1–23.

Robinson, Neil, et al. *Cyber-Security Threat Characterization: A Rapid Comparative Analysis.* Santa Monica, CA: RAND, 2011.

Taddeo, Mariarosaria, and Ludovica Glorioso. *Ethics and Policies for Cyber Operations: A NATO Cooperative Cyber Defense Centre of Excellence Initiative.* New York: Springer, 2017.

Weed, Scott A. *US Policy Response to Cyber Attack on SCADA Systems Supporting Critical National Infrastructure.* Maxwell AFB, AL: Air University Press, 2017.

Computer Network Attack (CNA)

Hostile activities that exert detrimental effects on a target's computer systems or on the information that it contains or uses. According to U.S. military doctrine expressed in Joint Pub 3–13, CNA is a component of the overarching category of computer network operations (CNO). This doctrinal scheme considers CNA distinguished from intelligence collection efforts targeting computer systems, which doctrine identifies as computer network exploitation (CNE). According to that definition, CNA entails solely activities that block access, degrade capabilities, or eradicate computer-based information, networks, and machines.

However, actions involved in CNE and CNA can be difficult to distinguish and frequently overlap, particularly since deriving intelligence through CNE is a prerequisite for maximizing the impact of a subsequent CNA action intended to exert damage. Cyber expert Bruce Schneider has suggested the distinction between CNE and CNA to be the equivalent of a marketing campaign, arguing that "cyber-espionage is a form of cyber-attack" (Schneider, 2014) and that shifting definitions have diplomatic rather that purely technical causes.

The cache of CNA came as a result of mounting infusion of computers into crucial systems, and it coincided with the continuation of this trend and with the increasing interconnection of computers through systems like the Internet from the end of the 20th century onward. First, however, came the manifestation of the advantages offered through computer-supported systems. The logistics, intelligence, and targeting demonstrated via U.S. technology in 1991 through the swift and dramatic victory of the U.S.-led international coalition that ejected Saddam Hussein's invading Iraqi forces from Kuwait. The conflict left many analysts globally struggling to come to terms with the recent and vivid impact of leveraging computerization to improve a range of vital systems.

In both the military and the civilian realm, the United States in the 1990s pursued computerization and connectivity as U.S. computer users led the way in an expansion of the Internet. Concurrently, analysts in the People's Republic of China (PRC) associated with the People's Liberation Army (PLA) were considering how the impact of the U.S. warfighting system, as used against Iraq, might be countered so that it could not be used either as a source of leverage against the PRC or of warfighting supremacy against the PLA. As with other technologically enabled systems, tampering with or eliminating the enabler can be expected to wreak cascading effects on the entities that relied on the enabler and the entities that had come to presume support from those now degraded entities.

This study resulted in renewed PLA interest in information operations in general and also to thought about CNA. Importantly, PLA analysts define CNA more broadly than do U.S. writers. Among U.S. planners, CNA target computer networks, and they do so by using computer technology. PLA documents have indicated that the neutralization or elimination of computer networks constitutes CNA, although the methods may involve computers launching digital actions, warfighting platforms using kinetic weapons, or a combination of the two approaches. The PLA's reported view of computer network attack reflects an effects-oriented focus. As a result, doctrinal thinking allows for different methods by which to neutralize networked targets. These can range from methods through the cyber domain itself to potentially including kinetic actions against elements of a network.

In keeping with the realities of warfare in the physical domains, effective intelligence of a target is reputed to be a vital prerequisite before launching a meaningful CNA. Since CNA fits within the context of information operations, it is most energetically touted as a warfighting tool by figures who foresee an adversary's reliance on computerized systems as significant and growing. Advocates have examined ways to maximize the impact of a CNA. One example would be to access and study electronic communications to more accurately map and prioritize candidate targets for later CNA.

Interest in conducting surveillance to understand how a target functions and how to most effectively degrade it implies a degree of continuity, or at least a relationship, between CNE and CNA. Although CNE may or may not exist as a precursor to subsequent CNA actions, a target's discovery of signs indicating CNE activity seem likely to raise concern in the minds of decision makers. Advanced persistent threats (APTs) have historically been most often directed toward espionage for economic or other strategic purposes. In these cases, exfiltration of data is both the ultimate objective and marks the most likely point at which an APT is discovered. However, occasionally operations have occurred that some experts identify as APTs, which result in data eradication or in limited physical destruction. These cases underscore the point that although CNA may be different from CNE, the assumption that CNE does not signal upcoming CNA is potentially faulty.

Analysis within the United States has considered whether this avenue gives an advantage to the United States as a country particularly invested in computerization and connectivity or if the prospect of these enabling technologies being threatened means that the country is especially vulnerable to an adversary's use of CNA against it. Notable CNA events include Stuxnet's targeting of physical systems that enriched uranium for Iran's nuclear program, the hack against Sony in 2014 since it involved the destruction of some of the company's data, and attacks against Ukrainian infrastructure such as the targeting of its electrical power grid in 2015.

Although the possible advantages of launching a CNA against an adversary was identified at least as early as the 1999 North Atlantic Treaty Organization air campaign against Serbia connected with Serb atrocities in Kosovo, debate among U.S. policy advisers about the character of CNA reportedly led to the rejection of the CNA option on the contemporary interpretation that CNA actions might be deemed war crimes. Other legal scholars since have continued the debate, and some have asserted that CNA might not even be deemed an "armed attack," and thus might fail to rise to the threshold at which retaliation by a targeted state would be accepted internationally.

The perspectives of some experts suggest that the focus on blocking or breaking computers and data may lead to a counterproductively narrow definition of CNA. This approach would, for example, include traffic analysis and other surveillance meant to exfiltrate information, although this would otherwise be defined as CNE. Furthermore, actions to insert misinformation into a system so as to grant it undue credibility when disseminating it might also come under a broader understanding of CNA. One example that analysts have already identified would be for a nation-state adversary in time of war to manipulate data related to command and control as well as to logistics used by the U.S. military. The cited effects would include extreme congestion at shipping and airport facilities due to confusion. Unanswered logistical needs, for food, ammunition, or equipment, would mean dire consequences for personnel. Experts have pointed to PLA documents indicating an interest in CNO, which would include attacks.

Nicholas Michael Sambaluk

See also: Advanced Persistent Threat; Computer Network Defense (CND); Malware; Snake, Cyber Operations against Ukraine, 2014; Sony Hack, 2014; Stuxnet; Ukraine Power Grid Cyberattack, December 2015

Further Reading

Dinstein, Yoram. "Computer Network Attacks and Self-Defense." *International Law Studies* 76 (2002), 100–119.

Merritt, Eric J. *Creating Network Attack Priority Lists by Analyzing Email Traffic with Predefined Profiles.* Wright-Patterson AFB, OH: Air Force Institute of Technology, 2012.

Schmitt, Michael N. "'Attack' as a Term of Art in International Law: The Cyber Operations Context." *4th International Conference on Cyber Conflict* (2012), 283–293.

Schneider, Bruce. "There's No Real Difference between Online Espionage and Online Attack: You Can't Hack Passively." *The Atlantic,* March 6, 2014.

Silver, Daniel B. "Computer Network Attack as a Use of Force under Article 2(4) of the United Nations Charter." *International Law Studies* 76 (2002), 73–97.

Wortzel, Larry M. *The Chinese People's Liberation Army and Information Warfare.* Carlisle, PA: Strategic Studies Institute, 2014.

Computer Network Defense (CND)

A term used primarily by the military but also sometimes in civilian sectors to refer to the protection of computer networks. It is defined by the U.S. Department of Defense (DoD) as the actions taken through the use of computer networks to protect, monitor, analyze, detect, and respond to unauthorized activity within information systems and computer networks.

CND allows an organization to defend and retaliate against network attacks such as intrusion, disruption, denial, degradation, exploitation, or other types of activities that would give access to information systems and their contents. CND actions not only protect systems from an external threat, but also from exploitation from within, and are considered a necessary function by the U.S. military in all its operations globally.

CND is a subcategory of the broader concept Computer Network Operations (CNO), which also includes Computer Network Attack (CNA) and Computer Network Exploitation (CNE). Together, all categories of CNO are therefore the use of capabilities to either protect, exploit, or attack electronic information and infrastructure. In other words, CNOs are actions taken to harness computer networks in order to gain information advantage and prevent the adversary from using network capabilities for the same purpose.

CNO were identified as one of the five core Information Operations (IO) capabilities in the 2006 document Joint Publications JP 3-13, which was a revised version of the 1998 publication that first introduced the doctrine of Information Operations for the U.S. military (the other four components were electronic warfare, psychological operations, operations security, and military deception). The document affirmed that IO were integral to the successful execution of military operations, with the key goal of achieving and maintaining information superiority for the United States and its allies. IO doctrine is considered to be the foundation of the U.S. military approach to the use of computer technologies.

However, in 2013, a new doctrinal document called JP 3-12 introduced the doctrine of Cyberspace Operations (CO), superseding the previous terminology related to IO. Cyberspace was there presented as a global domain within the information environment and one of the five interdependent domains with air, land, maritime, and space. Cyberspace Operations were defined as the employment of cyberspace capabilities with the primary purpose of achieving objectives in or through cyberspace. As for the term CND, it was replaced by Defensive Cyberspace Operations (DCO), encompassing passive and active cyberspace defense operations to preserve the ability to utilize cyberspace capabilities and protect data, networks, net-centric capabilities, and other designated systems.

At the end of the 20th century, the U.S. military progressively included computer-related operations as a core component of its strategic approach to defend the interests of the United States globally. This shift happened as the U.S. government as a whole was increasingly acknowledging the central role information and communication technologies were playing in economic and political activities, both in terms of opportunity and in terms of threats facing the country. Indeed, as government operations were becoming more dependent on computer networks, it also became necessary to protect and defend these vital systems against malicious actors and adversaries. In 1997, the DoD was made particularly aware of the vulnerability of computer networks during the military exercise known by the code name "Eligible Receiver," in which the National Security Agency (NSA) managed to infiltrate the electrical networks and emergency systems of nine major cities in the United States.

In order to respond to these evolving challenges and recognizing that computer-related activities were a critical aspect of the U.S. military operations around the globe, in 1998 the DoD established its first unit to combat cyber threats, known as Joint Task Force-Computer Network Defense (JTF-CND). In 2001, the unit was redesignated Joint Task Force-Computer Network Operations (JTF-CNO), and it was assigned to the U.S. Strategic Command in 2002. After many reorganizations over the years, the U.S. Cyber Command (USCYBERCOM) was created in 2010. It represents the latest evolution of the organizational designs for the enhancement of the U.S. military capabilities in cyberspace. USCYBERCOM is situated in Fort Meade, Maryland.

Karine Pontbriand

See also: Computer Network Attack (CNA); USCYBERCOM

Further Reading

Ballmann, Bastian. *Understanding Network Hacks: Attack and Defense with Python.* New York: Springer, 2015.

Computer Network Defense, Department of Defense Directive 8530.1. Washington, D.C.: Department of Defense, 2001.

Joint Publication 3-12 (R), Cyberspace Operations. Washington, D.C.: Department of Defense, 2013.

Joint Publication 3-13, Information Operations. Washington, D.C.: Department of Defense, 2006.

Kaplan, Fred. *Dark Territory: The Secret History of Cyber War.* New York: Simon & Schuster, 2017.

Klimburg, Alexander. *The Darkening Web.* New York: Penguin Press, 2017.

Covert Channels

A computer security attack marked by the transformation of data between otherwise incompatible processes. Since this skirts legitimate data transmission mechanisms, the transfer activity itself can go undetected by security assurance systems running on victim computers. As a result, it can be a valuable tool for hackers. Since the transfer of data on covert channels can include the insertion of malware, covert channels can be used as building blocks to form more complex computer security problems for a targeted system.

Two types of covert channels exist: channels that modify hard drives and other storage locations, and channels that manipulate timing. When a targeted system's patterns are studied and understood, a covert channel can be established that is reportedly difficult for computer security experts to subsequently uncover.

Predictably, covert channels also possess shortcomings. Installation is reported to be challenging, because that phase of an attack impacts system performance in ways that can be monitored and diagnosed. As is frequently the case with illicit communications, the unobtrusiveness of the transfer phase occurs because most activities are not the covert transfer itself. In covert channels, this involves a low signal-to-noise ratio, which means that illicit data transfer happens only slowly. Attackers must balance the preservation of their channel's secrecy with the throughput or capacity of their channel, since greater throughput accelerates the illicit movement of data but also raises the likelihood that the activity will betray the existence of the channel-to-system monitors.

Although covert channels were first hypothesized in the early 1970s, the contemporary experimental ARPANet invented with the support of the Defense Advanced Research Projects Agency (DARPA) grew into eventually becoming the Internet. The massive expansion in networks and communication provided new avenues for covert channel attacks, specifically "network covert channels" or "trapdoors." Cybersecurity experts have noted that systematic examination of network covert channels is complicated by the fact that the attacks tend to be specifically adapted to particular protocols or applications.

Computer science professionals have suggested "potential beneficial applications of covert channels" (Johnson, 2010), pointing to the possibility of blocking man-in-the-middle attacks in which an attacker eavesdrops on a conversation and potentially also manipulates data traveling between two legitimate parties. Researchers identify the analysis of information flow as a potentially valuable tool for defenders. Comparing the leverage of information flow to the game Bridge, "covert channels may be used for identification of opponents" (Evtyushkin, 2016, 118) and thereby help uncover adversaries. Researchers have hypothesized that applications could include eroding the anonymity of software designed to facilitate Internet browsing privacy.

Nicholas Michael Sambaluk

See also: Cyberweapons; Defense Advanced Research Projects Agency (DARPA)

Further Reading

Evtyushkin, Dmitry, et al. "Understanding and Mitigating Covert Channels through Branch Predictors." *ACM Transactions on Architecture and Code Optimization* 13, no. 1 (March 2016), 10:1–10:23.

Johnson, Daryl. *Covert Channels in the HTTP Network Protocol: Channel Characterization and Detecting Man-in-the-Middle Attacks.* Rochester, NY: Rochester Institute of Technology, 2010.

Murdoch, Steven J. *Covert Channel Vulnerabilities in Anonymity Systems.* Cambridge: Cambridge University Press, 2007.

CrowdStrike

A U.S.-based cybersecurity technology company founded in 2011 by George Kurtz and Dmitri Alperovitch. CrowdStrike provides endpoint security, threat intelligence, and pre- and postincident response services. CrowdStrike is well known for its countermeasure involvement in high-profile cyberattacks, such as the 2014 attack of Sony Pictures and the 2015 and 2016 attack of the Democratic National Committee.

CrowdStrike's primary service is a subscription cloud-based endpoint security platform that operates at the kernel level, approaching cybersecurity in a dualized manner. CrowdStrike uses the traditional defense and detection security model in which antimalware tools are used against known attacks in order to identify and block them. CrowdStrike also uses machine learning and crowd-sourced intelligence and behavioral analytics to detect and prevent attacks at the endpoint. Relevant data about each attack is transmitted to CrowdStrike for analysis, and commonalities of events across the entire sensor network are sought. This allows CrowdStrike to adapt to detect new attack modalities as they develop.

Diana J. Ringquist

See also: Malware; Sony Hack, 2014

Further Reading

Kassner, Michael. "CrowdStrike's Security Software Targets Bad Guys, Not Their Malware." *TechRepublic,* October 9, 2015, https://www.techrepublic.com/article/crowdstrikes-security-software-targets-bad-guys-not-their-malware.

Rid, Thomas, and Ben Buchanan. "Attributing Cyber Attacks." *Journal of Strategic Studies* 38, no. 1–2 (2015), 4–37.

Scott, James, and Drew Spaniel. *China's Espionage Dynasty: Economic Death by a Thousand Cuts.* Washington, D.C.: Institute for Critical Infrastructure Technology, 2016.

Crypto Phone

A mobile telephone built to prevent electronic eavesdropping by tools such as an International Mobile Subscriber Identity-catcher (IMSI-catcher). Secure calls require crypto phones on both sides of a conversation in order for the authentication system, which includes a three- or four-digit session key, to function. Electronic eavesdropping systems are able to detect and intercept a signal, meaning that the presence of a phone and the act of a call can be determined. Nonetheless, an attempted man-in-the-middle attack remains incapable of decrypting the phones' own encryption.

Interception attempts can impact the session key and thus reportedly alert crypto phone users to a surveillance effort, suggesting that surveillance efforts might actually prove counterproductive or alternatively that stymied surveillance actions might conceivably be one day used as a tool for authorities signaling to crypto phone users. Session keys are not used repeatedly, increasing the challenge involved in trying to conduct surveillance. Crypto phones therefore offer a degree of communications security to individuals and groups seeking to evade a potential eavesdropper's surveillance, particularly if the existence of a call does not itself compromise a user.

Nicholas Michael Sambaluk

See also: Encryption; International Mobile Subscriber Identity-Catcher (IMSI-Catcher)

Further Reading

Johanputra, Nirav, et al. *Emerging Security Technologies for Mobile User Accesses.* Report. San Jose State University, 2007.

Mulliner, Collin, et al. *SMS-Based One-Time Passwords: Attacks and Defence.* Technical Report. Berlin: Technische Universitat Berlin, 2014.

Cyber Caliphate

Also known as United Cyber Caliphate, Islamic Cyber Army, or Islamic State Hacking Division, an organization conducting activities in the cyber domain in promotion of the goals of the Islamic State of Iraq and Syria (ISIS). Although ISIS's cyber capabilities are considered relatively low, hackers appear to focus on projecting power by highlighting potential vulnerabilities within the U.S. national security apparatus through tactics such as defacement, distributed denial of service (DDoS), data theft, and disabling websites. The extent to which individuals claiming to operate on behalf of ISIS are sympathizers acting without guidance or active ISIS members conducting directed cyber operations on behalf of the organization is unknown.

The hacker most often affiliated with ISIS activities is Junaid Hussain, a British-born Muslim hacktivist who traveled to Syria in late 2013 to join ISIS and who was later killed in a drone strike in August 2015. Hussain was a founding member of the English language pro-Islamic State online propaganda and recruitment collective the FBI refers to as "The Legion" or "Raqqa 12." Although Hussain was considered the head of the Islamic State Hacking Division (ISHD), his hacking activities while in Syria are unclear. Varying accounts of Hussain's death and the targeting of other members of The Legion soon after led many ISIS fighters to abandon the use of the encrypted messaging app Surespot, as it was thought to be infiltrated and used by Western law enforcement and intelligence agencies to track militants.

Pro-ISIS accounts and websites frequently claim responsibility for hacks of local, state, and federal government agencies. Hackers sometimes post stolen data or employee information and allege it will be used to compile a list of human targets. In December 2014, the U.S. Department of Homeland Security and FBI issued a joint statement warning that ISIS sympathizers were using social media to compile targeting information on members of the military and their families to create "kill lists" and to identify individuals for potential recruitment into the Islamic State.

A January 2015 hack of U.S. Central Command (CENTCOM) social media accounts is most often affiliated with Cyber Caliphate. CENTCOM Twitter and YouTube accounts were hacked by individuals claiming affiliation to the Islamic State. The hackers posted messages and a video before the accounts were taken offline. Since 2015, Western intelligence and law enforcement have issued a number of warnings indicating an ISIS desire to expand cyberactivities to conduct attacks on critical infrastructure.

Many believe some of the activities attributed to Cyber Caliphate may be Russian hackers pretending to be affiliated with ISIS. The April 2015 attack on a French television channel, which occurred a few months after the Charlie Hebdo terrorist attack, was initially claimed by a group calling itself Cyber Caliphate. It was later attributed to the Russian group APT28. Martin Libicki notes false flag attacks are uncommon, but the attack on France's TV5 (TV5Monde) is one such example.

Melia T. Pfannenstiel

See also: Hacktivism; Operational Security, Impact of Social Media on; Surespot App; War on Terrorism and Social Media

Further Reading

Alkhouri, Laith, et al. "Hacking for ISIS: The Emergent Cyberthreat Landscape." *Flashpoint* (April 2018).

Hamid, Nafees. "The Hacker Who Became the Islamic State's Chief Terror Cyber Coach: A Profile of Junaid Hussain." *CTC Sentinel* 11, no. 4 (April 2018), 30–35.

Libicki, Martin C. *Cyberspace in Peace and War.* Annapolis, MD: Naval Institute Press, 2016.

U.S. Library of Congress, Congressional Research Service, *Information Warfare: DoD's Response to Islamic State Hacking Activities* by Catherine A. Theohary, Kathleen J. McInnis, and John W. Rollins, IN10486 (May 10, 2016), 1–3.

Cyber Pearl Harbor

A term first entering the mainstream discourse on cybersecurity in 1997 during the congressional testimony by Deputy Defense Secretary John Hamre. Cyber Pearl Harbor was referenced again in 2012 by Defense Secretary Leon E. Panetta and by Senator Harry Reid in 2015, leading to its development into a marketing reference.

The term was originally used as an analogy to describe the potential of a crippling cyberattack on the United States impacting command-and-control systems, operations and infrastructure, institutions and economies, and national morale. Although it is agreed that critical infrastructure has significant vulnerabilities, the likelihood and severity of exploitation of these vulnerabilities in a cyberattack is debated by experts, businesses, and politicians.

It is generally agreed that the concept of a Cyber Pearl Harbor has encouraged an awareness of the need to develop systems that use multipath systems that avoid collocation and encourage heterogeneity, redundancy, logical diversity, and independence within and between cyber system components and security strategies.

Diana J. Ringquist

See also: Computer Network Attack (CNA); Supervisory Control and Data Acquisition (SCADA)

Further Reading

Goldman, Emily O., and John Arquilla. *Cyber Analogies.* Monterey, CA: Naval Postgraduate School, 2014.

Stiennon, Richard. *There Will Be Cyberwar: How the Move to Network-Centric Warfighting Has Set the Stage for Cyberwar.* Birmingham, MI: IT-Harvest, 2015.

Wirtz, James J. "The Cyber Pearl Harbor." *Intelligence and National Security* 32, no. 6 (March 2017), 758–767.

CyberBerkut

A "hacktivist" group sympathetic to Russian influence in Ukraine in the context of the Crimean conflict starting in 2014. They take their name from their pro-Russian Ukrainian "Berkut" riot police notorious for their violent methods of repression against citizens during the EuroMaidan riots in Kiev in late 2013. CyberBerkut use various methods of attack such as malware, DDoS-style attacks, hacking security systems, blocking websites, and more to further their pro-Russian agenda on Ukrainian government targets and other intergovernmental entities like NATO. They are primarily known for their attempted malware attack on the Ukrainian elections of parliament in 2014, attempting to bring Ukraine's government more closely under Russian influence. Other activities ascribed to the group include attacks on the German government, European Union officials, the U.S. Embassy, and various U.S. private military companies. Other activities have targeted Ukraine's ability to mobilize its military. It is unclear if CyberBerkut is an independent Ukrainian organization or covertly supported by the Russian government.

Stephanie Marie Van Sant

See also: Distributed Denial-of-Service (DDoS) Attack; Hacktivism; Malware; Snake, Cyber Actions against Ukraine, 2014; Ukraine Power Grid Cyberattack, December 2015

Further Reading

Jasper, Scott. "Deterring Malicious Behavior in Cyberspace." *Strategic Studies Quarterly* (Spring 2015), 60–85.

Kostyuk, Nadiya, and Yuri M. Zhukov. "Invisible Digital Front: Can Cyber Attacks Shape Battlefield Events?" *Journal of Conflict Resolution* (November 2017).

Williams, Timothy J. *Cyberwarfare and Operational Art*. Ft. Leavenworth, KS: School of Advanced Military Studies, 2017.

Cyberweapons

Tools utilizing weaponized code to compromise the security services of information systems, including authorization, access control, nonrepudiation, confidentiality, availability, and integrity. Although there are many software, firmware, hardware, and network vulnerabilities that are known and sometimes remediable, the most coveted vulnerabilities are those that are not yet known to the vendors of these products or to the information technology (IT) professionals responsible for protecting them. Malware access tools leveraging these kinds of vulnerabilities are known as zero-day exploits. They are sometimes combined with other attack techniques such as social engineering (e.g., phishing) to facilitate the spread of malicious code. While cyberweapons damage the security services, they have also been used to cause physical damage to adversaries.

As is the case with cyberwar and cyberattack, different sources use the term cyberweapon in different ways. Two decades into the 21st century, a large array of cyberweapons have been identified, with diverse methods, effects, and appearing in massively varying sizes and levels of sophistication. According to a broad definition, the vast majority of cyberweapons and cyberattacks reflect low levels of technical sophistication. The term "script kiddies" took hold because of the youth and unoriginality that characterize a large proportion of the derivative forms of cyberattack. These may be launched by individuals, groups, or political entities, and the actors' motivations may include theft, espionage, interest in vandalism, political activism, or still other reasons. In contrast, the most sophisticated cyberweapons are typically presumed to have been created with the backing of nation-states, due to the time, financial, and expertise resources involved in their development.

Attribution is nonetheless more complicated in the digital realm than with physical weapons. The secrecy required to maintain the potency of many cyberweapons can be a liability and does not lend itself well to acting as a deterrent. These arms races therefore look different than those regarding physical weapons where missile tests and other military exercises put capabilities on full display. Experts note that this dynamic does not eliminate the potential for arms races in pursuit of cyberweapons, and in fact the most robust nation-state programs reportedly involve the stockpiling of both vulnerabilities to exploit against adversaries and also the stockpiling of patches to deploy at strategic times for one's own protection.

The early 21st century has already proven rife with assertions of attribution about modern cyberweapons, often giving rise to considerable controversy. Authorized disclosure of an event from the Cold War illustrates some of the potential of cyberweapons. During the Cold War, the Soviet Union often resorted to espionage and theft of Western high-tech intellectual property in efforts to keep pace with advances in the United States and elsewhere. When a Soviet spy defected and informed U.S. authorities of some Soviet espionage activities, U.S. policy makers authorized the Central Intelligence Agency (CIA) to write malicious code into industrial programs the Soviet Union was known to covet; the program was then lightly guarded so that Soviet spies could acquire the program for Soviet use. This code, applied to the trans-Siberian pipelines for energy transfers, interfered with normal operations and in 1982 led to an enormous nonnuclear explosion comparable to a 3-kiloton blast and visible from space.

Stuxnet, a cyberweapon known by name (in contrast to the nameless Trojan horse above), was deployed against nuclear centrifuges in Iran at a uranium enrichment facility named Natanz. Stuxnet provides a sophisticated example of the deployment of a cyberweapon. As the Industrial Control Systems (ICSs) and Programmable Logic Controllers (PLCs) that ran these centrifuges were appropriately air-gapped (e.g., without network connection), taking advantage of any vulnerability in the system would require an insider, either wittingly or unwittingly. By initially infecting the contractors who installed or serviced the ICS, the code was then in position to be carried inside the facility to the air-gapped centrifuges. The code utilized a zero-day exploit in an auto-run feature associated with USB drives. It is believed that the code was introduced via an infected USB drive at a computer in the nuclear facility at Natanz, transmitting the exploit code targeting the ICSs and PLCs. Malfunctions eroded the efficiency of the system, and inaccurate temperature and pressure readings impeded identification and understanding of the problem.

These examples illustrate the potential for cyberweapons to cause physical damage. Furthermore, while these appear to have been developed only by the application of resources available primarily to nation-states, some experts express concern that the launching and revealing of such cyberweapons opens the door to reverse engineering of the weapons and copycat attacks by substantially less sophisticated parties. With their temperature monitoring, valves, pumps, generators, turbines, public utilities, and other industrial facilities are frequently cited as potential targets for attacks, and the victims of ransomware attacks in 2017 notably included hospitals.

Since the deployment of a cyberweapon relies on exploitable vulnerabilities, some researchers work to uncover these vulnerabilities in order to expedite their patch process. Exploit code can be given as a proof of concept (PoC) to a vendor in order to disclose the presence of the vulnerability and demonstrate its dangerous nature. Although some individuals have been prosecuted for such research, some corporations offer "bug bounty" programs designed to economically reward those who discover exploitable problems with their software or hardware products. There are companies created to acquire zero-day exploits that can pay upwards of 1.5 million dollars for key zero days in high-value target software programs, though this is unusual. Presumably, these companies are doing so with the best interests of the victim software and hardware in mind. Various motivations may lead the developer of these exploits to look for less legitimate ways to benefit from exploit code.

The possibility that the use of a cyberweapon could lead to a hot war is a very real one. Therefore, it is a matter of critical national security that, where and when possible, vulnerabilities in software and hardware do not remain unknown and unpatched for long. Further, detection procedures for identifying that an attack is under way, and incident response procedures for when it has been identified, must be in place for when these weapons do hit.

Nicole Marie Hands

See also: Computer Network Attack (CNA); Computer Network Defense (CND); Exploit; Malware; Stuxnet; Vulnerability; Zero-Day Vulnerability

Further Reading

Ayala, Luis. *Cyber-Physical Attack Recovery Procedures: A Step-by-Step Preparation and Response Guide*. Berkeley, CA: Apress, 2016.

Libicki, Martin. *Cyberdeterrence and Cyberwar*. Santa Monica, CA: RAND, 2009.

Reed, Thomas. *At the Abyss: An Insider's History of the Cold War*. New York: Ballantine Books, 2007.

Zetter, Kim. *Countdown to Zero Day: Stuxnet and the Launch of the World's First Digital Weapon*. New York: Crown, 2014.

Distributed Denial-of-Service (DDoS) Attack

A relatively low-tech, basic attack against a computer system, network, or a website. Such an attack consists of a large number of computers simultaneously attempting to access a cyber resource. The raw number of requests serves to block or slow traffic flow to the resource, significantly degrading or completely stopping its availability to legitimate users. In part, a DDoS takes advantage of the underlying structure and governance of the Internet itself, which functions by transmitting information as small packets to be reassembled by the receiving system. A massive infusion of packets can overwhelm not just the receiver but also the systems within the network used to transfer packets. Because most communication protocols include mechanisms to detect network congestion, they can be used to enhance a DDoS, essentially by forcing the protocols to reduce data transmission rates to ease congestion.

Digital communication systems make it possible for even a single computer to send a high volume of information requests. When multiple computers are utilized in the same fashion, the effect is multiplied. Computers under the external control, usually unauthorized, of a single actor are typically referred to as "bots," making the network of such attackers a "botnet." A botnet of even a few dozen computers might suffice to slow or stop traffic to relatively small websites. Much larger botnets, including a few measuring more than 100,000 computers, have been discovered. Such botnets are usually created via malware, and the owners of individual machines typically do not know their systems have been compromised, as they experience a minor slowing of performance but no more overt clues about the underlying problem.

DDoS attacks require very little sophistication, which means that a variety of human users and robotic systems can participate in launching such attacks. DDoS attacks simply require a target and a minimal amount of coordination. Even if the target is protected by a firewall, they can still clog traffic outside of the network perimeter, beyond the cybersecurity efforts of the target. Such an attack might be likened to a medieval siege—while the target sits safe and secure inside a fortified castle, protected by high walls and a moat, the besieger can still cut off the flow of communications and resources to the castle by surrounding it. With the rise of DDoS effectiveness, there has been a corresponding rise in the size of available botnets, because these can be used to rapidly magnify the impact of DDoS activities. Also, a viable DDoS-rental market now exists, with the owners of botnets directing their cyber forces to attack targets on demand, for a relatively small payment of cryptocurrency.

Early DDoS attacks could potentially be foiled by packet filtering, which consisted of shutting down traffic that matched either the parameters of a DDoS attack or that originated from known DDoS sources. However, the advent of source address spoofing has made it easier for DDoS perpetrators to hide both their origin and intent, by masking their attacks as legitimate traffic flow. More advanced efforts to mitigate DDoS attacks rely on algorithms and behavioral heuristics, but even these approaches can still be overwhelmed by the brute force attack of a large botnet.

In 2004, a group of hackers created the Low Orbit Ion Cannon (LOIC), a simple mechanism for organizing voluntary contributions to DDoS. Praetox Technologies repackaged the LOIC as a network stress-testing program and released it as an open source free download, all but guaranteeing that it would be used in a malicious fashion. The hacktivist group Anonymous, in particular, has widely distributed and used the LOIC as a means to orchestrate DDoS attacks against individuals, organizations, and groups that have caused offense.

Just as DDoS attacks can be launched by contracting with a cyber mercenary force to provide them, so too can protection against DDoS be secured from independent providers. One key method is to use the Completely Automated Public Turing Test to Tell Computers and Humans Apart (CAPTCHA). This consists of requiring users to answer one or more questions, usually by identifying scrambled letters and numbers, before the system will accept a communication attempt. As malware programmers have attempted to defeat the most basic CAPTCHA systems, more advanced questions requiring a certain level of judgment, such as selecting all the images containing animals from a batch of pictures, have been developed. These are usually quite simple for human users, albeit a bit time consuming and occasionally frustrating. Other on-demand defense mechanisms include maintaining acceptable network address databases, installing data cookies for repeat users, or creating content delivery networks (CDNs).

DDoS have proven surprisingly effective, even as a tool of state-sponsored cyber attackers. When Russia and Estonia became embroiled in a diplomatic conflict in 2007, Russian hackers began attacking Estonian networks with ever-increasing DDoS attacks. Eventually, they all but knocked Estonia off the Internet—a major problem for a cyber-dependent country that had effectively become a cashless society. Russia's conflict with Georgia in 2008 came with similar attacks, facilitated by a custom-designed program, Georbot. Computer users who wished to support Russia in the dispute voluntarily loaded Georbot onto their devices, instantly assembling an enormous botnet for DDoS attacks directed by Russian intelligence services.

Paul J. Springer

See also: Anonymous; Botnet; Estonia, 2007 Cyber Assault on; Georgia, 2008 Cyber Assault on; Malware

Further Reading

Bartlett, Jamie. *The Dark Net: Inside the Digital Underworld.* New York: Melville House, 2015.

Coleman, Gabriella. *Hacker, Hoaxer, Whistleblower, Spy: The Many Faces of Anonymous.* London: Verso, 2014.

Kaplan, Fred. *Dark Territory: The Secret History of Cyber War.* New York: Simon & Schuster, 2016.

Libicki, Martin. *Cyberspace in Peace and War.* Annapolis, MD: Naval Institute Press, 2016.

Electronic Frontier Foundation (EFF)

An organization promoting digital rights. Formed in 1990, the founders were reportedly inspired by frustration dealing with accusations of digital misconduct by law enforcement entities, which lacked familiarity with the technological issues at hand. Although its activities include publicizing guidance and software that facilitate greater security in digital communication, such as *Protecting Yourself Online* in 1998 and *Pwning Tomorrow* in 2015, much of its efforts touch even more directly on law and policy regarding online security.

Part of the organization's efforts include publicizing activities that threaten digital security, such as politically motivated phishing during the 2016 U.S. elections. EFF also established its Pioneer Awards in 1992 as a platform for promoting figures and activities it identifies as advancing "the health, growth, accessibility, or freedom of computer-based communications" (EFF Pioneer Awards). Some recipients, such as Wikimedia head Jimmy Wales in 2006 and Internet pioneer Tim

Berners-Lee in 2000, relate to information availability broadly. Other recipients, such as AT&T whistleblower Mark Klein in 2008 and WikiLeaks informant Chelsea Manning in 2017, relate to more controversial events.

EFF frequently supports the legal defense efforts of individuals and of technologies that the organization considers to promote online civil liberties and to be targeted by governmental power. The people who first established EFF did so reportedly because of concern about governmental power exceeding the government's capability to coherently understand digital technology in the last decade of the 20th century. By 1998, EFF had supported the development of the "EFF DES cracker" decryption tool in order to demonstrate the obsolescence of an encryption system then still touted by the U.S. government.

The organization examines and reports on the security and encryption strength of digital communication technologies through its "Secure Messaging Scorecard." EFF has demonstrated a degree of reluctance about the scores it assigns to technologies being interpreted as constituting product endorsements. However, some figures in the digital civil liberties field have accused EFF of defending Internet companies rather than Internet users and of framing corporate interests as questions of censorship or privacy. EFF reportedly received $1 million from Google in 2011 as a result of what other digital civil liberty groups suggested was Google's attempt to pay funds from a class action settlement to sympathetic lobbying groups. Facebook provided $1 million to EFF in 2012. EFF's primary donor is the video game company Humble Indie Bumble, which supports several nonprofit organizations focused on digital information and human health.

Nicholas W. Sambaluk

See also: Klein, Mark; Manning, Chelsea /Manning, Bradley

Further Reading

Electronic Frontier Foundation. "EFF Pioneer Awards 2017." https://www.eff.org/awards/pioneer/2017.

Nhan, Johnny, and Bruce A. Carroll. "The Offline Defense of the Internet: An Examination of the Electronic Frontier Foundation." *Science and Technology Law Review* 15 (2012), 389–401.

Postigo, Hector. "Capturing Fair Use for the YouTube Generation: The Digital Rights Movement, the Electronic Frontier Foundation and User-Centered Framing of Fair Use." *Information, Communication & Society* 11, no. 7 (2008), 1008–1027.

Encryption

The process of protecting the confidentiality and integrity of messages and other data from surveillance or manipulation by unauthorized parties. Although cryptography dates to antiquity and the use of codes and cyphers carries millennia of history, encryption in modern contexts has increasingly come to be associated with the use of digital encryption of computer data and of digital communications. Cybersecurity experts consider encryption an important issue for two fundamental reasons, one being that from its genesis the Internet was not designed with a sense of the need to protect data or for security, since networked participants were originally assumed to be cooperative rather than hostile in their mind-sets.

The proliferation of the Internet accounts for the second reason; the prominence of social media comprising "Web 2.0" and the anticipated growth of the Internet of Things have resulted in vast amounts of data and communication vulnerable to illicit examination and therefore potentially in need of protection via encryption. Electronic commerce, and the opportunity to monetize or strategically leverage various forms of data, exemplify the need for security that encryption is commonly used to fulfill.

Encryption can provide varying degrees of privacy in online browsing. The Onion Router (TOR), created in the 1990s with the support of governmental entities such as the U.S. Naval Research Laboratory as well as watchdog groups like the Electronic Frontier Foundation (EFF), represents a notable system for encryption designed to protect the private browsing of Internet users who may face surveillance efforts.

Protected data can be accessed by authorized users through their use of decryption keys, which may be private keys shared by trusted users or alternatively public keys, depending on the encryption system. These families can be subdivided further into more specifically discernable types. Private keys systems, also known as symmetric key designs, utilize a shared key at both ends of a communication, and the key is used for decryption as well as the original encryption. Private keys are generally more straightforward to use once they are in place, although providing a shared key to both communicators poses potential weaknesses. Public key systems use asymmetric keys, wherein the sender and the receiver must utilize two different keys. Public key systems are inherently more complicated than private key systems. The concept of the public key system emerged only in the early 1970s and was refined during the 1990s to yield the Pretty Good Privacy (PCP) application purchased in 2010 and periodically updated by the cybersecurity company Symantec.

Many of the apps that appear as part of the Web 2.0 landscape provide a degree of messaging privacy to users. Although not necessarily comparable to the level of security that occurs in the sophisticated communications technologies developed by leading militaries or the level of security provided by specialized cryptophones, the privacy offers through encrypted communications apps permits a level of easily available privacy that had been heretofore nonexistent to the ordinary public. This opportunity has prompted concerns for the law enforcement and national security apparatus of several countries, and spokespeople have asserted that the use of encrypted communications by certain criminals or terrorist groups could hamper law enforcement efforts and endanger public safety.

One notable example involved the December 2, 2015, mass shooting in San Bernardino, California. Syed Farook and Tashfeen Malik, a husband-and-wife jihadi terrorist sleeper cell, murdered fourteen people and wounded twenty-two others. At the time, the attack was the deadliest terrorist incident on U.S. soil since the September 11, 2001, attacks. Following the attacks, the Federal Bureau of Investigation (FBI) wanted to examine the data on the iPhone 5 owned by San Bernardino County and used by Farook in his earlier capacity as a county employee. In February 2016, the FBI and the National Security Agency (NSA) proved unable to unlock the phone and access its data. As a result, the FBI secured a court order that phone provider Apple unlock the device, but Apple refused to comply on the grounds that cooperating with the court order would undermine the security of the encryption used more generally by iPhone users as a whole. A contentious legal battle ensued for weeks, during which U.S. public opinion split nearly in half about whether Apple's refusal was appropriate or unjustified. Continuing to rebuff federal efforts to force its compliance, Apple accused federal investigators of inadvertently complicating the task of unlocking the phone through an early effort to reset the password for Farook's cloud computing account.

On March 28, 2016, federal officials announced that they had unlocked the phone without Apple's assistance. Reports since emerged that the FBI paid professional hackers to deploy a zero-day vulnerability in order to break into the phone. Media demands to uncover the identity of the hacker and the sum paid by the government were rejected in federal court in September 2017 on the grounds of national security, although journalistic reports in March 2018 indicate that when Farook's phone could finally be accessed none of the examined data it held revealed any new information about the December 2015 attack or any follow-on plot. The San Bernardino attack therefore provided an intermittent public spotlight on the encryption issue for more than two years. Although demonstrating that federal authorities could energetically secure court orders and purchase hacking talent to overcome encrypted data, Apple arguably set a high-profile precedent for major technology companies to muscularly resist compliance while using the privacy of other users' systems as its justification.

Although only tangentially connected with the San Bernardino case, cloud computing also raises an area of cybersecurity concerns that involve encryption. By definition the geographic dispersion implicit in cloud computing means that physical security mechanisms must be at least complemented by encryption. Reaction has already included interest not only in arranging "backdoor" access for law enforcement to defeat encryption but also initiatives by some countries to prohibit domestic use of foreign servers.

Nicholas Michael Sambaluk

See also: Crypto Phone; Electronic Frontier Foundation (EFF); National Security Agency (NSA); Surespot App; Telegram App; The Onion Router (TOR); Web 2.0; Zero-Day Vulnerability

Further Reading

Batten, Lynn Margaret. *Public Key Cryptography: Applications and Attacks.* Piscataway, NJ: IEEE Press, 2013.

Holden, Joshua. *The Mathematics of Secrets: Cryptography from Caesar Ciphers to Digital Encryption.* Princeton, NJ: Princeton University, 2017.

Kessler, Gary C. *An Overview of Cryptography (Updated Version, 3 March 2016).* Daytona Beach, FL: Embry-Riddle Aeronautical University, 2016.

Saper, Nathan. "International Cryptography Regulation and the Global Information Economy." *Northwestern Journal of Technology and Intellectual Property* 11, no. 7 (Fall 2013), 673–688.

Estonia, 2007 Cyber Assault on

A surge of distributed denial-of-service (DDoS) attacks that stymied Estonian commercial, governmental, and private networks in late April and early May 2007. Although no direct evidence of Russia's involvement or culpability for the attack is known, it is the likely origin. The Estonian attacks demonstrated how difficult tracing and punishing malicious actors is in the ungoverned space of the Internet. Further, the attacks displayed the economic, administrative, and national security vulnerabilities of highly connected states. The furor is likely the result of efforts by the Estonian government to remove Soviet symbols and narratives from the public space in places like Tallinn. In addition, the attacks provided Russian hackers with the first of many opportunities to refine their skills. In response to the attacks on a North Atlantic Treaty Organization (NATO) member, a joint, multinational center stood up in Estonia. NATO member states also faced the reality of threats from cyber domains. Although not the first coordinated cyberattack against a state actor, the Estonian attacks marked a distinct shift in how the national security elements of states view network-based threats.

In the course of these cyberattacks, sophisticated hackers commandeered hundreds of thousands of computers worldwide to overload the country's network capacity. Critical systems related to emergency response, banking and finance, and transportation failed under the density of network activity. The initial response from officials was to take down their own networks, ceding victory to the hackers. Only after appealing to international network service providers to identify, isolate, and disconnect the hackers did the attacks dissipate. Even then, only some IP addresses identified specific individuals, few of whom faced any consequences, and most were lost in the anonymity of the web. Indirect and anecdotal evidence implicate Russia as the origin of the cyberattack. The Russian government officially denied any involvement in the matter. A few years after the attack members of the pro-Putin youth organization Nashi claimed responsibility. Cyber expects believe it to be a reasonable claim but unlikely without significant assistance from the Russian state.

Expert consensus identified Russian involvement and motivation. In 2007 the Estonian national government decided to relocate a Soviet monument from a prominent location in Tallinn to the outskirts of town. The monument, a bronze Soviet soldier, honored the sacrifices of the Soviets during World War II to stop and roll back Nazi incursion to the Baltic states. To most Estonians, however, it symbolized the legacy of Soviet occupation and suffering. The Russian government was irate at the pronouncement, and ethnic Russians living in Estonia rioted. Moscow threatened that moving the monument would be "disastrous for Estonians." The ensuing riots saw more than 800 arrests, 153 documented injuries, and one death. Workers removed the statue and transported it shortly after the unrest began. More broadly, the Estonian cyberattacks were one of the Russian government's first substantive usage of kinetic information warfare. Over the next decade, Russia was accused of unleashing multiple, comprehensive, and destructive cyberattacks against neighbors and adversaries, including Georgia in 2008, Ukraine in 2014, and the United States in 2016. Each attack can be traced to specific Russian connections, but there is no direct evidence to implicate the Russian state.

The 2007 Estonian cyberattacks were a wake-up call regarding the vulnerability of well-connected states generally, and specifically also to the lack of preparation that NATO gave to cyber threats. Consequently, NATO aided the establishment of the Cooperative Cyber Defense Centre of Excellence as a fusion center for cybersecurity encompassing both government and industry partners. In addition, the attacks illustrated the seriousness of cyber threats and the iterative process Russia used to become more efficient with cyber tactics and techniques. The downside is that despite the long ramp-up of cyber operations from Moscow, the West, NATO, and the United States in particular were unprepared for the physical and institutional damage Russia learned how to enact.

Estonia's experience in 2007 as a target of focused and effective cyberattack revealed to the world the seriousness of electronic and cyber war. The attacks crippled the governmental functions, and economic activity of a well-developed and technologically advanced democratic state. It illustrated how unprepared most states were to respond to cyber threats and that NATO had not adequately calculated the cyber domain in the alliance's strategy. The best explanation to target Estonia is twofold: first as a retaliatory measure against removing Soviet artifacts and reshaping history; and second, the attacks clearly acted as a test case for the further refinement of Russian information attacks. Although the international community cannot formally hold Russia accountable for these attacks, experts consider the identity of the aggressor to be obvious. One result of the Estonian attack was that NATO members and other Western and European U.S. allies and partners were forced to accept the serious nature of cyberattacks and acknowledge the resolve of the Russian state to use them. This lesson nonetheless failed to reign in subsequent aggression connected to Russia. In the years that followed, Russia unleashed more cyberattacks and more sophisticated attacks in conjunction with other foreign policy measures that threatened Europe, the Baltics, and NATO.

Andrew Akin

See also: Georgia, 2008 Cyber Assault on; Snake, Cyber Actions against Ukraine, 2014; Ukraine Power Grid Cyberattack, December 2015

Further Reading

Connell, Michael, and Sarah Vogler. *Russia's Approach to Cyber Warfare.* Arlington, VA: CNA, 2017.

Herzog, Stephen. "Revisiting the Estonian Cyber Attacks: Digital Threats and Multinational Responses." *Journal of Strategic Security* 4, no. 2 (Summer 2011), 49–60.

Kozlowski, Adrzej. "Comparative Analysis of Cyberattacks on Estonia, Georgia and Kyrgyzstan." *European Scientific Journal* 3 (February 2014), 237–245.

Shackelford, Scott J. "Estonia Three Years Later: A Progress Report on Combating Cyber Attacks." *Journal of Internet Law* (February 2010).

Singer, P. W., and Allan Friedman. *Cybersecurity and Cyberwar: What Everyone Needs to Know.* Oxford: Oxford University Press, 2013.

EternalBlue Exploit

An exploit taking advantage of a flaw in Microsoft Windows software. This enabled ransomware called WannaCry, which was used in May 2017 and which highlighted the widespread vulnerability of infrastructure such as the health sector to illicit activities in the cyber domain.

In August 2016, hackers known as the "Shadow Brokers" revealed, through leaked government documents, the existence of a long list of flaws in computer systems and exploits developed by the National Security Agency (NSA) to use them in order to gain access to a wide range of hardware and software. Among them was EternalBlue, which targets a flaw in Microsoft Windows, and which NSA programmers considered so dangerous they debated warning the company. Instead, they reportedly kept the flaw secret, simultaneously also leaving the tool to use it vulnerable to theft. They are believed to have employed the exploit from its development in 2011 to the Shadow Broker data dump in 2016. Upon becoming cognizant of the flaw, Microsoft in March 2017 created a patch. The company deemed the problem sufficiently threatening that it included a fix for Windows XP, despite this being an obsolete system that had otherwise stopped receiving support from the manufacturer in 2014.

Malicious hackers, however, weaponized the exploit, which targets Windows Server Message Block 1.0 network file sharing protocol. In May 2017, WannaCry, a ransomware program, spread among computers that had not updated with the Microsoft patch, encrypting users' files and demanding payment in bitcoin. Those who did not pay saw their files destroyed after 72 hours. The most effective response found was to wipe the infected computers and restore from untainted backups. In the next week, more than 300,000 computers were affected, including the National Health Service in Great Britain, the tax bureau of Brazil, and the Russian Interior ministry, along with 150 more countries. Microsoft considered the outbreak a "wake-up call" about the U.S. government's hoarding of knowledge about vulnerabilities that could cripple the national infrastructure, as well as the damaging potential for these weaknesses to be weaponized by bad actors and unleashed on the public. This attack was also an illustration of the potential of hacking and malware to paralyze the operations of the health care system through seizure of patient records and disabling of automated systems on which hospitals rely.

Margaret Sankey

See also: Exploit; National Security Agency (NSA); Vulnerability

Further Reading

Comptroller and General Auditor. *Investigation: WannaCry Cyber Attack and the NHS.* London: National Audit Office, 2018.

The WannaCry Ransomware. Port Louis, Mauritus: CERT-MU, 2017.

"WannaCry" Ransomware Attack: Technical Intelligence. London: Ernst & Young, 2017.

Zimba, Aaron, Zhaoshun Wang, and Hongsong Chen. "Multi-Stage Crypto Ransomware Attacks: A New Emerging Cyber Threat to Critical Infrastructure and Industrial Control Systems." *ICT Express, SI: CI & Smart Grid Cyber Security* 4, no. 1 (March 2018), 14–18.

Exploit

In computer security contexts, a noun referring to software or command sequences that utilize a software vulnerability in order to gain unauthorized access to a system. Commonly, once this access has been achieved, subsequent activities include efforts to escalate the privileges of a compromised user so that the compromised account is considered by the system to be a "root" administrator with maximum authorities and access. Other exploits are developed with the purpose of gaining these privileges upon compromise of an account. Once both access and privileges are in the hands of a hacker, however, the intruder's ability, both to survey and to impact, a compromised system is considerable.

The continual unveiling of new software, the proliferation of apps, and the increasing ubiquity of connectivity among devices that inspires predictions of an Internet of Things (IoT) combine to vastly expand the attack surface that may include software vulnerabilities ripe for exploitation. Developing an exploit starts with the discovery of a vulnerability in software. Experts note that this process is sometimes accelerated by the use of automated tools that probe for weaknesses in software, although vulnerabilities are also hunted manually.

In some instances, distinct variants of an exploit may each utilize a vulnerability, demonstrating that the relationship between exploits and known vulnerabilities is not a one-to-one ratio. Sophisticated malware may easily include a number of different exploits, and the most complex identified to date, such as Stuxnet, may include one or multiple exploits tailored to target "zero-day vulnerabilities," whose weaknesses had not previously been known in the computer science community.

Some experts have suggested computer network defense frameworks that include modeling exploits. Such a model would define the required vulnerability, access to it, and any prerequisite privileges related to use of an exploit. Attacks might be understood through understanding linkages between successive exploits.

Other researchers have conceptualized other approaches to foreclosing the development and deployment of exploits. Experts demonstrated automatic patch-based exploit generation, identifying patches and then reverse engineering the patches to understand the vulnerability that has recently been detected and secured, and then developing a fresh exploit can be derived that will target systems which have not yet been patched. Since patching is not an activity that occurs simultaneously worldwide, and since many systems have historically not been kept updated with patches, the emergence of a patch inadvertently publicizes a window of vulnerability for unpatched systems. Cybersecurity experts who note the possibility of reverse engineering patches as a way to establish exploits point out that "a patch distribution scheme that staggers roll out of a patch over a large time window is insecure since the first user to receive a patch may be able to create an exploit before most users have received the patch" (Brumley, 2008, 6).

Nicholas Michael Sambaluk

See also: Computer Network Attack (CNA); Cyberweapons; Malware; Vulnerability; Zero-Day Vulnerability

Further Reading

Ball, Justin R. *Detection and Prevention of Android Malware Attempting to Root the Device.* Wright Patterson AFB, OH: Air Force Institute of Technology, 2014.

Brumley, David. *Analysis and Defense of Vulnerabilities in Binary Code.* Thesis. Pittsburgh: Carnegie Mellon, 2008.

Cullum, James, Cynthia Irvine, and Tim Levin. *Performance Impact of Connectivity Restrictions and Increased Vulnerability Presence on Automated Attack Graph Generation.* Monterey, CA: Naval Postgraduate School, 2007.

Singer, P. W., and Allan Friedman. *What Everyone Needs to Know about Cybersecurity and Cyberwar.* Oxford: Oxford University Press, 2014.

Facial Recognition Software

Technology that uses digital images and frames from video sources to identify a person. A camera captures images and searches a database to match the face to a name and other background information. The greatest advantage of this software over other biometric identification technology is its ability to function on large scales without cooperation from test subjects.

This software is most often deployed as a security measure in combination with other biometrics such as fingerprinting. Law enforcement agencies and militaries around the world use this technology to monitor sensitive locations and locate persons of interest. Social media platforms, namely Snapchat and Facebook, also use this technology to tag or apply filters to images. The technology relies on three primary techniques: traditional, three-dimensional (3D), and skin texture analysis. Each of these approaches concentrates on different aspects of the face and attempts to assign measurable values to facial features for precise identification. The most accurate systems use a combination of these three methods.

The traditional method records facial characteristics such as the shape of the eyes, cheekbones, chin, mouth, and nose from several angles. These features then feed into an algorithm that classifies the face using one of two methods: geometric or photometric. Geometrics focus on distinctive features, and photometrics condense images into statistics and compare those numbers to a template. The 3D method uses specialized sensors or sets of cameras to generate a 3D map of a person's face. Changes in lighting and angle have less effect on this method, but changes in expression can still impact accuracy. This approach has generally proven more accurate than the traditional method alone. The skin texture analysis method relies on the unique pattern of lines, spots, and marks on a person's skin. First, a picture of the skin, called a "skinprint," is broken into small sections, and each feature receives a mathematical value that can be compared with a database. Skin texture analysis is the only method to have successfully distinguished between identical twins.

Other techniques such as thermal imaging, while better able to surmount problems such as facial hair and disguises, are generally less effective because of limited databases. Entities such as the U.S. Army Research Laboratory (ARL) have attempted to fix this issue by creating artificial intelligence programs that translate thermal images into traditional databases. This approach, known as the "cross-spectrum synthesis method," is still in experimental stages.

The development of this technology has received sharp criticism from various sources. Some activist organizations such as the Electronic Frontier Foundation (EFF), Big Brother Watch, and the American Civil Liberties Union (ACLU) have expressed concern about this software. They cite expansion of government surveillance as a danger to privacy and a possible threat to persecuted groups of people. Others have warned that individuals could abuse this technology as well. For instance, some users of Russian social media platforms used facial recognition to harass women purportedly involved in Internet pornography. This software has also proved less effective in identifying persons of color, leading to reservations about using the technology as evidence in criminal investigations.

James Tindle

See also: Electronic Frontier Foundation (EFF); Facebook, Use of

Further Reading

Fretty, Douglas A. "Face-Recognition Surveillance: A Moment of Truth for Fourth Amendment Rights in Public Places." *Virginia Journal of Law & Technology* 16, no. 3 (Fall 2011), 430–463.

Introna, Lucas D., and Helen Nissenaum. *Facial Recognition Technology: Survey of Policy and Implementation Issues.* New York: Center for Catastrophe Preparedness and Response, 2009.

Welinder, Yana. "A Face Tells More than a Thousand Posts: Developing Face Recognition Privacy in Social Networks." *Harvard Journal of Law & Technology* 26, no. 1 (Fall 2012), 166–239.

Gaza War, 2008–2009

An Israeli military incursion into the Gaza Strip from December 2008 to January 2009, which catalyzed concerted social media campaigns on both sides. Israel is generally thought to have lost the global public relations battle in a campaign (Operation Cast Lead) that it mounted in hopes of eliminating rocket artillery attacks mounted by Hamas militants within Gaza who fired into Israel. The Israel Defense Force (IDF) air and artillery bombardment began on December 27, 2008, and was followed by a ground invasion from January 3, 2009, until a unilateral ceasefire two weeks later.

In 2005, Prime Minister Ariel Sharon oversaw the evacuation of IDF forces and Israeli settlers from the 365 square kilometers of the Gaza Strip southwest of Israel and bordering the Mediterranean Sea on the west and Egypt on the south. Although this left the Palestinian Authority (PA), the self-governing authority for the Palestinian communities in both the Gaza Strip and West Bank, in control within the territory, the IDF still maintained a tight military belt around Gaza. Palestinian militants continued to launch sporadic rocket attacks against Israeli urban areas, and tensions increased when Hamas won a plurality in January 2006 elections in Gaza and Israel subsequently cut off economic aid to the PA. Smuggled shipments of rockets and other weapons triggered Operation Cast Lead two years later. Both sides sought to advantageously leverage social media during the conflict.

Although Israel's uses included an effort to promote civilian safety within Israel, much of its effort aimed at bolstering the country's reputation in international eyes. Israel used a Web 2.0 application called "Red Alert" to warn Israeli civilians of incoming rockets from militant forces in Gaza, a harbinger to later uses of social media outside war zones to push warnings of dangers such as tornadoes. Israeli forces posted video clips of air strikes onto YouTube to illustrate their use of restraint, discrimination, and proportionality when attacking targets that their enemy consciously placed amid civilian sites. Israel also worked to promote a positive image of the country and its purposes in the conflict by recruiting recent Israeli immigrants to post favorable messages on social media sites such as Twitter. Other social media activity, particularly present on Twitter, YouTube, and Flickr, showed Israeli citizens under bombardment from Gaza and seeking shelter against rocket attacks.

Hamas directed some of its social media efforts to undercutting Israeli morale while the organization focused most of its efforts on creating sympathy for Palestinian civilians and damaging Israel's reputation on the international stage. Messages written in Hebrew were sent via social messaging to known Israeli military personnel, and hackers sympathetic to Hamas infiltrated Israeli media outlets and disseminated messages meant to damage Israeli morale. Various Twitter handles, including #GazaUnderAttack, #PrayForGaza, #StopIsrael, and especially #FreePalestine, constituted part of a social media propaganda campaign that highlighted Palestinian casualties for a global audience in Web 2.0. Some Hamas fighters wore body cameras so that they could take combat footage that could be collected for use in future propaganda and recruitment purposes.

The legacy of the war remains controversial both within Palestinian and Israeli communities. Outside the general humanitarian crisis, Hamas, for one, stands accused of using civilian locations, such as schools, as launching points for their rocket attacks. The group also stands accused of using civilians as human shields throughout the conflict. Outside of Gaza, an investigation into the IDF's use of force and alleged misconduct, conducted by Richard Goldstone, a former Balkan Wars criminal prosecutor and South African jurist, was released in September 2009. Highly critical of the IDF's conduct, especially in the use of disproportionate force toward Palestinian civilians and infrastructure, the Goldstone Report found that the war resulted in 1,300–1,450 Palestinian deaths, with 40 percent of those women and children. Goldstone personally retracted his assessment that the IDF committed the equivalent of war crimes in 2011 citing disputed evidence. However, the coauthors of the original report rejected his retraction and stand by their original findings. The 2008–2009 Gaza War, alongside the Goldstone Report, continues to divide and generate controversy in Israeli society, Arab communities, the United Nations, and other human rights organizations.

Kate Tietzen

See also: Cyberweapons; Twitter, Use of

Further Reading

Filiu, Jean-Pierre. *Gaza: A History*. New York: Oxford University Press, 2014.

Mansour, Camille. "Reflections on the War on Gaza." *Journal of Palestine Studies* 38, no. 4 (Summer 2009), 91–95.

Mishal, Shaul, and Avraham Sela. *The Palestinian Hamas: Vision, Violence, and Coexistence*. New York: Cambridge University Press, 2000.

Slater, Jerome. "Just War Moral Philosophy and the 2008–09 Israeli Campaign in Gaza." *International Security* 37, no. 2 (Fall 2012), 44–80.

el Zein, Hatem, and Ali Abusalem. "Social Media and War on Gaza: A Battle on Virtual Space to Galvanise Support and Falsify Israeli Story." *Athens Journal of Mass Media and Communications* (April 2015), 109–120.

Georgia, 2008 Cyber Assault on

A focused and continuous cyberattack against the government and economic sectors of the Republic of Georgia occurring between August 8 and 12, 2008. This coincided with a Russian territorial incursion into South Ossetia and Georgia. Cybersecurity experts agree that the Russian state engaged in further testing and refinement of their information warfare capabilities in a joint effort against Georgia, but no Russian officials were implicated in the affair.

The attack overwhelmed Georgian government networks and defaced key political websites of the Georgian political actors. The August 2008 military intervention crowned months of escalating tension between Russia and Georgia over unresolved political issues relating to ethnic minority populations, which remained unsolved from the Soviet Union's dissolution. A deliberate and sustained cyberattack that accompanied a kinetic military action was a new phenomenon, and experts argue that in hindsight it should have garnered more attention from the international community as a harbinger of Russia's use of gray zone or hybrid warfare. Fortunate for Georgia in 2008 was the low degree of network capability and reliance. Although the world is far more aware of the potent combination of cyberattacks leveraged with other elements of foreign policy projection, Georgia remains Russia's test case.

Beginning in July 2008, two types of cyberattacks hit Georgia, focusing on Georgian government and key economic and communication systems. The attacks were continuous, dynamic, and coincided with a joint air and land military campaign into South Ossetia and Georgia. The first type, a distributed denial of service (DDoS), floods the targeted servers with massive amounts of network traffic, inhibiting them from processing commands and causing them to crash. The second type involves modification of code that is supposed to control the website's appearance. Illicit modifications by hackers can deface the site. Meanwhile, DDoS attacks forced multiple Georgian government websites, national financial institutions, and telecom functions to shut down. Although the malicious activity began a few weeks before the August invasion of Russian troops, the attacks morphed over time to respond to defensive measures implemented by Georgian security. Further, the cyberattacks forced key elements of the Georgian national telephone grid offline almost continuously, crippling the ability for pro-Georgian forces from communicating with international supporters or media outlets. The attacks appear to follow a highly effective and supporting script for the military operations against Georgia.

The Russian military incursion to Georgia and South Ossetia was the latest chapter in a long and complicated relationship. The small provinces of Abkhazia and South Ossetia are semiautonomous entities within the geographical boundaries of the Georgian state. Both territories contain sizable populations of ethnic minorities, some of whom claim Russian heritage. After the Soviet Union dissolved, a law allowed former Soviet citizens to apply for Russian citizenship and relocate to the Russian Federation, if desired. Consequently, Russia maintained a keen interest in any regional power disputes that might foster a mass migration across Russian borders, and in extending protections to ethnic Russians living in the near abroad.

A number of specific policies undertaken by both the Russian Federation and the Republic of Georgia had already escalated residual tensions since the collapse of the Soviet Union. The Russians issued massive numbers of Russian passports to Georgian citizens living in South Ossetia and held large-scale war games that summer which left substantial numbers of troops, aircraft, and support staff in striking distance from Georgia. Georgian president Mikhail Saakashvili ordered the forward deployment of military assets and troops as a show of force and precaution during the Russian exercises. Pro-Russian militias initiated skirmishes over the summer before the August invasion.

The response to the cyberattacks was more opaque than the Georgian and international reaction to Russia's contemporary military incursion into the country. First, there is no direct link to implicate the Russian state or its security services in the cyberattacks. The surge in network traffic that overwhelmed Georgian websites originated from across the globe, likely from infected "zombie" computers that users unwittingly allowed. Consequently, tracing the root of and initiation of the attacks is impossible. Second, the attacks outpaced cyber defenses mounted against them. Most online systems went dark in an effort to mitigate further damage and data theft.

The Georgian government directed some critical information and data be transferred to allied states. Some of those third-party states experienced DDoS attacks after downloading Georgian government data. A mitigating factor that worked in favor of Georgia was that in 2008 it was still one of the least well connected and networked countries on earth. The failure of networked technology to form the basis of both the government and economic backbone acted as a firewall to prevent the full devastation of a cyberattack from crippling the state's infrastructure and rendering the military useless.

The 2008 cyberattack against Georgia was an element of a larger military strike aimed at the Georgian state. Although Russia never accepted responsibility for the cyberattacks and no definitive evidence of them being ordered by Russian officials exists, the international community attributes this attack to the Russian government. Fortunately, the Georgian cyber infrastructure was in its infancy at the time of the attacks, which prevented serious damage to key elements of national defense. The attacks demonstrated Russia's intent to use force to resolve lingering political questions from the Soviet collapse. Most significantly, the attacks demonstrated the effectiveness of using cyberwar as an element of a larger military fight.

Andrew Akin

See also: Distributed Denial-of-Service (DDoS) Attack; Estonia, 2007 Cyber Assault on; Snake, Cyber Actions against Ukraine, 2014; Ukraine Power Grid Cyberattack, December 2015

Further Reading

Connell, Michael, and Sarah Vogler. *Russia's Approach to Cyber Warfare.* Arlington, VA: CNA, 2017.

Hollis, David. "Cyberwar Case Study: Georgia 2008." *Small Wars Journal* (January 2011).

Kozlowski, Adrzej. "Comparative Analysis of Cyberattacks on Estonia, Georgia and Kyrgyzstan." *European Scientific Journal* 3 (February 2014), 237–245.

Libicki, Martin C. *Cyberspace in Peace and War.* Annapolis, MD: Naval Institute Press, 2016.

Shakarian, Paulo. "The 2008 Russian Cyber Campaign against Georgia." *Army History* (November–December 2011), 63–68.

Singer, P. W., and Allan Friedman. *Cybersecurity and Cyberwar: What Everyone Needs to Know.* Oxford: Oxford University Press, 2013.

GhostNet

An extensive social malware program identified by Canadian cybersecurity analysts and believed to have originated in the People's Republic of China (PRC). In 2008, the Dalai Lama approached computer experts at the University of Cambridge and the Information Warfare Monitor, a think tank with personnel from the Munk Center for International Studies at the University of Toronto. In the run-up to the Beijing Olympics, Tibetan activists were increasingly being surveilled, and the Dharamsala, India, headquarters of the Dalai Lama noticed that people with whom they'd recently had contact were being arrested and confronted with printouts of their online conversations by the Chinese authorities.

A ten-month investigation revealed a widespread system the researchers dubbed "GhostNet," classified as an advanced persistent threat (APT), with command and control originating at internet protocol (IP) addresses in Hainan, China. The operation functions through careful social engineering, in which the recipient opens an e-mail attachment designed to look urgent and harmless, and which then allows the originator to take full control of the computer and its contents, even using that machine to send further messages to new targets.

Although the researchers found that a number of embassies and government agencies had been affected, including those of India, Thailand, South Korea, Taiwan, and the Canadian finance department, only the Dalai Lama has allowed public disclosure of the mechanism through which it occurred. The researchers have been very careful not to directly accuse the Chinese government of this hacking. Experts have instead meticulously ensured that they pointed to the IP address, allowing speculation that the activities might have been the work of private Chinese citizens rather than an operation by a nation-state. With respect to the geographic origins, GhostNet appears not to be an isolated case, since many notable and long-lasting cyber actions have been traced to the boundaries of the PRC. The GhostNet reportedly continues to operate, with the most protections against it being expensive and extensive training for computer users to resist the e-mail lures and practice an "antisocial" response to request information and help that are often antithetical to the purpose of the institutions targeted, making the GhostNet difficult to fend off and extremely dangerous.

Margaret Sankey

See also: Advanced Persistent Threat; Operation Aurora; Operation Shady RAT; People's Liberation Army Unit 61398; People's Liberation Army Unit 61486/"Putter Panda"

Further Reading

Information Warfare Monitor. *Tracking GhostNet: Investigating a Cyber Espionage Network.* Toronto: Munk Centre for International Studies, 2009.

Nagaraja, Shishir, and Ross Anderson. "The Snooping Dragon: Social-Malware Surveillance of the Tibetan Movement." *Technical Report, Number 746.* Cambridge, UK: University of Cambridge Computer Laboratory, 2009.

Singer, P. W., and Allan Friedman. *Cybersecurity and Cyberwar: What Everyone Needs to Know.* Oxford: Oxford University Press, 2014.

Government Communications Headquarters (GCHQ)

Government Communications Headquarters (GCHQ) is the intelligence and security authority for the United Kingdom, based in Cheltenham, England. It was started in 1919 after World War I previously as part of the Royal Navy and the Army as the Government Code and Cypher School and served during World War II to help Allied code-breaking efforts. The aim of the organization is to gather intelligence as well as protect all communications of the United Kingdom's government. It is made up of two smaller agencies, namely the Composite Signals Organisation and the National Cyber Security Centre. The Composite Signals Organisation handles the gathering of information while the National Cyber Security Centre is entrusted with the security of all communication within the United Kingdom. The GCHQ was the center of much scrutiny in 2012 when Edward Snowden leaked classified information from the agency. It also partners with other corresponding national intelligence authorities in English-speaking countries such as the United States, Canada, Australia, and New Zealand. The GCHQ falls under the Foreign and Commonwealth Office of British Government, which is overseen by the British foreign secretary. They have stations all over the world.

Stephanie Marie Van Sant

See also: National Security Agency (NSA); WikiLeaks, Impact of

Further Reading

Aldrich, Richard J. "Counting the Cost of Intelligence: The Treasury, National Service and GCHQ." *The English Historical Review* 128, no. 532 (2013), 596–627.

Anderson, Ross, and Michael Roe. "The GCHQ Protocol and Its Problems." *International Conference on the Theory and Applications of Cryptographic Techniques* (1997), 134–148.

The Great Cannon

A cyber tool using malicious JavaScript to hijack bystander web traffic and repurpose it into a distributed denial-of-service (DDoS) attack. Developed by the People's Republic of China (PRC), the Great Cannon is distinct from the "Great Firewall" because it is focused on an offensive capability rather than strictly the blocking of Chinese access to websites or web searches unapproved by the PRC government.

Use of the Great Cannon was first identified in March 2015 when GreatFire.org came under DDoS attack via compromised servers of the Chinese search engine Baidu. GreatFire provided services to Chinese Internet users interested in avoiding the strictures of the Great Firewall, a tool PRC developed since the late 1990s to restrict domestic access to unauthorized sites critical of the regime or advocating democracy. The Great Cannon hijacks third-party "bystander" online traffic originating outside China but using a Baidu infrastructure server; in 1–2 percent of cases, the computer is turned unwittingly into a participant in a DDoS attack against an entity targeted by the PRC. To date, the targets of the Great Cannon have been websites that host censorship-evading tools, such as GitHub and GreatFire. In this sense, the Great Cannon can be understood as an offensive counterpart to the Great Firewall.

When announcing its discovery of the Great Cannon, Citizen Lab specifically noted that "the incorporation of Baidu in this attack suggests that the Chinese authorities are willing to pursue domestic stability and security aims at the expense of other goals, including fostering economic growth in the tech sector" (Marczak et al., 2015, 10). Frequently, PRC cyberespionage activities reflect the alignment of economic interests and espionage, so the Great Cannon represents a divergence where proactive information control is prioritized above long-term potential economic impacts of indirectly compromising the global reputation of the country's leading Internet search engine.

Ostentatious use of the Great Cannon also indicates a belief by the regime that the use of "loud" or easily identifiable offensive tools in the cyber domain is either in PRC's interest or will not harm it. Citizen Lab asserted that PRC's "conducting such a widespread attack clearly demonstrates the weaponization of the Chinese Internet to co-opt arbitrary computers across the web and outside of China to achieve China's policy ends" (Marczak et al., 2015, 12). Thus, while still representing a nondestructive capability, the Great Cannon may contribute to a watershed in the leveraging of the Internet for strategic purposes.

Nicholas Michael Sambaluk

See also: Computer Network Attack (CNA); Cyberweapons; Distributed Denial-of-Service (DDoS) Attack

Further Reading

Cheng, Dean. *Cyber Dragon: Inside China's Information Warfare and Cyber Operations.* Santa Barbara: Praeger, 2016.

Fritz, Jason R. *China's Cyber Warfare: The Evolution of Strategic Doctrine.* Lanham, MD: Lexington, 2017.

Inkster, Nigel. *China's Cyber Power.* New York: Routledge, 2016.

Hjelmvik, Erik. "China's Man-on-the-Side Attack on GitHub." March 31, 2015, http://www.netresec.com/?month=2015-03&page=blog&post=china%27s-man-on-the-side-attack-on-github.

Klimburg, Alexander. *The Darkening Web: The War for Cyberspace.* New York: Penguin, 2017.

Marczak, Bill, et al. "China's Great Cannon." The Citizen Lab: University of Toronto Munk School of Global Affairs. April 2015, https://citizenlab.ca/2015/04/chinas-great-cannon.

Hacktivism

A word combining, and referring to, an amalgamation of "hacking" and "activism." While more apparent in the past few years, the term "hacktivism" is actually believed to have been coined in 1996 by a member of the hacking group "Cult of the Dead Cow." Previously, hacking of networks or computers could be attributed to a number of factors. Those conducting criminal hacking (cybercrime) were thought to have a financial motivation behind their actions. Some hackers simply are motivated by notoriety, while others may be hacking in order to pilfer state or organizational secrets (cyberespionage). Finally, hacking or attacks in cyberspace could be to cause mass panic or disruption (cyberterrorism). A new motivation has been added to that list, which is hacking due to a particular political or social agenda. There is not much new in terms of techniques or tactics that were previously used to hack in cyber. The difference lies especially in the motivation for attacking the attack. Put simply, hacktivism is the use of computers and networks in a rebellious manner to promote a political or social agenda. Although the agenda can be social change, hacktivist activity typically comes down to free speech, human rights, or awareness about freedom of information.

Although other forms of cyberattacks can have a number of targets, hacktivists seem to focus predominantly on government entities at the local, state, or government level. The crowning achievement for hacktivism was the release of numerous classified memos and messages from the State Department to the website WikiLeaks in 2012. That incident highlights the target and the motivation for hacktivism, although several examples exist as well. For example, hacktivists were thought to be responsible for attacks on Michigan's state website in order to focus attention on the water crisis in the Michigan city of Flint. When the Michael Brown case became national news in Ferguson, Missouri, the hacking group Anonymous threatened to release the officer's identity if the police department and the state refused to do so. Other examples include targeting North Carolina's state websites because the state passed a law that said people had to use the public bathroom that matched their birth certificate gender. Another example involved a police shooting in Baton Rouge. This time hackers took their frustrations out on the city's websites as a way to voice their frustration with the system.

Hacktivism seems to be a way to voice frustration and anger over the bureaucratic system by attacking the physical and cyber systems. Since in the hacktivist mind-set, the bureaucratic system moves at a glacier pace, the hacktivist can have a more immediate impact on systems that move at the speed of light. These individuals have had success in their attempts to take on the system. Hacktivists have been able to freeze government servers and websites, release e-mails, and even access highly sensitive data. The cause sometimes goes outside the government system. Although these attacks may have good intentions, they are unlawful and considered a form of criminal trespassing. More importantly, they can deny common citizens access to government services, increase wait times, and cause increased taxes because targeted government entities consequently are compelled to invest in improved firewalls and antivirus software.

Hacktivists are also thought to have helped orchestrate the antiauthoritarian movements across North Africa and the Middle East in 2011, popularly known as the Arab Spring. During "Operation Tunisia," recruited volunteer hackers helped take down government websites, which allowed dissidents the time to organize and have a lasting impact. Hacktivists have focused their attention on corporations and unlawful individuals as well, including pedophiles. Although these attacks have had an impact, it is the recruiting and training that bares some explanation.

Most of the hacktivist attacks have been done using distributed denial-of-service (DDoS) techniques, or some have used vulnerability scanners on network systems. Both of these techniques are commonly found in dark web or underground hacking forums where tools and techniques can easily be shared. The ease of conducting the attack and the appeal of the target make hacktivism a growing form of cyberattack. It has seen rapid growth as more and more look for a way to rapidly affect the system than trying to speed up the wheels of justice.

Melvin G. Deaile

See also: Distributed Denial-of-Service (DDoS) Attack; Doxing; Social Media and the Arab Spring; WikiLeaks, Impact of

Further Reading

Cumming, Penny. "Hactivism: Will It Pose a Threat to Southeast Asia and, If So, What Are the Implications for Australia?" *Indo-Pacific Strategic Digest* (2017), 269–280.

Deseriis, Marco. "Hactivism: On the Use of Botnets in Cyberattacks." *Theory, Culture & Society* 34, no. 4 (October 2016), 131–152.

Fish, Adam, and Luca Follis. "Gagged and Doxed: Hactivism's Self-Incrimination Complex." *International Journal of Communication* 10 (2016), 3281–3300.

Howlett, William IV. *The Rise of China's Hacking Culture: Defining Chinese Hackers.* Master's thesis. San Bernardino: California State University, 2016.

Intellipedia

A collaborative online information-sharing tool established in 2005 for use by U.S. intelligence entities. Formally announced in April 2006 by the Director of National Intelligence (DNI), it operates on MediaWiki, the same software used by Wikipedia. Intellipedia is available for employees from the seventeen agencies of the intelligence community who have the appropriate security clearance to create and modify content. The users can also add their reflections on the topics and thus contribute to the building of a base of knowledge for the whole community. Intellipedia operates at three classification levels: unclassified, secret, and top secret. The idea was inspired by a 2004 paper written by a CIA officer, D. Calvin Andrus, who suggested the creation of a collaborative encyclopedia that could be allowed edits by any user. Intellipedia was then created in an effort by the intelligence community to share more information across agencies.

Intellipedia records surpassed a quarter of a million documents by the start of 2014, and the tool has received considerable praise for its value and has been considered very useful by the intelligence community. In 2009, the two CIA officers who implemented it, Sean P. Dennehy and Don Burke, won a Homeland Security Service to America Medal for the development of Intellipedia.

Karine Pontbriand

See also: Computer Network Defense (CND); National Cybersecurity Center; National Security Agency (NSA)

Further Reading

Andrus, D. Calvin. "The Wiki and the Blog: Toward a Complex Adaptive Intelligence Community." *Studies in Intelligence* 49, no. 3 (2005).

Burke, Cody. "Freeing Knowledge, Telling Secrets: Open Source Intelligence and Development." *CEWCES Research Papers,* Paper 11, 2007.

International Mobile Subscriber Identity-Catcher (IMSI-Catcher)

The International Mobile Subscriber Identity-catcher (IMSI-catcher) is a piece of electronic equipment used for spying, listening in on, or "tapping" mobile phone calls. It allows the user to monitor and intercept mobile phone data of a desired subject. They are used by law enforcement, security, and intelligence agencies all over the world to track suspects as well as locate for the purpose of search and rescue. During use the IMSI-catcher poses as an artificial mobile phone tower and communicates with the target device between an actual tower and the target device's user.

IMSI-catcher activities fit the definition of "man-in-the-middle" attacks. Use of 3G and LTE mobile networks make the use of IMSI-catchers difficult, as the authentication process is more complicated. Attacks can also be thwarted by IMSI-catcher "detector apps"; however, these might have limited effectiveness. One complication of IMSI-catchers is that when they are used, multiple IMSIs can be picked up, and the user must know which he or she intends to monitor. Due to these issues, IMSI-catchers remain controversial as they raise questions regarding citizens' right to privacy but conversely offer the opportunity to secure telecommunications.

Stephanie Marie Van Sant

See also: Crypto Phone; Encryption

Further Reading

Dabrowski, Adrian, et al. "IMSI-Catch Me If You Can: IMSI-Catcher-Catchers." *Annual Computer Security Applications Conference* (2014).

Winner, Kristi. "From Historical Cell-Site Location Information to IMSI-Catchers: Why Triggerfish Devices Do Not Trigger Fourth Amendment Protection." *Case Western Reserve Law Review* 68, no. 1 (2017), 243–274.

iSIGHT

An Apple brand webcam technology expanding the opportunity for convenient webchatting, although its lack of security magnifies the risk of surveillance targeting personal computer users. iSIGHT was first introduced in 2003 as an external item and within five years was no longer sold except as an integrated component of Apple's array of portable computers. It uses internal microphones and noise suppression capability and weighs barely 2 ounces.

Although the advantages of the iSIGHT include its capabilities as a compact and integrated camera-microphone system, these characteristics and its presence since 2008 as an integral component of Apple computers have raised cybersecurity concerns. The iSeeYou vulnerability in particular has been identified as an avenue by which a remote access tool can be used by a third party to covertly leverage the camera and microphone to record an Apple computer user. As with other vulnerabilities connected to unsecure webcams and microphones, the result is a considerable degree of susceptibility to illicit surveillance.

Nicholas Michael Sambaluk

See also: Malware; Vulnerability

Further Reading

Daley, John. "Insecure Software Is Eating the World: Promoting Cybersecurity in an Age of Ubiquitous Software-Embedded Systems." *Stanford Technology Law Review* 19, no. 533 (Spring 2016), 533–546.

Santos, Eunice. *BRI: Cyber Trust and Suspicion.* Arlington, VA: Air Force Research Laboratory, 2017.

Teng, Che-Chun, et al. "Firmware over the Air for Home Cybersecurity in the Internet of Things." *Network Operations and Management Symposium* (2017).

Klein, Mark

A whistleblower who reportedly disclosed detailed information about cooperation between the telecommunications industry and the U.S. National Security Agency (NSA) and its efforts to monitor U.S. traffic. Prior to releasing information that began a wave of journalistic coverage about NSA activities, Klein had been a technician for the AT&T for over twenty years, the first half of his career working in East Coast facilities in New York and spending the second half on the West Coast in California. Klein's revelations sparked legal actions by the Electronic Frontier Foundation (EFF) against AT&T in 2006, and the EFF celebrated Klein's actions while also highlighting its own perspective of the events by naming Klein among the recipients of its EFF Pioneer Award for 2008. The following year, Klein published a book entitled *Wiring Up the Big Brother Machine . . . and Fighting It* for which best-selling author-journalist James Bamford wrote the introduction.

Nicholas Michael Sambaluk

See also: Electronic Frontier Foundation (EFF); National Security Agency (NSA)

Further Reading

Stiennon, Richard. *There Will Be Cyberwar: How the Move to Network-Centric War Fighting Has Set the Stage for Cyberwar.* Birmingham, MI: IT-Harvest, 2015.

Toxen, Bob. "The NSA and Snowden: Securing the All-Seeing Eye." *Queue* 12, no. 3 (2014), 1–12.

Wood, D. M., and S. Wright. "Before and after Snowden." *Surveillance and Society* 13, no. 2 (2015), 132–138.

Libicki, Martin C.

(1952–), American scholar known for research on domestic and national security implications of information technology, specifically cybersecurity and cyberwar. He is the Maryellen and Richard L. Keyser Distinguished Visiting Professor in Cyber Security Studies at the U.S. Naval Academy. Since 1998 he has also worked for RAND Corporation in various capacities, including senior policy analyst, professor at the Pardee RAND Graduate School, and adjunct senior management scientist.

Libicki is the author of three published books and the author or coauthor of more than 60 monographs on topics ranging from information warfare to strategic cooperation. As a RAND researcher, he helped create post-9/11 information systems for counterinsurgency and counterterrorism through participation in several research projects for the U.S. Department of Justice, FBI, CIA, and Department of Defense. Notable projects include the development of an information technology strategy for the U.S. Department of Justice, information security analysis capabilities for Defense Advanced Research Projects Agency's (DARPA) Terrorist Information Awareness program, and assessment of the CIA's In-Q-Tel research and development project. Since 2010, Libicki has contributed to a number of projects for the U.S. Air Force focused on vulnerabilities to air superiority and cyberspace, particularly pertaining to relative Chinese capabilities. While advocating greater understanding of the role of cyberactivities in conflict, Libicki has worked to avoid what he sees as overplaying the impact of cyberwar or of cyberweapons such as the Stuxnet malware discovered in 2010.

Prior to joining RAND, Libicki spent twelve years at the National Defense University, three years on Navy staff as program sponsor for industrial preparedness, and was a policy analyst for the Government Accountability Office's Energy and Minerals Division. He has a PhD from the University of California, Berkeley (1978), and an undergraduate degree from MIT.

Melia T. Pfannenstiel

See also: Computer Network Defense (CND); Stuxnet

Further Reading

Libicki, Martin C. *Conquest in Cyberspace: National Security and Information Warfare.* New York: Cambridge University Press, 2007.

Libicki, Martin C. *Cyberdeterrence and Cyberwar.* Santa Monica, CA: RAND, 2009.

Libicki, Martin C. *Cyberspace in Peace and War.* Annapolis, MD: Naval Institute Press, 2016.

Malware

Malicious software intended to damage or undermine targeted computers, networks, or related infrastructure. Malware uses code known as exploits to take advantage of flaws, known as vulnerabilities, in a targeted system's software. During the first two decades of the 21st century, proliferating forms of malware appeared and were adapted increasingly toward economic exploitation or surveillance purposes, usually by criminal elements and sometimes infamously by entities condoned or sponsored by nation-states.

Among early malware in the 1970s and 1980s, the typical form was the Trojan (software that contained malicious code in what was ostensibly a harmless program), and the most common motivation was computer vandalism. Over time, additional forms of malware appeared, although Trojans remained particularly common and reportedly represented two-thirds of all identified malware as recently as 2011. During the early years of the 21st century, the most popular motivations for launching malware became criminal profit. By 2016, the annual economic damage to the U.S. economy through malware was believed to lie between approximately $50 and $100 billion. This was a result of criminals recognizing important factors, including that the wealth of data processing and commercial activity occurring via computers represented a massive set of potential targets, and furthermore that the standard architecture of computers and networks presupposed an environment of trust in a way that the attack surface was not only lucrative but also vulnerable.

The proliferation of malware by the early 21st century included increasing awareness of successive waves of computer viruses and worms. A virus is a self-replicating computer program, necessarily present without a user's knowing consent and potentially carrying destructive code known as a payload. Although viruses, like Trojans, actually date to the 1970s, their spread was limited by the small number of computers then in existence and also the factors involved in the propagation of early malware. Different viruses operate in various ways, and while some are written to hide a computer's infection by tampering with the computer's logs, others may seek to compromise the antivirus software that might have been installed to recognize and eliminate viruses. Viruses require user activity in order to spread. In contrast, computer worms propagate and run themselves on victims' devices, meaning that the victim does not need to take any unwary or unintentional action as is the case with a virus. Although viruses and worms can be distinguished by the processes by which they spread, different kinds of both forms of malware abound, and the diversity of their payloads means that viruses cannot be thought of as "less" or "more" dangerous relative to worms.

Concealment is particularly important for forms of malware that are intended to operate stealthily. For these, rootkits and backdoors are popular and useful features. Rootkits allow malware to gain access to a targeted operating system so that records can be manipulated in ways that hide the entry and existence of malware from the user. Rootkits can also be designed to conceal the copying and exfiltration of information being taken from a targeted system. Parallel to their physical counterpart, backdoors are methods for accessing a target by avoiding standard authentication procedures. Increased access and elevated administrative privileges are standard objectives of hackers, because greater access to information increases the ability to survey a device or network, and elevated administrator privileges facilitate manipulation of devices and systems.

Backdoors are therefore important tools for nonstate actors, but also for other entities such as governmental institutions interested in surveillance, espionage, manipulation, or deletion of targeted data. Malware is often designed to evade detection, by techniques that include mapping or "fingerprinting" targeted environments to understand their appearance prior to a malware infiltration, but also various processes that disable or confuse antivirus systems, which typically search for previously identified examples of malware so that they can be deleted or quarantined.

Because malware uses exploits against software vulnerabilities, a common approach to combating malware is to find and address vulnerabilities with software solutions called patches. This has prompted software companies to develop updates that carry patches to users. The continual process of applying patches, while important, also struck many users as inconvenient and resulted in widespread consumer reluctance to patch software; some companies such as Microsoft have reacted by forcing the adoption of patches and by ceasing to continue supporting older versions of software with new patch development.

Adopting and maintaining antimalware or antivirus software on devices is another complementing approach to security. Maintaining even the most up-to-date malware protection cannot guarantee against intrusion, however. Frequently, through efforts such as phishing, malware is unknowingly allowed into a system by a human user who does not recognize its true nature; antivirus programs may identify and warn, but they cannot prevent such intrusions from occurring. Furthermore, antivirus programs have historically been written to recognize and protect against known malware, which can be identified because they use previously identified exploits against previously discovered vulnerabilities.

Newly discovered vulnerabilities are known as zero-day vulnerabilities, for which zero-day exploits can be created to support malware that cannot be recognized or blocked by antivirus programs. Identification of zero-day vulnerabilities and development of zero-day exploits vary in cost, but those relevant to particularly popular software such as contemporary Microsoft Windows or Apple iOS operating systems reportedly run in the hundreds of thousands of dollars or higher. Malware is of interest to a variety of actors, from individuals to nation-states, and for a range of purposes, meaning that potential purchasers of zero days often involve governmental organizations and well-financed criminal groups. Interest by some governments, in buying stockpiles of zero days or of trying to persuade software creators to quietly establish backdoors through which law enforcement can operate, have resulted in considerable controversy and concerns about privacy and about the implications of built-in vulnerabilities, which might subsequently be discovered by third parties either independently or as a result of leaked government information.

Grayware refers to unwanted applications that can negatively impact computer performance without necessarily being considered malware. These include adware that bombards a user with unwanted advertisements, and in the process of operating the adware may work to incapacitate antivirus software, which would otherwise hinder its own functioning. As a result, however, the adware may have left the device unprotected and vulnerable against other attack. Spyware covertly collects and exfiltrates information about a targeted system. In support of distributed denial-of-service (DDoS) attacks, some hackers have used malware to control background operations of victim computers so that they can be formed into bots to participate in DDoS activities. A victimized computer hosting a bot may suffer slightly degraded performance while being used in a DDoS attack, but a user might not otherwise recognize that the device has suffered an intrusion.

Whereas most malware is meant to run quietly so that the intrusion can continue for an extended period, other malware deliberately announces itself. This is notably the case with ransomware, which locks a device so that the legitimate user cannot access the machine or its data unless some conditions have been met. Typically, this is a monetary payment of an extortion demand, so that the device can be unlocked when the ransomed money has been received by the hacker. Isolated cases of ransomware date to the late 1980s, but waves of ransomware attacks grew continually during the 2010s. The first half of 2018 reflected a more than twofold increase over the first half of the preceding year, which itself was marked by the start of the wave of WannaCry ransomware using the EternalBlue exploit. Forensic experts traced this to North Korea by the end of 2017, and the malware exploit was alleged to have been developed as a result of leaked information from a National Security Agency stockpile.

Likely the most infamous malware to date is Stuxnet, which was developed to target the computer systems running equipment used by Iran to enrich uranium. The probable development time frame of the malware and its operation coincided with international concern about the continuation of that country's nuclear development program and its parallel interest in the development of long-range ballistic missiles. Stuxnet was identified in 2010 after its operation was significantly accelerated, and it spread beyond the initially compromised Iranian machines, thus capturing the attention of different groups of cybersecurity experts worldwide. Stuxnet was unique in that, to date, very few examples exist of malware that actively triggered physical damage rather than degradation or inoperability of computers and electronic devices. The code appears to have been carefully written so that, even if it were introduced to a variety of machines, it would exert harm only onto the kinds of nuclear enrichment systems used by Iran. Stuxnet was also particularly unusual in its being constructed with the inclusion of an unprecedented four separate zero-day exploits. Cumulatively, this forensic information and the political context has led experts to consider Stuxnet to be the product of one or more nation-states rather than being malware developed by criminals or nongovernmental organizations. A decade after the Stuxnet actions, it remains a nearly unique form of malware in its complexity, damage to physical targets, attention to precision, and its probable cost of development.

Nicholas Michael Sambaluk

See also: Code Red Worm; Distributed Denial-of-Service (DDoS) Attack; EternalBlue Exploit; Exploit; MSBlast; SQL Slammer; Stuxnet; Vulnerability; WannaCry Virus; Zero-Day Vulnerability

Further Reading

Council of Economic Advisers. *The Cost of Malicious Cyber Activity to the US Economy.* Washington, D.C.: Government Printing Office, 2018.

Han, Rui. *Data-Driven Malware Detection Based on Dynamic Behavior Features.* Dissertation. Coral Gables, FL: University of Miami, 2017.

Libicki, Martin C. *Cyberspace in Peace and War.* Annapolis, MD: Naval Institute, 2016.

Singer, P. W., and Allan Friedman. *Cybersecurity and Cyberwar: What Everyone Needs to Know.* Oxford: Oxford University Press, 2014.

"WannaCry" Ransomware Attack: Technical Intelligence Analysis. London: Ernst & Young, 2017.

Zetter, Kim. *Countdown to Zero Day: Stuxnet and the Launch of the World's First Digital Weapon.* New York: Crown, 2014.

Mandiant

A Virginia-based cybersecurity firm founded in 2004 by former U.S. Air Force officer Kevin Mandia. The firm was originally named Red Cliff Consulting but was rebranded in 2006. Mandiant provides cybersecurity consulting for organizations with regard to incident response, computer forensics, network and application security, and training.

The firm rose to fame on February 8, 2013, when it publicly released a report on a Chinese advance persistent threat (APT) actor that it named APT1. Entitled *APT1: Exposing One of China's Cyber Espionage Units,* the report documented security breaches attributed to the People's Liberation Army (PLA) Unit 61398, based in Shanghai, and unveiled cyberattacks compromising at least 141 companies in the United States and in other countries, with the attacks going as far back as 2006. According to the report, APT1 managed to steal, over several months or years, broad categories of intellectual property including technology blueprints, proprietary manufacturing processes, business plans, etc.

The main affirmation of the report was that APT1 had been able to conduct a lasting and wide-ranging cyberespionage campaign because it was government sponsored. In effect, the report stated that the Chinese government was tasking the PLA to commit systematic cyberespionage and data theft against organizations around the world, therefore demonstrating the implication of the Chinese government in attacks targeting (although not only) several American corporations. However, the report received some criticism, notably from the Chinese government, which affirmed that cyber operations were difficult to attribute and that China was itself victim of numerous cyberattacks from the United States. The public, mainly the cybersecurity community, also accused Mandiant of exploiting fear and the difficulties of some companies for publicity.

It has been argued that this report has had an important impact on global cybersecurity policy, especially with regard to the relationship between China and the United States in cyberspace. Indeed, the Chinese cyber threat gradually gained a dominant place within the American strategic discourse around cybersecurity. In 2014, a federal grand jury in the Western District of Pennsylvania indicted five Chinese military officers, allegedly linked to Unit 61398, of cyberespionage activities. The officers were accused of computer hacking, economic espionage, and other crimes directed at six American companies across different industries. This indictment was viewed as an attempt by the U.S. government to deter China from its aggressive behavior in cyberspace.

Mandiant profited generously from the publication of the report. The firm gained notoriety and generated new clients by raising awareness about cyber threats. On December 30, 2013, Mandiant was purchased by the larger cybersecurity firm FireEye, based in California, in a deal worth more than $1 billion.

Karine Pontbriand

See also: Advanced Persistent Threat; APT1 Report; People's Liberation Army Unit 61398

Further Reading

Mandiant. *APT1: Exposing One of China's Cyber Espionage Units,* 2013.

Singer, Peter W., and Allan Friedman. *Cybersecurity and Cyberwar: What Everyone Needs to Know.* Oxford: Oxford University Press, 2014.

Springer, Paul J. *Cyber Warfare: A Reference Handbook.* Santa Barbara, CA: ABC-CLIO, 2015.

Manning, Chelsea/Manning, Bradley

Manning (1987–), a former private first class in the U.S. Army who downloaded terabytes of classified U.S. military reports and files, as well as more than 250,000 State Department cables from the Secret Internet Protocol Router Network (SIPRNet) onto CD-ROM discs, which were disseminated through Julian Assange's WikiLeaks website.

While serving in Iraq in 2010, Manning's work in intelligence provided him access to classified materials. Correspondence with hackers during his deployment through his online handle "bradass87" indicated a growing hostility toward American foreign policy, compounded by a personal struggle with gender dysphoria and an impending discharge from the U.S. Army. After classified materials were published through WikiLeaks and major news outlets, Manning used instant messaging to confide in a hacker called Adrian Lamo, who turned Manning in to the FBI. Manning was court martialed on multiple charges under the Uniform Code of Military Justice, including providing aid to the enemy (UCMJ, Article 104) and violations of the Espionage Act (UCMJ, Article 134).

Manning pled guilty in 2013 to ten of the charges while denying having aided the enemy, a potential capital offense. Prosecutors pursued the additional charges without seeking the death penalty. Manning was sentenced in August 2013 to thirty-five years imprisonment, reduction in rank, forfeiture of pay, and a dishonorable discharge. Days after sentencing, Manning petitioned for a legal name change from Bradley Edward Manning to Chelsea Elizabeth Manning, which was granted in April 2014. On January 17, 2017, President Barack Obama commuted Manning's sentence to seven years, and Manning was released from Fort Leavenworth federal prison on May 17, 2017.

Melia T. Pfannenstiel

See also: Assange, Julian; Snowden, Edward; WikiLeaks, Impact of

Further Reading

Sheldon, John. "Toward a Theory of Cyber Power." In *Cyberspace and National Security,* ed. Derek Reveron. Washington, D.C.: Georgetown University Press, 2012, 207–224.

Singer, P. W., and Allan Friedman. *Cybersecurity and Cyberwar: What Everyone Needs to Know.* Oxford: Oxford University Press, 2014.

U.S. Library of Congress, Congressional Research Service, *Criminal Prohibitions on Leaks and Other Disclosures of Classified Defense Information*, by Stephen P. Mulligan and Jennifer Elsea, R41404 (March 7, 2017), 21–22.

MonsterMind

A program reported to have been designed by the National Security Agency (NSA) as both a deterrent and an automated cyberweapon. In the fall of 2014, a year after his initial disclosures, Edward Snowden met with James Bamford in Moscow and disclosed that the program provided his motivation to leave the NSA. Described as a "digital Star Wars missile defense system" or as a "hackback," the program is intended to have algorithms constantly analyzing metadata in order to detect malicious or unusual patterns and respond to them with a "kill" of the denial of service or worm attack as well as launch a retaliatory counterattack. Through this program, monitoring and response would be automated and out of human hands. The United States and other nations have predetermined and preplanned responses to cyber and kinetic attacks, but analysts conclude that MonsterMind is international conflict on a tripwire.

Although similar to the Einstein 2 and Einstein 3 programs, which monitor government systems for attacks, and the ADAMS (Anomaly Detection at Multiple Scale) program of the Defense Advanced Research Projects Administration (DARPA), Snowden was opposed to MonsterMind on the grounds that it would be possible for an attacker to fool it using an innocent proxy, so that the automatic retaliation by the United States might start a conflict with an uninvolved country, or cause collateral damage, most often citing as an example a cyberattack on a hypothetical Russian hospital. A hostile actor is likely to use a botnet, taking control of and using other's computers to launch the attacks, making it likely that the retaliation will cut off or harm unwitting people and businesses. Additionally, in order to analyze the scale of metadata necessary, the program would need to access all traffic flows, raising questions of privacy, Fourth Amendment rights, and the soundness of the algorithms used to identify patterns. Defenders of the program argue that its existence is a deterrent to potential bad actors, and that the NSA's algorithms are sophisticated enough to avoid targeting innocent parties, and that scanning that volume of traffic is necessary to defend the United States.

Margaret Sankey

See also: Defense Advanced Research Projects Agency (DARPA); National Security Agency (NSA); Snowden, Edward

Further Reading

Bamford, James. "Edward Snowden: The Untold Story." *Wired*, August 2014, http://hiveware.com/doc/Edward_Snowden_The_Untold_Story.pdf.

Pasquinelli, Matteo. "Anomaly Detection: The Mathematization of the Abnormal in the Metadata Society." Presented at the Transmediale, Berlin, January 2015.

Templeton, Graham. "Snowden Went Too Far by Revealing the NSA's MonsterMind Cyber Weapon." *ExtremeTech,* August 14, 2014, https://www.extremetech.com/extreme/187992-snowden-went-too-far-by-revealing-the-nsas-monstermind-cyber-weapon.

Zetter, Kim. "Meet Monstermind, the NSA Bot That Could Wage Cyberwar Autonomously." *Wired*, August 13, 2014, https://www.wired.com/2014/08/nsa-monster mind-cyberwarfare.

MSBlast

A worm created through the reverse engineering of a Microsoft software patch that posed problems for Microsoft users throughout the second half of 2003. Known to have impacted at least 100,000 computers worldwide, estimates have ranged to its infecting one hundred times that many computers. The vulnerability was identified by Polish cybersecurity experts, and a patch was developed and released in late May 2003. However, hackers believed to be in the People's Republic of China reverse-engineered the patch to create the MSBlast worm within weeks.

Although concerted monitoring by Internet service providers and Microsoft's own release of a software repair tool in January 2004 curbed the worm's impact, derivative malware continued to appear throughout the entire period in which MSBlast itself was being contained. At least one of the subsequent worms was crafted by a U.S. teenager, who was prosecuted. The absence of internal firewalls meant that, once the worm compromised a single machine on a business or university network, it quickly spread to the network's other computers. One of the telltale signs of the worm's infecting a machine was a message on the affected monitor chiding Microsoft founder Bill Gates to "stop making money and fix your software!" The experience pointed to the shortcomings of operating systems and other software that required continual patching. The advent of MSBlast, like the emergence of Code Red and Nimda two years earlier, marked another sign of malware creation abilities emanating from the People's Republic of China.

Nicholas Michael Sambaluk

See also: Code Red Worm; Exploit; Malware; Nimda; Vulnerability

Further Reading

Bailey, Michael, et al. "The Blaster Worm: Then and Now." *IEEE Privacy and Security* (July/August 2005), 26–31.

Hoogstraten, John van. *Blasting Windows: An Analysis of the W32/Blaster Worm*. Bethesda, MD: SANS Institute, 2003.

Wagner, Arno, and Bernhard Plattner. "Flow-Data Compressibility Changes during Internet Worm Outbreaks." Zurich, Switzerland: Swiss Federal Institute of Technology, 2009.

Multinational Cyber Defense Capability Development (MN CD2)

A Dutch-led cyber defense development initiative supported by five North Atlantic Treaty Organization (NATO) member states (Canada, Denmark, the Netherlands, Norway, and Romania) and undertaken by the NATO Communications and Information Agency. The group officially inaugurated the effort in March 2013, prompted by a shared awareness among participating countries of the prevailing opinion that, unique to the cyber domain, defense is a more complex and expensive challenge than offense. Members pool their resources to help address the problem of cost in developing credible capability in defending against cyberattack, although participants have noted that interoperability is another intended benefit anticipated through the cooperative approach.

Two years into its existence, MN CD2 expanded its attention to supporting and assessing progress on cyber defense capabilities. Also complementing MN CD2 is a malware information-sharing plan and a cyber defense education and training effort by NATO members. In a sense, these efforts extend from earlier cooperative efforts, such as the nonbinding international law study by some NATO members of the *Tallinn Manual*. MN CD2 aims to help NATO members build more robust cyber defense capabilities and to gauge progress. Semiautomated response reportedly represents a notable avenue of MN CD2 effort, given the ease with which offensive actions can be launched and the risk of manual defense efforts becoming inundated and overwhelmed during a crisis.

Nicholas Michael Sambaluk

See also: Computer Network Defense (CND); *Tallinn Manual*

Further Reading

Caton, Jeffrey L. *NATO Cyberspace Capability: A Strategic and Operational Evolution.* Carlisle, PA: Strategic Studies Institute, 2016.

Jordan, Frederic, and Gier Hallingstad. "Towards Multi-National Capability Development in Cyber Defense." *Information & Security* (2011), 81–89.

Pintat, Xavier. *From Smart Defense to Strategic Defense: Pooling and Sharing from the Start.* Brussels, Belgium: NATO Parliamentary Assembly, 2013.

National Cybersecurity Center

A subsidiary of the U.S. Department of Homeland Security (DHS) that was created in March 2008 under the Obama administration via National Security Presidential Directive 54 and Homeland Security Presidential Directive 23 (NSPD-54/HSPD-23). It is tasked with protecting the nation's cyber infrastructure from attack. Located in Colorado Springs, Colorado, the appointment of entrepreneur and civilian technology expert Rod Beckstrom as its first director signaled that it would be a public-private partnership and clearinghouse for more enforcement and intelligence-oriented agencies like the DoD, FBI, and NSA. However, Beckstrom's appointment was followed by his resignation within months.

Although the organization hosts a monthly meeting of directors of federal cyber centers and operates an ongoing cyber event conference call to respond to acute threats, the center was not mentioned in the June 2009 Cyber Security Policy Review. Reorganization within DHS in the fall of 2009 entailed the conglomeration of the U.S. Computer Emergency Readiness Team and of a national telecommunications coordinating office into a new National Cybersecurity and Communications Integration Center (NCCIC); the reorganization also reinforced connections between the National Cybersecurity Center and the DHS Office of Intelligence and Analysis, as well as with external DHS partners.

With the support of the state of Colorado, especially Governor John Hickenlooper, the center has received substantial financial underwriting and physical location space. In April 2011, it began an initiative to implement a National Strategy for Trusted Identities in Cyberspace, exploring cost-effective and business-friendly methods to foil identity theft and exploitation of cyber vulnerabilities, offering grants to pilot programs in private business, state, and local government entities. As it currently exists, the center focuses on raising awareness of cyber issues and policy implications, creating jobs in the cybersecurity field and providing workforce development, often in partnership with the University of Colorado, Colorado College, Pike's Peak Community College, and the U.S. Air Force Academy. The center hosts a variety of events, including conferences and workshops, aimed at engaging with business leaders. The National Cybersecurity Center thus works to be an economic engine for the region as well as serving as a national resource for cybersecurity.

Margaret Sankey

See also: Computer Emergency Response Team (CERT); Computer Network Defense (CND)

Further Reading

Harknett, Richard, and James Stever. "The New Policy World of Cybersecurity." *Public Administration Review* 71, no. 3 (June 2011), 455–460.

Heilman, Wayne. "National Cybersecurity Center Opens in Colorado." *Government Technology*, January 22, 2018, http://www.govtech.com/security/National-Cybersecurity-Center-Opens-in-Colorado.html.

"National Cybersecurity Center." Accessed October 22, 2018. https://cyber-center.org.

National Security Agency (NSA)

The National Security Agency (NSA) is an intelligence agency of the Department of Defense that was created on November 4, 1952, to conduct data collection and monitoring for intelligence and counterintelligence purposes. Although it maintains staff in various countries, unlike the Central Intelligence Agency (CIA), the NSA does not perform its data gathering using foreign human espionage, but specializes in signal intelligence (SIGINT) and surveillance of electronic media such as the Internet communications, telephone calls, and radio signals of various organizations and individuals. As of 2013, the NSA has been organized into many directorates that are designated by an alphabetic character, including (among many others): F directorate, with activities covering clandestine electronic monitoring throughout the world, conducted in cooperation with the CIA and NSA; G directorate, with activities that include management of platforms like the U-2 spy plane; I directorate, with activities for protecting the nonrepudiation, confidentiality, and integrity of national security and telecommunication systems; J directorate, cryptological activities; Q directorate, with security and counterintelligence responsibilities; R directorate for research; and S directorate for signal intelligence, including intercepts from satellites.

In addition to foreign intelligence gathering, the NSA (the largest U.S. intelligence agency) is tasked with the protection of U.S. communication networks and information systems. Following the terrorist attacks on September 11, 2001, the Patriot Act was passed, which provided new powers to the NSA in its responsibilities to combat domestic terrorism and conduct necessary surveillance. In 2004, new authorization was provided for mass surveillance of Internet records and e-mails. The agency has been reported to eavesdrop on domestic communications including (according to a 2010 article in *The Washington Post*) the collection and storage of 1.7 billion e-mails and other communications each day. The extent of the controversial activity was revealed largely through the unauthorized disclosures of a former NSA contractor, Edward Snowden.

The gathering of data in an attempt to stop terrorists both within and outside the country has been highly controversial. In December 16, 2005, *The New York Times* reported that the NSA had been intercepting phone calls made to people outside the country without obtaining warrants from the U.S. Foreign Intelligence Surveillance Court, under the Foreign Intelligence Surveillance Act (FISA). FISA was passed in 1978 during the tenure of President James E. Carter, and provided that the president may authorize, through the attorney general, electronic surveillance without a court order for the period of one year, provided that it is only to acquire foreign intelligence information. However, proponents of the surveillance program argue that the president's constitutional powers override the limitations of FISA, especially when the rules are applied to government equipment. One surveillance program that was directed by President Bush included the NSA's relaying of conversations, including cell phone calls, from monitoring stations to Army Signal Intelligence Officers; the calls included conversations of both U.S. citizens and foreign nationals.

Section 215 of the Patriot Act had been interpreted to provide the NSA, as the agent of the administration, to gather data on a range of communications, including domestic telephone calls. In 2013 the NSA obtained a court order requiring Verizon to provide "metadata" (such as the numbers of both parties on a call, location data, call duration, unique identifiers, and the time and duration of all calls) for a three-month period. Verizon had more than 120 million U.S. subscribers. The government's interpretation of the Patriot Act provisions asserted that the entirety of U.S. communications may be considered "relevant" to a terrorism investigation if it is expected that even a tiny minority may relate to terrorism. The value of the NSA's massive data collection is undetermined, and as may be expected, there have been a number of lawsuits over the years challenging the NSA and the administration, including *Hamdan v. Rumsfeld, American Civil Liberties Union v. NSA,* the *Center for Constitutional Rights v. Bush,* and *Electronic Frontier Foundation (EFF) v. NSA.* In 2015 the courts ruled that the interpretation of Section 215 of the Patriot Act was wrong and that the NSA program that has been collecting Americans' phone records in bulk was not lawful. The ruling provided that Section 215 could be clearly interpreted to allow government to collect national phone data. Congress's replacement law, the USA Freedom Act, enables the NSA to continue to have bulk access to citizens' metadata but with the stipulation that the data will now be stored by the companies themselves. Telecommunications companies maintain bulk user metadata on their servers for at least 18 months, and will be provided to the NSA upon the agency's request. The ruling did not impact on other NSA activities that include techniques used by the NSA to collect and store Americans' communications or data directly from the Internet.

NSA director general Keith B. Alexander stated on June 27, 2013, that the NSA's bulk phone and Internet intercepts had been instrumental in preventing fifty-four terrorist "events," including thirteen in the United States. Although officials conceded to the Senate that these intercepts had not been vital in stopping any terrorist attacks, they were very important in identifying and convicting four San Diego men for sending funds to terrorists in Somalia. The NSA was reportedly unable to assist the Federal Bureau of Investigation (FBI) during February 2016 when the FBI sought to unlock an iPhone that had been used by a domestic terrorist prior to his assault on a county building in San Bernardino, California, in December 2015. Since the founding of the U.S. Cyber Command (USCYBERCOM) in June 2009, the director of the NSA has also led USCYBERCOM as well. The latter is a unified command of the U.S. Department of Defense organized to strengthen the country's military cyberspace capabilities.

Nicholas W. Sambaluk

See also: Electronic Frontier Foundation (EFF); Encryption; Government Communications Headquarters (GCHQ); Rogers, Michael S.; Snowden, Edward; USCYBERCOM

Further Reading

Hayden, Michael V. *Playing to the Edge: American Intelligence in the Age of Terror.* New York: Penguin, 2016.

Kaplan, Fred. *Dark Territory: The Secret History of Cyber War.* New York: Simon & Schuster, 2016.

Singer, P. W., and Allan Friedman. *Cybersecurity and Cyberwar: What Everyone Needs to Know.* Oxford: Oxford University Press, 2014.

Nimda

A multivector computer worm unleased at the start of the 21st century and targeting computers using a variety of then popular operating systems, including Windows 95, 98, and XP. Nimda derives its name from the reversal of the letters for "admin." Although the timing of its appearance led to initial concern that it was derived by nonstate terrorists, it was instead a sophisticated product of a leading state in the cyber domain.

As a multivector malware, Nimda proved especially rampant because it was able to spread through several contemporarily popular methods, including by e-mail and compromised webpages as well as via network shares. Rather than relying on zero-day vulnerabilities, Nimda frequently made use of vulnerabilities for which patches already existed but which had not been adopted by large numbers of computer users. Previous forms of malware had typically relied on a single vector, which had considerably simplified the efforts involved in protection. The use of multiple vectors meant that Nimda persisted for several months before it was finally eliminated.

The first public warnings about the Nimda worm occurred in the immediate days following Al Qaeda's coordinated terrorist attacks against the World Trade Center and the Pentagon in September 2001. Within less than half an hour, Nimda set a new record as the fastest-spreading malware up to that time. The economic impact was estimated to be in the billions of dollars, due in large part to the impact the malware directly exerted on the financial sector. Speculation soon followed that Nimda had been a digital component in the terrorist campaign targeting the United States, but this concern was subsequently replaced by forensic signs identified by the information security company F-Secure, which pointed to the People's Republic of China as a state creator of the malware.

The acute concern prompted by the worm notwithstanding, analysts later reported that "the Internet community was better trained to stop the spread of worms with the prior outbreaks of Code Red and Nimda" (Sullivan, 2005, 13) as a result of the attention and effort that Nimda had caused in response.

Nicholas Michael Sambaluk

See also: Advanced Persistent Threat; Code Red Worm; Exploit; Malware; Vulnerability

Further Reading

Halle, Ann M. *Cyberpower as a Coercive Instrument.* Maxwell AFB, AL: School of Advanced Air and Space Studies, 2009.

Mell, Peter, and Miles C. Tracy. *Procedures for Handling Security Patches: Recommendations of National Institute of Standards and Technology.* Gaithersburg, MD: National Institute of Standards and Technology, 2002.

Shaw, Darryl S. *Cyberspace: What Senior Military Leaders Need to Know.* Carlisle Barracks, PA: U.S. Army War College, 2010.

Sullivan, Matthew W. *National Security Agency (NSA) Systems and Network Attack Center (SNAC) Security Guides versus Known Worms.* Thesis. Wright-Patterson AFB, OH: Air Force Institute of Technology, 2005.

NIPRNet

Created in the 1980s by the Defense Information Systems Agency (DISA) as a replacement for the MILNET functions of the original ARPANET computer network sponsored by the Defense Advanced Research Projects Agency, the Nonsecure Internet Protocol Router Network (NIPRNet), often called the "nipper" net, is the U.S. Department of Defense's private Internet protocol network for protected and often encrypted, but not secure, functions. NIPRNet, which is the largest private network in the world, handles military personnel, financial, scheduling, communications, and educational functions on which the Department of Defense relies for day-to-day operations. Encryption is generally deemed beneficial in the context of these unclassified communications. Classified and secure information is held and accessed on the Secure Internet Protocol Router Network (SIPRNet), which is a far more restrictive network that can be accessed only through dedicated terminals that are themselves located in secured areas.

The rapid and continuing expansion of computerization across society is paralleled by enormous increased use of computers and networks by the armed forces, resulting both in significantly increased traffic and additionally in substantially larger numbers of devices within the network. Responding to this situation, the Department of Defense in 2010 launched a ten million dollar effort to thoroughly map the NIPRNet in order to catalog the more than ten million devices using it at that point. This would also identify the presence of unauthorized users.

The military faces problems similar to those of civilian workplaces, with employees regularly using work terminals and Internet access to pursue personal entertainment, conduct business, and use social media, although with the added complication of these being done over a restricted private network, into which contact with the open Internet can introduce viruses and other malicious threats. Further, the restrictions and firewalls in place on NIPRNet make it difficult for the military to install and use the full functionality of commercial, off-the-shelf software programs for everyday operations and Internet usage consistent with mission requirements. Characterized as the "soft, chewy outside" of the military's presence online, NIPRNet's functioning remains a balance between the necessity of protecting sensitive data and the collaborative and open structure of the original ARPANET and World Wide Web today.

Margaret Sankey

See also: Computer Network Defense (CND); SIPRNet; U.S. Military Use of Social Media

Further Reading

Grant, Rebecca. "Battling the Phantom Menace." *Air Force Magazine*, April 2010.

"Information Warfare: Lost in the NIPRNET." Strategy Page, January 23, 2010, https://www.strategypage.com/htmw/htiw/articles/20100123.aspx.

Kastenberg, Joshua. "Changing the Paradigm of Internet Access from Government Information Systems: A Solution to the Need for the DOD to Take Time-Sensitive Action on the Niprnet." *Air Force Law Review* 64 (2009), 175–210.

Koelsch, Bernard. *Solving the Cross-Domain Conundrum*. Carlisle, PA: Army War College, 2013.

Nth Order Attacks

Nth order attacks are the deliberate targeting of ancillary systems in an attempt to degrade, divert, or disable more vital or more secure systems and undermine confidence in their functioning. In the case of cyberattacks, a bad actor may choose to aim at a vulnerable part of the overall target, for example, scrambling the easy-to-access schedule for garbage pickup, rather than manipulating the hardened traffic-signal system in the process of attacking a city's population. Both result in angry, frustrated people with a reduced confidence in their leadership and increased dissatisfaction with city services, accomplishing the attacker's goals with less risk. A key example of an Nth order attack was the Russian-originated 2007 assaults on Estonian cyber infrastructure and consumer-directed programs. Nth order attacks leverage the necessity of easily accessed systems to the functioning of a free and open society by exploiting those systems and forcing the operators of them to restrict access or spend money and effort to secure them, making them a favored tactic in asymmetric warfare and strategies of systemic disruption.

Margaret Sankey

See also: Computer Network Attack (CND); Distributed Denial-of-Service (DDoS) Attack; Estonia, 2007 Cyber Assault on; Malware

Further Reading

Bilar, Daniel. "On Nth Order Attacks." Presented at the 4th International Conference on Cyber Conflict, Tallinn, Estonia, June 2012, http://www.ccdcoe.eu/sites/default/files/multimedia/pdf/CyCon_2012_Proceedings_0.pdf#page=169.

Czosseck, Christian, and Kenneth Geers, eds. *The Virtual Battlefield: Perspectives on Cyber Warfare.* Amsterdam: IOS Press, 2009.

Oghab 2

Iranian counterespionage entity established in December 2005 and charged with the protection of the country's nuclear facilities. Oghab 2 may be nested under the Ministry of Intelligence and Security (MOIS). Oghab 2 was reportedly organized after a pair of foreign intelligence agents were captured at secret nuclear development sites in 2005. Iran's nuclear program could be potentially damaged via several alternative threat vectors, including physical sabotage and the cyber domain as well as kinetic strikes, and the agency was therefore tasked with preventing a very broad set of possible attacks and intrusions.

The founding of Oghab 2 in late 2005 also happens to coincide with the time frame in which experts estimate work began on the creation of the sophisticated Stuxnet malware that would be deployed against Iranian nuclear enrichment technologies. Stuxnet and a family of related software eluded detection and wrought carefully calibrated damage of the gas centrifuges needed for enrichment as well as causing wastage of the gaseous form of uranium the Iranian program sought to enrich. Enrichment employees noticed the material attrition but were unable to pinpoint the cause. Stuxnet was conclusively identified only in the spring of 2010, indicating that the enormous care taken by its creators had succeeded in delaying the malware's discovery. Although the exact extent of the damage to Iran's nuclear program is a subject for debate, the long deployment of Stuxnet apparently caused significant consternation in Iran's counterespionage community.

By the time that Stuxnet was discovered, the organization had already suffered leadership turnover due to the uncovering of a suspected spy ring in Iran in 2007. The agency's new leadership oversaw an expansion of the organization's staff to an estimated 10,000 agents but was unable to stem an extended campaign of assassinations of Iranian nuclear experts that analysts believe to be the result of Israeli Mossad activities. In this context, the revelation that lagging centrifuge performance had been the result of a major security breach was a further humiliation for Oghab 2.

In mid-2011, Iran reportedly created a cyber unit for the dual purpose of parrying future digital realm efforts to damage the country's nuclear program and also to launch cyberattacks beyond the county's borders. Some have suggested that the Shamoon malware that impacted 30,000 computers at Saudi Aramco and Qatar's RasGas in August 2012 originated in Iran, and thus conceivably from Oghab 2 or its new cyber-oriented counterpart.

Nicholas W. Sambaluk

See also: Gas Centrifuge; Operation Olympic Games; Shamoon; Stuxnet

Further Reading

Anderson, Collin, and Karim Sadiadpour. *Iran's Cyber Threat: Espionage, Sabotage, and Revenge.* Washington, D.C.: Carnegie Endowment for Peace, 2018.

Library of Congress. *Iran's Ministry of Intelligence and Security: A Profile.* Washington, D.C.: Federal Research Division, 2012.

Zetter, Kim. *Countdown to Zero Day: Stuxnet and the Launch of the World's First Digital Weapon.* New York: Crown, 2014.

The Onion Router (TOR)

Popularly known by the acronym "TOR," a network of digital relays that allows users to browse the Internet anonymously. Individuals often use TOR to avoid censorship or monitoring and to keep web communication private. TOR software is free to the public, and volunteers support the network. The software does not mask the fact that someone is using TOR, but it does hide the user's information. Some domains, such as Wikipedia, may block a TOR user's ability to edit information on their site. These restrictions do not reveal the user's identity.

When a computer accesses the Internet using this software, TOR adds layers of encryption to the user's data and sends it through a series of randomly selected virtual relays. Each relay decodes a layer of data to discover where to direct the information next, like peeling an onion, and the final relay translates the last of the encryption. The data then arrives at its destination with no trace of its path or origin, making the source of the information appear to be the final relay instead of the actual sender. This rerouting of information through layers of encryption is what gives The Onion Router its name.

Development of "onion routing" began in the mid-1990s as a way to safeguard U.S. intelligence communications. The first, or "alpha," version of the TOR software launched in 2002, and the beta version came out in 2004. Organizations like the Naval Research Laboratory and the Electronic Frontier Foundation (EFF) provided support and development. The TOR Project officially opened in 2006, and much of the project's funding thereafter has come from the U.S. government.

Individuals take advantage of TOR's features for a variety of activities. One of the most common legal applications of TOR in the United States is protecting victims of domestic violence and the social workers supporting them. News organizations commonly use TOR to shield the identity of sources and whistleblowers, while some people merely want to avoid advertisements and cyber spying.

This software also has use for illegal conduct such as bank fraud, money laundering, leaking of classified or sensitive documents, identity theft, and other activities. In conjunction with services like Bitcoin and Silk Road, TOR has earned a reputation as a dark corner of the web where illicit sales of counterfeit currency, controlled substances, sex workers, weapons, and other contraband occur regularly. Those wishing to evade online censorship use TOR to bypass Internet restrictions and monitoring in many countries.

Despite its history spanning a quarter century, TOR remains a developing technology. Agencies within the U.S. government, as well as those from other countries, often attack the service and try to decode the encryption when combating illegal activities utilizing the technology. Cracking TOR security has also been a pursuit of academic research. The TOR Project welcomes such tests and continues to improve its software and make the browser available to more people around the world.

James Tindle

See also: Electronic Frontier Foundation (EFF); Encryption

Further Reading

Bartlett, Jamie. *The Dark Net: Inside the Digital Underworld.* Brooklyn, NY: Melville, 2014.

Komal, Nakil, and Sonkar Shriniwas. "A New Approach towards The Onion Router Network Using an Attack Dependent on Cell-Counting." *International Journal of Emerging Technology and Advanced Engineering* 3, no. 7 (July 2013), 500–507.

McCoy, Damon, et al. *Shining Light in Dark Places: Understanding the Tor Network.* Research paper. Boulder, CO, and Seattle, WA, 2009.

Operation Ababil

A cyberattack on American financial institutions occurring in successive stages, beginning September 18, 2012, and ending July 23, 2013. A secretive group called the Izz Ad-Din Al Qassam Cyber Fighters claimed responsibility for the attacks. The group's chosen name is also that of an early-20th-century Syrian Islamic scholar who participated in resistance movements against French, British, and Jewish-Zionist presence in the Levant. Throughout all phases of the operation, the Qassam Cyber Fighters (QCF) communicated threats and demands to the American government via the online text storage site Pastebin .com. QCF consistently claimed that its attacks were in response to the YouTube posting of an anti-Islamic short film *Innocence of Muslims.* QCF demanded that the U.S. government remove the video. YouTube maintained that the video did not violate its guidelines and did not remove it except in some Middle Eastern nations.

Operation Ababil targeted major U.S. financial institutions including the New York Stock Exchange, Bank of America, JP Morgan Chase and Associates, Wells Fargo and Company, and HSBC. The method of attack was distributed denial of service (DDoS), aiming to overwhelm an online service with artificially massed traffic in order to trigger temporary collapse of the targeted site. Such attacks either use volunteers to click on links designed to bombard the target, or they employ botnets (networks of virus-infected computers controlled by hackers). The specific vehicle used by QCF is known as "Brobot." QCF's hacking techniques were effective enough to disrupt major banks' online customer services, preventing users from accessing their accounts for short periods. The attacks were successful despite QCF's prior warnings on Pastebin.

Operation Ababil's initial phase lasted from September 18, 2012, to October 23, 2012, when it was suspended for the Eid al-Hadha holiday, according to QCF's Pastebin page. The group announced a new round of attacks on December 10, 2012, which it suspended January 29, 2013, while continuing to issue warnings. A third phase lasted from March 6 to May 6, as QCF continued to urge the U.S. government to completely remove all vestiges of the film from the Internet. A final limited attack occurred July 23, 2013.

QCF claimed to be composed of young volunteers from throughout the Muslim Middle East. Some American officials argued that the attacks emanated from the Iranian government in retaliation for economic sanctions imposed by the United States. Senator Joseph Lieberman presented this idea on C-SPAN in late September 2012. On October 11, 2012, U.S. secretary of defense Leon Panetta asserted that Iran was actively undertaking efforts at cyber warfare. Both the Iranian government and QCF denied Iranian involvement. In 2016, the U.S. Department of Justice (DOJ) charged seven Iranians in connection with the attacks and linked the suspects to the Iranian government. The DOJ also announced that 95 percent of the botnets associated with Operation Ababil had been neutralized.

Daniel Campbell

See also: Distributed Denial-of-Service (DDoS) Attack

Further Reading

Guiton, Clement. *Inside the Enemy's Computer: Identifying Cyber Attackers.* Oxford: Oxford University Press, 2017.

Mazanec, Brian M. *The Evolution of Cyber War: International Norms for Emerging Technology Weapons.* Lincoln, NE: Potomac Books (University of Nebraska Press), 2015.

Mazanec, Brian M., and Bradley A. Thayer. *Deterring Cyber Warfare: Bolstering Strategic Stability in Cyberspace.* New York: Palgrave Macmillan, 2015.

Radzowill, Yaroslav. *Cyber-Attacks and the Exploitable Imperfections of International Law.* Leiden, The Netherlands: Brill, 2015.

Operation Aurora

An extended intrusion by Chinese hackers targeting dozens of organizations including several second-tier defense industry companies that provide components as subcontractors. The intrusion, which was first announced by Google on January 12, 2010, appears to have begun in mid-2009 and continued at least through the end of that year, with reports of further incidents occurring as late as February 2010.

The action itself was dubbed Operation Aurora by Dmitri Alperovitch, then a senior executive for the cybersecurity company McAfee, in reference to part of its code. Experts at Symantec traced the activities to a group that they in turn referred to as "Elderwood" because of a source code variable that the attackers had used. The Elderwood group has been connected with the cyber domain espionage activities of the People's Republic of China (PRC).

Operation Aurora used spear phishing to lure unsuspecting targeted workers at Google to follow an e-mailed link to a website that contained malicious JavaScript code carrying an exploit called Trojan.Hydraq. This utilized a zero-day vulnerability impacting several contemporary versions of the popular Internet Explorer Browser (6, 7, and 8) on machines running Windows 7, Vista, Windows XP, and other software. Hackers then used a remote access tool (RAT) to collect information about the user's activities and files, and hackers proceeded to send e-mail messages to further victims in order to expand the impact of the exploit.

Experts noted the sophistication of the attack, which not only utilized an exploit against a zero-day vulnerability but also allowed attackers to access Google source code. This created the opportunity to create fresh vulnerabilities to espionage among customers and partners of the victim system based on their trust of the source code. Reportedly, the potentially compromised source code involved the Gaia system that was built to enable users to consolidate access to several Google services through a single password.

The array of targets ranged widely, although experts came to identify patterns. Corporations involved in the defense, electronics, energy, transportation, and manufacturing industries were all represented among the victims. Software company Adobe Systems, aerospace corporation Northrump Grumman, the financial firm Morgan Stanley, Dow Chemical, and Yahoo were among the more prominent. Three dozen of the companies targeted are headquartered in the United States, indicating that industrial espionage and economic motivations played an important part in the making of the operation. Many more companies were also victimized, numbering perhaps in the thousands.

However, other targets point to a more diversified set of objectives. Activists working on civil rights issues in the PRC appeared to also be prime targets. This complemented the unearthing of technical evidence including domain names, IP addresses, and malware signatures that associated Operation Aurora with the activities of Elderwood.

Experts have furthermore suggested a link between Operation Aurora and another large hacking effort, called Operation Shady RAT, which dates to 2006, reflecting the activities of an advanced persistent threat. Specifically, researchers from the security firm VeriSign connected IP addresses used by the instigators of Operation Aurora to those used in distributed denial-of-service (DDoS) attacks against U.S. and South Korean targets from the time that Operation Aurora itself appears to have begun. VeriSign researchers identified patterns between the two actions and deduced linkage between the actors in both sets of activities.

Some experts have asserted linkage between the hacks and some of the leading centers of computer science education in the PRC. In particular, suspicion has been directed at the Lanxiang Vocational School, whose officials have denied any such linkage with malicious action in the cyber domain and who have even rejected the assertion that they possess the technical sophistication prerequisite in the launching of such an attack that exploits a zero-day vulnerability and accesses a victim's source code. Other experts outside China have suggested a relationship between Operation Aurora and Shanghai-based Jiaotong University. Students there have demonstrated considerable prowess in computer technology dating at least to IBM-sponsored computer competitions in the late 1990s, and Jiaotong University is located in the same city as People's Liberation Army (PLA) Unit 61398, which experts have linked with a variety of cyberespionage activities. Other cyber professionals, such as those at HBGary, have argued that evidence linking the malware to Chinese language coders does not necessarily implicate the PRC government itself.

Operation Aurora appears to have come to a surprise to many. Spontaneous but anonymous expressions of support for Google occurred in Beijing, although these were quickly condemned by the government there as illegal. No official government response has emerged in PRC. Prior to the discovery of Operation Aurora, Google had agreed to comply with constrictive rules PRC had set in the formation of the Google.cn search engine for the country. Following the announcement of the espionage effort, Google reversed this policy of cooperating with rules set by the PRC government. PRC official sources have in turn suggested that Google's reaction was directed by the U.S. government.

Nicholas Michael Sambaluk

See also: Advanced Persistent Threat; GhostNet; Operation Byzantine Hades; Operation Shady RAT; People's Liberation Army Unit 61398; Zero-Day Vulnerability

Further Reading

Fritz, Jason R. *China's Cyber Warfare: The Evolution of Strategic Doctrine.* Lanham, MD: Lexington, 2017.

Inkster, Nigel. *China's Cyber Power.* New York: Routledge, 2016.

Lindsay, Jon, Tai Ming Cheung, and Derek Reveron. *China and Cybersecurity: Espionage, Strategy, and Politics in the Digital Domain.* Oxford: Oxford University Press, 2015.

Rid, Thomas. *Cyber War Will Not Take Place.* Oxford: Oxford University Press, 2013.

Rosenzweig, Paul. *Cyber Warfare: How Conflicts in Cyberspace Are Challenging America and Changing the World.* Santa Barbara, CA: Praeger, 2013.

Segal, Adam. "From 'Titan Rain' to 'Byzantine Hades': Chinese Cyber Espionage." In *A Fierce Domain: Conflict in Cyberspace: 1986 to 2012*, ed. Jason Healey. Alexandria, VA: Cyber Conflict Studies Association, 2013.

Shakarian, Paulo, et al. *Introduction to Cyber-Warfare: A Multidisciplinary Approach.* Rockland, MD: Syngress, 2013.

Springer, Paul J., ed. *Cyber Warfare.* Santa Barbara, CA: ABC-CLIO, 2015.

Operation Buckshot Yankee

A fourteen-month-long campaign to prevent damage after a malicious computer worm was discovered operating within classified U.S. Strategic Command (USSTRATCOM) computer networks in October 2008. The operation was a major turning point in U.S. cybersecurity and led directly to the creation of the subordinate organization, U.S. Cyber Command (USCYBERCOM) in 2009.

In October 2008, the National Security Agency (NSA) discovered a piece of malware known as "Agent.btz" on both the Secret Internet Protocol Router Network and the Joint Worldwide Intelligence Communications System, both of which are used by the U.S. State Department and the Department of Defense to transmit secret, classified information to U.S. officials around the world. Agent.btz was capable of capturing information and then transmitting it back to a third party through a remote server. The program was discovered when it began sending beacons in an attempt to get instructions from that remote server.

Although the point of origin for the infection was not discovered, analysts concluded it to have most likely been the result of an American soldier, contractor, or other official in the Middle East who used a USB thumb drive in a public computer connected to the Internet, which became infected, then placed that drive into a computer connected to the classified government network, spreading the infection from there. No foreign nation has been officially blamed for the infiltration, although some analysts suspect that Russia could be responsible.

Operation Buckshot Yankee was the effort to repair the damage from this attack as well as strengthen American cyber defenses. To begin, NSA operators devised a plan to neutralize the infection by writing a program that could respond to the malware's beacon and order it to shut itself down. After this plan was implemented successfully, the agency then had to find and neutralize Agent.btz, which had already spread to many computers throughout the government and military, by reformatting hard drives of individual computers. Then the NSA conducted an electronic spying campaign to seek out variants of the malware elsewhere, although senior government officials rejected proposals to search for the malware on non-government systems and computers of other countries.

The incident, and the operation to counter it, spurred the United States to make significant changes in the realm of cybersecurity. To prevent further attacks, USSTRATCOM banned the use of thumb drives and took steps to disable autorun features that had enabled Agent.btz to spread. The rate of new infections declined by 2009, and official sources claim there is no evidence that any secret information fell into enemy hands. General Keith Alexander, then director of the NSA, used the incident as the foundation for his argument that a separate joint command to handle cyber threats was necessary. USCYBERCOM was thus created in June 2009 and began full operations in October 2010, under the command of General Alexander.

Michael Hankins

See also: Advanced Persistent Threat; Exploit; Malware; National Security Agency (NSA)

Further Reading

Healey, Jason. *A Fierce Domain: Conflict in Cyberspace, 1986–2012.* Alexandria, VA: Cyber Conflict Studies Association, 2013.

Kaplan, Fred. *Dark Territory: The Secret History of Cyber War.* New York: Simon & Schuster, 2016.

Lynn, William J., III. "Defending a New Domain: The Pentagon's Cyberstrategy." *Foreign Affairs* (September/October 2010).

Operation Byzantine Hades

The code name given to a long-term campaign of cyberespionage carried out by elements of the People's Republic of China (PRC) through specialized cyber warfare units in the People's Liberation Army (PLA). The campaign lasted nearly a decade and had an extremely broad target list, making it one of the most sophisticated and dangerous advanced persistent threats (APTs) documented to date. Although many of the specific details of Byzantine Hades remain classified, certain aspects of the campaign have been released.

As is the case for most APTs, particularly those perpetrated by the PLA, Byzantine Hades commenced with a series of social engineering attacks. These types of intrusions involve manipulating a human computer user into accidentally introducing malware into a target system, supplying a password or other credentials, or otherwise allowing unauthorized access into a network. In the specific case of Byzantine Hades, Facebook proved to be a particularly fruitful attack vector. After compromising specific user accounts, the Byzantine Hades perpetrators posted legitimate-looking URLs, provoking further victims to click the supplied links and inadvertently introduce malware onto their computers. This approach relied primarily on the unfortunate habits of most social media users, specifically of moving rapidly between various forms of content with little thought given to operational security. Because many users browse their Facebook pages on the same computers (and networks) as they use for their professional activities, once the malware was introduced it had almost unrestricted access to spread throughout the system.

Byzantine Hades proved to be such a large campaign that it was subsequently broken into a series of subsections, each with its own targeted systems and objectives. Byzantine Candor was the largest, and probably the most damaging, of the Byzantine Hades attacks. Byzantine Candor specifically targeted the U.S. Department of Defense (DoD), major economic nodes in the United States, and centers of political influence. When it was first discovered, it was mislabeled Operation Titan Rain III, on the assumption that it was part of the earlier Titan Rain campaigns. Further forensic analysis demonstrated that it was part of a previously unknown campaign, and that it had penetrated a far larger number of systems than originally assumed. The perpetrators of Byzantine Candor managed to infiltrate a substantial number of sensitive networks, and to steal information related to military technology, the parameters of oil and gas exploration contracts, and political activity associated with anti-PRC nonstate organizations. This information was then likely passed on to the relevant PRC entities that could benefit most from its exploitation. For example, PRC energy companies had a substantial negotiating advantage when seeking oil and gas leases because they knew not only what their Western competitors had offered but also the limits of how high those companies might raise their bids.

While frustrating, the Byzantine Candor attacks also presented certain opportunities for U.S. cyber operators. Forensic analysis discovered that most of the Byzantine Candor hop points were based in the United States, but the attacks were initiated elsewhere. By launching cyber counterattacks, it became possible to compromise the routers being used to launch the attacks, effectively allowing the National Security Agency (NSA) to backtrack to the original actors. Not only did this allow for a positive attribution of the attacks, but it actually included obtaining the billing records for the attackers' computer accounts and Internet access. In a stunning display of poor tradecraft, those billing records included the actual physical address and unit designation of the PLA unit carrying out the attacks. NSA operators managed to first observe the attackers' Internet traffic in real time, and then to insert data into the stream. This led to the compromise and penetration of the computer workstations of the PLA members carrying out the attacks, including the most likely team leader. NSA operators could see the upcoming targets, the code being written and used, and the specific techniques used for computer exploitation. They could also see the personal photos of the hostile actors, including numerous shots of those individuals in their PLA uniforms. The information served as the basis for the indictment of five PLA members on charges of cybercrimes in 2014. Although the PRC has steadfastly refused to consider extradition for those members, and even denied any responsibility for the attacks, the resolution of the Byzantine Candor case has proven quite embarrassing for the PLA. It has also forced the PRC to question how many of its own systems have been penetrated by the West, and to what extent.

Other subsections of Byzantine Hades have not been described in such detail as Byzantine Candor, but have at least been identified in public sources. Byzantine Raptor appeared primarily focused upon the DoD, although it had limited attempts to penetrate congressional networks and individual systems. Byzantine Anchor targeted weapon systems, information systems, and the National Aeronautics and Space Administration (NASA). Byzantine Trace focused almost entirely on British targets, primarily the Ministry of Defence and the Ministry of Affairs. Byzantine Foothold attempted to penetrate U.S. Transportation Command (TRANSCOM), the DoD organization tasked with the movement of military personnel and equipment around the globe. It also attacked U.S. Pacific Command (PACOM), the joint military command responsible for the Pacific region and hence the command structure most concerned with PRC military activities. At the very least, TRANSCOM networks were compromised and thousands of files exfiltrated to servers in the PRC. Those files included Trusted Authority certificates and administrator passwords, which suggest further attacks will attempt to exploit those new resources. The full extent of the penetration remains classified and might not even be clear to U.S. cyber defense experts.

Paul J. Springer

See also: APT1 Report; Operation Titan Rain; People's Liberation Army Unit 61398

Further Reading

Ambinder, Marc. "Inside the Black Box: How NSA Is Helping Companies Fight Back against Chinese Hackers." *Foreign Policy,* March 13, 2013, http://foreignpolicy.com/2013/03/07/inside-the-black-box.

Department of Justice. "U.S. Charges Five Chinese Military Hackers for Cyber Espionage against U.S. Corporations and a Labor Organization for Commercial Advantage." May 19, 2014, http://www.justice.gov/opa/pr/2014/May/14-ag-528.html.

Lemos, Robert. "'Byzantine Hades' Shows China's Cyber Chops." *CSO News,* April 21, 2011, https://www.csoonline.com/article/2128120/social-engineering/-byzantine-hades—shows-china-s-cyber-chops.html.

Libicki, Martin. *Cyberspace in Peace and War.* Annapolis, MD: Naval Institute Press, 2016.

Operation Olympic Games

A large-scale cyberattack reportedly initiated by the United States, partnering with Israel, against Iran's nuclear program. The operation began in 2006, under the President George W. Bush administration and continued throughout the majority of President Barak Obama's administration. Before that time, the United States had attempted to slow the Iranian nuclear program through physical sabotage but had achieved little effect. General James Cartwright, then Commander, U.S. Strategic Command (USSTRATCOM) and later Vice Chairman of the Joint Chiefs of Staff, joined with other intelligence officials in proposing a plan to attack Iranian nuclear facilities through cyberattacks.

The first step in this plan involved gaining detailed knowledge of the computer system inside the Iranian nuclear facility at Natanz. This was achieved by planting code in the Natanz facility that could produce a digital map of the computers controlling the facility's centrifuges. The rouge program could then broadcast that map back to the National Security Agency (NSA), who could then use the map to create a cyberweapon designed specifically to interrupt the centrifuges. The creation of this worm, later named Stuxnet, was achieved through collaboration between the United States, especially the NSA, and Israeli intelligence, including Unit 8200. U.S. officials indicated that the reason for the collaboration was in part due to Israeli expertise but also a desire to discourage Israel from conducting preemptive conventional strikes against Iran on their own.

Before the worm was deployed, it was reportedly first tested on replicas of the Natanz facility, constructed in Tennessee using the same model of centrifuge used by Iran that had been previously captured from Libyan dictator Colonel Muammar el-Qaddafi in 2003. After proving successful in tests, it was deemed ready for deployment against the Natanz facility in 2008 and is possibly the first major cyberattack designed to cause offensive physical destruction, rather than simply slowing target computer systems or intelligence gathering.

Planting Stuxnet in Natanz could not be done remotely, as the facility was protected by an "electronic moat," or "air gap," in which the facility's computers were completely physically separated from the outside. The first version of the worm was planted in 2008 using engineers and maintenance workers employed at Natanz, some of whom were spies, others who were unaware that they carried the bug. The early version of the worm spread using the Step 7 project files that programmed the Siemens-designed computers. In June 2009, an updated version of Stuxnet was introduced into the facility using employees from five companies that Natanz used as contractors, including Foolad Technic, Behpajooh, Neda Industrial Group, and Control Gostar Jahed.

This updated version of Stuxnet could be spread using USB flash drives, spreading throughout the computer system via the Windows autorun feature and an exploit in Microsoft's print spooler service. Stuxnet was designed to work slowly, causing centrifuges to spin out of control and cease to function or even cause destruction of their parts. The goal was that Iranian engineers would suspect either faulty equipment or employee incompetence rather than a cyberattack. Over about five months almost 20 percent of the centrifuges at Natanz ceased to function. In addition to this, Iranian workers began to shut down still-functional equipment as part of their search for a source of the problem.

The Obama administration continued the program and expanded it further, introducing more updated versions of the worm. In the summer of 2010, Stuxnet spread outside of the isolated environment of Natanz after it infected an employee's computer that was connected to the centrifuges. When that employee later left the facility and connected his computer to the Internet, Stuxnet quickly spread across the globe and was discovered and analyzed by a wide variety of international security organizations. American officials blamed the unintended spread of the worm on an Israeli programming error. Because Stuxnet was designed to attack Iranian centrifuges in specific ways, when infecting computers outside that network, it remained dormant. Despite the release of the worm onto the Internet, the Obama administration did not slow the operation, and it continued to cause destruction of more centrifuges. In total approximately 1,000 centrifuges were destroyed as part of Operation Olympic Games.

Another potential element of Operation Olympic Games includes another worm, called Flame, targeting Iran's oil industry and primarily intended to gain intelligence in preparation for larger-scale cyberattacks. Flame was also a collaboration between the Israeli military with American intelligence agencies including the Central Intelligence Agency (CIA) and the NSA. Flame contains some of the same code as Stuxnet and was developed around the same time, although Flame might possibly predate Stuxnet. Flame's capabilities include keystroke logging, activating microphones and cameras, capturing screen shots, extracting location data from images, and using Bluetooth functionality to send data and receive commands.

Operation Olympic Games was intended to remain classified, although in 2011, after his retirement, General Cartwright discussed much of the program with *New York Times* reporter David Sanger. In 2016 he pled guilty to denying that he leaked classified information. In 2017 Cartwright was pardoned by President Obama.

Further investigation by the documentary filmmaker Alex Gibney claims to have information from unnamed sources within the NSA and CIA claiming that Operation Olympic Games was intended to be followed up with a much larger cyberattack against Iran called Nitro Zeus. This plan would potentially have been able to disable Iran's air defense network, communications capability, and parts of its power grid, but was not implemented after the 2015 deal in which Iran agreed to limit its uranium production.

Operation Olympic Games was the largest and most sophisticated state-sponsored cyberattack in history at the time it occurred. It most likely set back the Iranian nuclear program by several years. However, some analysts have suggested that the unprecedented and offensive nature of the operation may invite other nations to become more aggressive with their own cyberattack operations.

Michael Hankins

See also: Gas Centrifuge; Malware; National Security Agency (NSA); Shamoon; Stuxnet

Further Reading

Kaplan, Fred. *Dark Territory: The Secret History of Cyber War.* New York: Simon & Schuster, 2016.

Kello, Lucas. *The Virtual Weapon and the International Order.* New Haven, CT: Yale University Press, 2017.

Klimberg, Alexander. *The Darkening Web: The War for Cyberspace.* New York: Penguin, 2017.

Sanger, David. *Confront and Conceal: Obama's Secret Wars and Surprising Use of American Power.* New York: Broadway Press, 2012.

Zetter, Kim. *Countdown to Zero Day: Stuxnet and the Launch of the World's First Digital Weapon.* New York: Crown, 2014.

Operation Orchard

Computer network operation disrupting Syrian air defenses in support of September 2007 Israeli air strikes that destroyed the Syrian Al Kibar nuclear reactor at Dair Alzour. Operation Orchard highlights the utility of cyber as a means to acquire intelligence as well as cyber capabilities used alongside traditional military force.

Although many details surrounding Operation Orchard remain unclear, most scholars and analysts trace its origin to the Israeli intelligence agency Mossad's infiltration of a senior Syrian government official's laptop computer. While visiting London in 2006, the Syrian official left his laptop computer in a hotel. When he left the hotel room, Mossad installed a Trojan horse onto the laptop to monitor his communications and search for evidence of Syria's nuclear program. When the hard drive files were analyzed, a photo and other files provided evidence of covert collaboration between North Korea and Syria to advance the Syrian nuclear program. The photo appeared to be taken at Dair Alzour, a military installation in the Syrian desert. Israeli intelligence identified the two men in the photo as Chon Chibu, a ranking official in North Korea's nuclear program, and Ibrahim Othman, the director of the Syrian Atomic Energy Commission. Other files on the computer revealed construction plans for the Al Kibar nuclear facility and evidence the Syrian government possessed a type of pipe used for fissile material.

After midnight on September 6, 2007, seven Israeli F-15Is dropped several bombs on the Al Kibar facility. The Syrian air defense network never fired on the fighter jets, and the Syrian military was unaware of the bombing until it was under way. Prior to the bombing, Israelis conducted a computer network attack to penetrate the Syrian military computers. The Israelis replaced data streams in the network that provided fake imagery of the Syrian border. By disabling the air defense, Israeli fighters were able to successfully level the Al Kibar facility and escape Syrian airspace without confrontation.

In response to pressure from Congress, in April 2008 the George W. Bush administration disclosed supporting evidence that the Al Kibar facility under construction was intended to produce weapons-grade plutonium and confirmed it was bombed by Israel in September 2007. At the time, media reports also suggested

South Korean intelligence officials believed approximately ten North Koreans were killed in the Israeli bombing. An IAEA investigation of the undeclared facility at Dair Alzour also concluded the design of the building, infrastructure within it, and positive environmental tests were consistent with a nuclear reactor facility.

Syria did not respond to the cyberattack or the bombing of the reactor site in Operation Orchard. Possibly to avoid a Syrian counterattack, Israeli government officials refused to publicly acknowledge Israel's role in the attacks until March 2018.

Melia T. Pfannenstiel

See also: Computer Network Attack (CNA); Cyberweapons

Further Reading

Libicki, Martin C. *Cyberspace in Peace and War.* Annapolis, MD: Naval Institute Press, 2016.

Lin, Christina Y. "The King from the East: DPRK-Syria-Iran Nuclear Nexus and Strategic Implications for Israel and the ROK." *Korea Economic Institute* 3, no. 7 (October 2008), 1–14.

Singer, P. W., and Allan Friedman. *Cybersecurity and Cyberwar: What Everyone Needs to Know.* Oxford: Oxford University Press, 2014.

U.S. Library of Congress, Congressional Research Service. *North Korea's Nuclear Weapons Development and Diplomacy,* by Larry A. Niksch, RL33590 (January 5, 2010), 20–24.

Operation Rolling Tide

Extending from August 2013 to February 2014, the U.S. Navy's first official operation designed specifically to counter hostile cyberactivity targeting friendly command and control systems. Intended to "accelerate the Navy's ability to prevent, constrain, and mitigate cyber-attacks and critical vulnerabilities" (U.S. Navy, 2015, 1), the operation was catalyzed by concern following intrusion of the Navy-Marine Corps Intranet system by Iranian hackers. The Iranian intrusion was reportedly discovered to be more substantial than had been first believed when it was originally disclosed in September 2013, and one U.S. government official noted, " 'the thing got into the bloodstream' " (Gorman and Barnes, 2014). Mapping Navy networks became one aspect of this effort. Officials explained that restoring Navy networks required time to coordinate with various commanders across the service as a result of the impact that changes had on command and control procedure.

The intrusion also fit within a larger contest of retaliations in cyberspace between Iran and the United States ever since the discovery in 2010 of the Stuxnet malware that targeted Iranian nuclear supervisory control and data acquisition (SCADA) and the widespread accusation of the United States or Israel for creation of Stuxnet code. Before hacking into the Navy's unclassified system upon which 800,000 users in more than 2,000 locations rely, Iran sought to retaliate against other systems, including the U.S. banking system and the Saudi Arabian oil industry.

This also coincided with the Navy's establishment of a cyber command for that service. Operation Rolling Tide is believed to have helped reinforce the message within the Navy Cyber Command that it is an operational (rather than a supporting) entity, and to have emphasized leaders' messages against the ongoing presumption that the cyber domain is a permissive environment. The operation was heralded in the Navy as a success, both in terms of restoring security and control over the Navy's unclassified systems and also in terms of deriving meaningful lessons learned. The lessons learned were subsequently applied in the preparation of the Navy's subsequent $3.5 billion Next Generation Enterprise Network-Recompete (NGEN-R) program.

Nicholas Michael Sambaluk

See also: Malware; Oghab 2; Shamoon; Stuxnet; Supervisory Control and Data Acquisition (SCADA); Vulnerability

Further Reading

Gilday, Michael M. "Statement by Vice Admiral Michael M. Gilday, Commander U.S. Fleet Cyber Command before the Senate Armed Services Committee Subcommittee on Cybersecurity." May 23, 2017, https://www.armed-services.senate.gov/imo/media/doc/Gilday_05-23-17.pdf.

Gorman, Siobhan, and Julian E. Barnes. "Iran's Infiltration of a Navy Computer Network Sure to Get Attention in Vetting of Next Head of Embattled National Security Agency." *The Wall Street Journal,* February 18, 2014, https://johnib.wordpress.com/2014/02/18/irans-infiltration-of-a-navy-computer-network-sure-to-get-attention-in-vetting-of-next-head-of-the-embattled-national-security-agency.

Stiennon, Richard. *There Will Be Cyberwar: How the Move to Network-Centric War Fighting Has Set the Stage for Cyberwar.* Birmingham, MI: IT-Harvest, 2015.

U.S. Navy. *Information Sys Security Program,* 2015. https://www.globalsecurity.org/military/library/budget/fy2016/navy-peds/0303140n_7_pb_2016.pdf.

Operation Shady RAT

An advanced persistent threat targeting more than seventy government agencies, nongovernmental organizations (NGOs), and private sector firms in fourteen countries between July 2006 and August 2011. Shady RAT was discovered in 2011 by Dmitri Alperovitch and a team of threat researchers at the cybersecurity firm McAfee. Shady RAT's name is derived from Remote Access Tool, software that provides a hacker full remote control of a device. The Shady RAT attacks are often attributed to a Chinese espionage unit 61398, referred to in a February 2013 report published by the cybersecurity consulting firm Mandiant (now FireEye).

Much of the available information on the associated cyber intrusions is detailed in a report published by McAfee in August 2011. Shady RAT refers to the collective of targets impacted by the security breaches between 2006 and 2011. Some affected organizations included the United Nations, International Olympic Committee, World Anti-Doping Agency, Hong Kong and New York Associated Press offices, U.S. Department of Energy, U.S. Department of Defense, and defense contractors. The intruders focused efforts on a broad range of sectors including intelligence gathering on energy firms, industry research and development, as well as acquiring sensitive military data and satellite communications. Forty-nine of the more than 70 firms targeted in Shady RAT are based in the United

States, although McAfee did not publicly disclose many of the affected private companies.

Similar to other advanced persistent threats (APTs), spear phishing and faked e-mails were typically used to cause malware to be downloaded. Malware was implanted when an e-mail attachment was opened, which created a backdoor communication channel for an outside web server. After the malware was downloaded, it communicated with a command and control server. This allowed the remote intruders to access the device. The remote access to the device then provided those individuals the ability to spread malware to other devices in the network and steal data. The attacks were sophisticated both in the precision they targeted specific individuals within the organization that held appropriate levels of access to the desired information and the ability of the malware within the e-mail attachments to elude antivirus protection and remain hidden in the networks and operating systems.

The McAfee researchers responsible for uncovering Shady RAT found a server through which stolen data from dozens of organizations and companies were cached for distribution. They noted a mistake made by the intruders led them to the command and control server and all Internet protocol (IP) addresses the server had controlled since 2006 were recorded in a log.

A number of other known cyber intrusions between 2006 and 2011 commonly attributed to China were conducted concurrent to Shady RAT. GhostNet (2007–2009) was an advanced persistent threat directed toward the Dalai Lama and Tibetan independence movement. Operation Aurora (2009–2010) targeted Google, including Chinese dissident Gmail accounts, and dozens of other firms. Aurora prompted Google's exit from China and is considered one of the first instances of a private company disclosing a large security breach. Byzantine Hades (2002–2010) focused on a range of U.S. government agencies and international organizations such as the International Monetary Fund (IMF) and World Bank. The breach of the RSA cryptosystem (2010–2011) compromised the security of defense contractors such as Lockheed Martin.

In particular, an operation between November 2009 and February 2011 referred to as Night Dragon, seemed to complement the information sought in some Shady RAT intrusions. Multinational firms in the oil and energy sector were targeted for specific information pertaining to oil exploration and bidding. The perpetrators gained insights into what competitors were prepared to bid on drilling rights and the negotiation positions of specific firms. Martin Libicki notes law firms are common targets in cyber operations because they typically store high-value information on computers but often lack the resources to protect it.

China is often implicated in cyber intrusions due to the selection of targets, which are often of political and economic interest to the Chinese government. In the case of Shady RAT, an indication was the targeting of the Olympic Committee and World Anti-Doping Agency prior to the 2008 Beijing Olympics. Despite considerable evidence of Chinese government involvement in data breaches, the theft of intellectual property, and cyberespionage activities, positive attribution using open-source data remains difficult.

Melia T. Pfannenstiel

See also: Advanced Persistent Threat; APT1 Report; Operation Aurora; Operation Byzantine Hades; Operation Titan Rain

Further Reading

Libicki, Martin C. *Cyberspace in Peace and War.* Annapolis, MD: Naval Institute Press, 2016.

Lindsay, Jon R., and Tai Ming Cheung. "From Exploitation to Innovation: Acquisition, Absorption, and Application." In *China and Cybersecurity: Espionage, Strategy, and Politics in the Digital Domain,* ed. Jon Lindsay, Tai Ming Cheung, and Derek S. Reveron. Oxford: Oxford University Press, 2015, 51–56.

Lindsay, Jon, Tai Ming Cheung, and Derek S. Reveron, eds. *China and Cybersecurity: Espionage, Strategy, and Politics in the Digital Domain.* Oxford: Oxford University Press, 2015.

Nakashima, Ellen, "Report on Operation Shady RAT identifies widespread cyber spying." *Washington Post,* August 3, 2011, https://www.washingtonpost.com/national/national-security/report-identifies-widespread-cyber-spying/2011/07/29/gIQAoTUmqI_story.html?utm_term=.f732c41307aa.

Singer, P. W., and Allan Friedman. *Cybersecurity and Cyberwar: What Everyone Needs to Know.* Oxford: Oxford University Press, 2014.

Operation Titan Rain

Code-name designation for cyberespionage activities by China beginning in 2003. Titan Rain is sometimes narrowly applied to activities targeting the U.S. Department of Energy laboratories between 2003 and 2005. Most cyber scholars and experts use Titan Rain in reference to the collective cyberespionage activities of Chinese origin directed toward the U.S. Department of Defense, NASA networks, and various private contractors between September 2003 and August 2005. Titan Rain is considered the first publicly known state-sponsored advanced persistent threat (APT) traced to the Chinese government.

In the series of intrusions associated with Titan Rain, hackers operating from China penetrated multiple sensitive but not secret computer systems from the Department of Defense Non-classified Internet Protocol Router Network (NIPRNet). No classified systems were reported to have been breached, but unauthorized access to files containing sensitive industrial technology and defense operations came from agencies such as the U.S. Army Information Systems Engineering Command, the Naval Ocean Systems Center, the Missile Defense Agency, and Sandia National Laboratories. The targets and high-level sophistication of procedures used in Titan Rain indicated the attacks were conducted by Chinese military intelligence, rather than a nonstate actor operating from within China.

Frequent acts of Chinese cyberespionage have been directed at U.S. government websites and private defense contractors since 2003. After Titan Rain was exposed in 2005, cyber exploitation activities originating from China continued to target government agencies, military universities, and the private sector. Department of Defense officials claimed in 2006 that the Pentagon received millions of daily scans, in what appeared to be an attempt to penetrate the Global Information Grid.

Although considered separate from Titan Rain, between 2006 and 2007 a number of publicly reported cyber intrusions occurred that are attributed to the

Chinese government. Some examples include the targeting of Department of State's Bureau of East Asian and Pacific Affairs (June–July 2006) that resulted in the U.S. Embassy in Beijing loss of connectivity for two weeks; the Department of Commerce Bureau of Industry and Security (June–October 2006); the U.S. Naval War College in November 2006, which caused a shutdown of the campus network; and the June–September 2007 intrusion of computers in the office of the Secretary of Defense, including the computer used by Secretary Robert Gates. Defense contractors such as Raytheon, Lockheed Martin, Boeing, and Northrup Grumman also experienced data breaches between 2006 and 2008 attributed to China.

Melia T. Pfannenstiel

See also: Advanced Persistent Threat; APT1 Report; NIPRNet; Operation Byzantine Hades; Operation Shady RAT; People's Liberation Army Unit 61398

Further Reading

Carr, Jeffrey. *Inside Cyber Warfare.* Sebastopol, CA: O'Reilly, 2010.

Libicki, Martin C. *Cyberspace in Peace and War.* Annapolis, MD: Naval Institute Press, 2016.

Lindsay, Jon, Tai Ming Cheung, and Derek S. Reveron, eds. *China and Cybersecurity: Espionage, Strategy, and Politics in the Digital Domain.* Oxford: Oxford University Press, 2015.

U.S. Library of Congress, Congressional Research Service. *Botnets, Cybercrime, and Cyberterrorism: Vulnerabilities and Policy Issues for Congress,* by Clay Wilson, RL32114 (January 29, 2008), 14–15.

People's Liberation Army Unit 61398

A cyber operations unit of the People's Liberation Army (PLA) of the People's Republic of China (PRC). The PRC has developed an increasingly aggressive approach to technological advancement and economic competition in the 21st century. One key element of this strategy is an elaborate and extensive attempt to make major leaps forward in virtually every area of economic activity through widespread cyberespionage. Because the Chinese government maintains a very tight relationship with Chinese corporations, government-backed espionage efforts often share the fruits of their labors with development companies. By utilizing government agencies to conduct cyberespionage, the Chinese government can maintain a close hold on the target lists, the capabilities of its most advanced cyber warriors, and the risks assumed by individual attacks. Also, a victim considering some form of retaliation through the cyber domain will be forced to engage against a national force rather than a rival company, giving pause to almost any potential response.

Spearheading the cyberespionage campaign is an extremely secretive military cyber unit. Dubbed "People's Liberation Army Unit 61398," this organization appears to be based in a 12-story building in Shanghai. It is also often referred to as the Third Office of the PLA General Staff Department, Third Department, Second Bureau. This longer, more bureaucratic nomenclature has the added effect of anonymizing the organization, burying it in nondescript modifiers that offer few clues as to its function.

The existence of Unit 61398 became public knowledge in 2013, when Mandiant Corporation released a report entitled "APT-1." The acronym in the title refers to "Anonymous Persistent Threat Number One," the original designator applied by Mandiant when it became aware of the unit's activities. Mandiant was retained to investigate a series of debilitating attacks against the computer networks of the *New York Times*. The cyber assault against the newspaper commenced after a series of articles illustrating the vast wealth accumulated by the outgoing premier of the PRC, Wen Jiabao. In a Communist country so ostensibly dedicated to raising the standard of living for all of its citizens, such a fortune in the hands of a powerful politician seemed at the very least suspicious. The articles apparently infuriated leaders within the PRC and provoked a major sabotage campaign.

As Mandiant began tracing the source of the attacks against the newspaper, it discovered a much larger campaign against hundreds of Western targets. Mandiant experts were able to document the activities of the unit in real time, literally watching cyberattacks as they commenced against a wide variety of targets. Upon investigating the originating point of the attacks, Mandiant researchers managed to pinpoint the exact physical location from which they emerged, and to identify the unique operators involved in the attacks. Interestingly, the vast majority of attacks coming from Unit 61398 occur during normal business hours throughout the workweek, with a corresponding drop-off during nights and weekends.

Unit 61398 appears devoted to the theft of virtually any intellectual property with economic value. As such, it targets almost any company in the West, including those outside the electronics and telecommunications sectors. The unit seems to focus primarily on English-speaking companies, although its attacks have not been exclusively limited to such targets. Most Unit 61398 attacks rely on social engineering rather than brute-force tactics. This more sophisticated approach requires an attacker to trick victims into providing confidential information such as login credentials for secured networks. Once access is obtained, unit members seek to upload malware that will allow remote control of computer systems, followed by attempts to escalate access to more privileged information. Once a network is completely compromised and mapped, the cyber intruders begin transferring information back to Unit 61398 servers. This process can take months or years but has proven extremely effective at eluding antivirus protections.

After the Mandiant report was released, the U.S. Department of Defense made strong references to Chinese hacking activities in a report to Congress. In 2014, the federal government indicted five purported members of Unit 61398 for espionage offenses. Unsurprisingly, the PRC government has denied all of the accusations and refused to extradite any citizens indicted for the attacks. Although some scholars have suggested that Unit 61398 might simply be a false front to cover for the activities of other organizations or individuals, there is substantial evidence to back the claim that it is directly responsible for the attacks documented by Mandiant. This evidence includes the enormous volume of fiber-optic cables entering the building, the presence of supporting organizations normally associated only with prestigious military units, and the military precision of the attacks originating from the location's servers.

The unit has been associated with some of the largest and longest cyber intrusions detected by Western agencies. These include Titan Rain, a campaign against U.S. Department of Defense organizations and associated contractors that commenced in 2003. Operation Shady RAT and Operation Night Dragon were a series of attacks against American businesses that commenced in 2006 and lasted for at least five years. Operation Aurora targeted high-technology companies in the West and ran for most of 2009. The GhostNet attacks targeted servers in more than one hundred countries and appears to have focused primarily on Tibetan independence movements and the activities of the Dalai Lama. In each case, the attackers managed to exfiltrate terabytes of information before being discovered.

Paul J. Springer

See also: APT1 Report; GhostNet; Malware; Operation Aurora; Operation Shady RAT

Further Reading

Austin, Greg. *Cyber Policy in China.* Cambridge, UK: Polity, 2014.

Inkster, Nigel. *China's Cyber Power.* London: Routledge, 2016.

Libicki, Martin. *Cyberspace in Peace and War.* Annapolis, MD: Naval Institute Press, 2016.

Lindsay, Jon R. *China and Cybersecurity: Espionage, Strategy, and Politics in the Digital Domain.* New York: Oxford University Press, 2015.

Mandiant Corporation. "APT-1: Exposing One of China's Cyber Espionage Units." Alexandria, VA: Mandiant Corporation, 2013. https://www.fireeye.com/content/dam/fireeye-www/services/pdfs/mandiant-apt1-report.pdf.

People's Liberation Army Unit 61486/"Putter Panda"

A cyberespionage organization believed to have been established by the People's Liberation Army (PLA) of the People's Republic of China (PRC) in 2007, focused particularly on espionage deriving information about foreign satellite technology on behalf of PRC's space surveillance apparatus.

The organization was first identified by researchers at the cybersecurity firm CrowdStrike in 2012, and CrowdStrike's intelligence team studied the organization for two years before publicly announcing its discovery and evidence of Unit 61486's activities in June 2014. This announcement came in the wake of a U.S. indictment of five PLA personnel connected with the activities of another cyberespionage organization, PLA Unit 61398. PRC officials denied the accusations in the U.S. indictments, although the indictments themselves were of limited effect against personnel working overseas.

CrowdStrike's publicized findings about PLA Unit 61486, occurring a month after the U.S. indictments against members of PLA Unit 61398, illustrated reasons for suspicion that cyberespionage was accepted and even supported at the nation-state level rather than happening only independently. Given that CrowdStrike had been tracking Unit 61486's activities for two years prior to their announcing the unit's existence, it is conceivable that the timing of the security firm making the announcement was related to the U.S. indictment handed down in absentia against other PRC cyberespionage actors and particularly to the official PRC denunciation of the U.S. indictment.

Unit 61486 was part of the 12th Bureau of the 3rd General Staff Department prior to the PLA's reorganization that began in late 2015. The 3rd General Staff Department has been likened to the U.S. National Security Agency, and Unit 61486 appears to be invested in espionage that supports PRC's space and satellite activity. Mistakes by personnel in the unit helped divulge the character of the unit, and these included the use of a former student's e-mail address by a unit member when deploying a remote access tool (RAT). The e-mail address corresponded to his account when he studied at Shanghai's Jiaotong University, which experts had earlier accused of feeding the PLA's cyberespionage organizations with computer science graduates.

Experts have pointed to evidence of cooperation between PLA Unit 64186 and Unit 61398. This includes activity corresponding to the working hours kept in Chinese time zones, which suggests that activities are officially supported and directed and are not the actions of individuals driven by individual motivations in their personal time. The domains used to control the malware were purchased by a then thirty-five-year-old Chen Ping, and these correspond to the physical location in Shanghai associated with Unit 61486. Shanghai is home not only to Jiaotong University but also to Unit 61398.

CrowdStrike reportedly dubbed the organization "Putter Panda" after identifying a pattern in its targeting golf-playing conference goers. Frequently, the unit compromised and then leveraged foreign websites as platforms for their own attacks. In at least one instance, victims were lured with a fake advertisement for a supposed yoga studio in Toulouse, France. Clicking on the link opened a path by which the victim's network could be surveilled. At the time that the unit's existence was publicly announced, the satellite industry garnered annual revenue of nearly $200 billion, and espionage in this sector represented a lucrative opportunity as well as a strategic shortcut. CrowdStrike cofounder George Kurtz noted that "targeted economic espionage campaigns compromise technological advantage, diminish global competition, and ultimately have no geographic borders" (CrowdStrike, 2014, 1).

Unit 61486 appeared to specialize in space and satellite information collection. Specifically, space and satellite technology of European entities, European aerospace companies, and Japanese and European telecommunication companies were prioritized targets. Unit 61486 utilized pngdowner malware that used comparatively simple C++ source code, although CrowdStrike observed that related versions of its RAT indicate "that despite the simple nature of the tool, the developers have made some attempts to modify and perhaps modernize the code" (CrowdStrike, 2014, 33).

Nicholas Michael Sambaluk

See also: Advanced Persistent Threat; CrowdStrike; People's Liberation Army Unit 61398

Further Reading

Cheng, Dean. *Cyber Dragon: Inside China's Information Warfare and Cyber Operations.* Santa Barbara, CA: Praeger, 2016.

CrowdStrike Global Intelligence Team. *CrowdStrike Intelligence Report: Putter Panda,* June 2014, https://cdn0.vox-cdn.com/ . . . /crowdstrike-intelligence-report-putter-panda.original.pdf.

Fritz, Jason R. *China's Cyber Warfare: The Evolution of Strategic Doctrine*. Lanham, MD: Lexington, 2017.

Inkster, Nigel. *China's Cyber Power*. New York: Routledge, 2016.

Klimburg, Alexander. *The Darkening Web: The War for Cyberspace*. New York: Penguin, 2017.

Lindsay, Jon, Tai Ming Cheung, and Derek Reveron. *China and Cybersecurity: Espionage, Strategy, and Politics in the Digital Domain*. Oxford: Oxford University Press, 2015.

Rid, Thomas

(1975–), a scholar focusing on information technologies and known for offering arguments opposed to what he considers hype about the rise of cyberwar. Currently a professor of strategic studies at John Hopkins School of Advanced International Studies, Dr. Rid taught earlier at King's College in London. He grew up in Germany and obtained his doctorate from Humboldt University of Berlin in 2006.

Among his works on the cyber domain and cybernetic history is an article in 2011 and a subsequent book in 2013 titled *Cyber War Will Not Take Place*. These rebutted a 1993 RAND Corporation report titled *Cyber War Is Coming*, and later works such as *There Will Be Cyberwar* by Richard Stiennon and Richard Clarke's *Cyber War*, which argued that cyber warfare would approach Armageddon levels of destruction analogous to those made possible by nuclear weapons in the Cold War. Rid argued that cyber would never approach the level of true warfare, which is characterized by violence and bloodshed. He pointed out that not one person has lost their life to a cyberattack. Although cyber was certainly a contested domain, Rid noted, these operations mainly consisted of sabotage, espionage, and subversion. Although these three operations have become more sophisticated because of cyber, the likelihood of them causing wide-scale death and destruction was unlikely.

Melvin G. Deaile

See also: Cyber Pearl Harbor; Stiennon, Richard; Stuxnet

Further Reading

Rid, Thomas. *Cyber War Will Not Take Place*. Oxford: Oxford University Press, 2013.

Rid, Thomas. *Rise of the Machines: A Cybernetic History*. New York: Norton, 2016.

Rid, Thomas, and Ben Buchanan. "Attributing Cyber Attacks." *Journal of Strategic Studies* 38, no. 1 (2015), 4–37.

Rogers, Michael S.

(1959–), "dual-hatted" commander of the U.S. Cyber Command (USCYBERCOM) and of the National Security Agency (NSA) from April 2014 to May 2018. A four-star Navy admiral, he assumed command in the wake of revelations that Edward Snowden had disclosed massive droves of classified material. In one of his more controversial decisions, he broke down bureaucratic divisions between offensive and defensive cyber operators.

Born in Chicago in 1959, Rogers was educated at Auburn University and commissioned into the U.S. Navy in 1981, working first in naval gunnery and then cryptology. This gave Rogers experience with the nascent cyber field, and by 2003 he contributed to computer network attacks during the initial stages of Operation Iraqi Freedom. He then filled two sequential positions as a director of intelligence, first for the Pacific Command and then for the Joint Chiefs of Staff. He subsequently commanded the U.S. Fleet Cyber Command and the U.S. 10th Fleet, which focuses on cyber warfare. His promotion to admiral—the first officer in the Information Warfare Community to reach this rank—highlighted the increasing importance of cyber to the Navy and to the U.S. military at large.

The organizational structure by which the NSA and USCYBERCOM shared a commander attracted controversy. Rogers's successor, four-star Army general Paul M. Nakasone, nonetheless continues to date to serve in the dual-hatted role.

Heather Pace Venable

See also: National Security Agency (NSA); USCYBERCOM

Further Reading

Kaplan, Fred. *Dark Territory: The Secret History of Cyber War.* New York: Simon & Schuster, 2016.

Klimberg, Alexander. *The Darkening Web: The War for Cyberspace.* New York: Penguin, 2017.

Russian Business Network (RBN)

A cybercrime entity based in St. Petersburg, Russia, whose business model included illicit mercenary activities at the edge of cyber warfare. First established in 2006, RBN steadily drifted from serving as a conventional Internet service provider toward instead hosting a range of illicit activities, including traffic in child pornography, phishing, and the propagation of malware. Associated with the Russian underworld, RBN was also directly involved in aggressive cyber actions and cybercrime. Identity theft ranked high among its criminal activities, and cybersecurity analysts note that RBN identity theft efforts frequently involved RBN's malware simulating legitimate antispyware and antimalware services. Other RBN malware masqueraded as innocent try-for-free web products. Some experts have suggested that at the organization's height its activities accounted for more than half of all cybercrime globally.

RBN achieved infamous notoriety by renting out its capabilities, such as the creation of botnets and the launching of distributed denial-of-service (DDoS) actions. Through these actions (which peaked with the cyber actions against Estonia in 2007 and against Georgia in 2008), RBN demonstrated the viability of an organization that included professional hackers to act in the role of a digital mercenary force. With respect to the Estonian and Georgian actions, RBN's suspected activities aligned closely with the desires of the Russian government. RBN's profit motive suggests the organization, despite its actions being apparently in concert with the priorities of the Russian state, was not an example of simple hacktivism. However, given the paucity of potential revenue to be gained by targeting Estonian,

Georgian, and Kirgiz websites, cybersecurity analysts have concluded that RBN participation in DDoS actions has taken place alongside the activity of conventional hacktivists. This conclusion has been further supported by some anecdotal evidence presented through Internet journalism.

Russian authorities purportedly disassembled RBN's cybercrime capabilities, with reports of its elimination appearing as early as November 2007. Another more decentralized system of organized cybercrime subsequently took its place. Reports have suggested that "syndicates, such as the now infamous (and defunct) Russian Business Network (RBN) are often tolerated because they provide services that the state needs and income to government cronies" (Connell and Vogler, 2017, 17). RBN's actions as a cyber mercenary entity show how groups that lack cyber capabilities can nonetheless rent them on the dark web.

Nicholas Michael Sambaluk

See also: Botnet; Distributed Denial-of-Service (DDoS) Attack; Estonia, 2007 Cyber Assault on; Georgia, 2008 Cyber Assault on; Malware

Further Reading

Armin, Jart. *Tracking the Russian Business Network.* Cambridge: Cambridge University Press, 2007.

Connell, Michael, and Sarah Vogler. *Russia's Approach to Cyber Warfare.* Arlington, VA: CNA Analysis and Solutions, 2017.

Korns, Stephen W., and Joshua E. Kastenberg. "Georgia's Cyber Left Hook." *Parameters* (Winter 2008–2009), 60–76.

Kozlowski, Andrzej. "Comparative Analysis of Cyberattacks on Estonia, Georgia and Kyrgyzstan." *European Scientific Journal* (February 2014), 237–245.

Medvedev, Sergei A. *Offensive-Defensive Theory Analysis of Russian Cyber Capability.* Thesis. Monterey, CA: Naval Postgraduate School, 2015.

Sandworm

The name of a cyber hacking group attributed to Russia that has been active since at least 2014, perhaps even earlier. The group has also been identified by the names Voodoo Bear and Telebots. The group's targets include the North Atlantic Treaty Organization and the European Union, as well as various telecommunications and infrastructure entities across Europe. Some analysts have concluded Sandworm to be associated in some way with the Russian government, citing connections between the group's activities and Russian geopolitical initiatives.

In particular, cyberattacks on Ukraine have been attributed to Sandworm actors. Analysts from iSight Partners have additionally linked Sandworm to the trojan "BlackEnergy3," a surveillance malware tool. Furthermore, Sandworm has demonstrated a pattern of employing phishing malware and fake ransomware in its attacks. These have resulted in energy blackouts and information theft. Arguably the most prominent of these activities was the Christmas attack on Ukraine's power grid in December 2015, which left several hundred thousand people briefly without electricity.

Stephanie Marie Van Sant

See also: CyberBerkut; Distributed Denial-of-Service (DDoS) Attack; iSIGHT; Malware; Russian Business Network (RBN); Ukraine Power Grid Cyberattack, December 2015

Further Reading

Connell, Michael, and Sarah Vogler. *Russia's Approach to Cyber Warfare.* Arlington, VA: CNA Analysis Solutions, 2017.

Park, Donghui, et al. "Cyberattack on Critical Infrastructure: Russia and the Ukrainian Power Grid Attacks." *The Henry M. Jackson School of International Studies* (October 2017).

Weedon, Jen. *Beyond 'Cyber War': Russia's Use of Strategic Cyber Espionage and Information Operations in Ukraine.* Tallinn, Estonia: NATO CCD COE, 2015.

Script Kiddies

A generally pejorative term describing a particular subgroup of hackers who nonetheless possess little to no understanding of the mechanics or protocols of the Internet, how vulnerabilities are discovered, or how exploits are written. Due to their lack of skills and knowledge, script kiddies download and use the numerous and readily available malware toolkits from the Internet in order to automate attacks.

These malware toolkits allow the user to exploit software and system vulnerabilities using techniques such as cross-site scripting, denial-of-services attacks, buffer overflows, cache poisoning, and SQL injections, and make releasing a virus or worm simple. Script kiddies are usually motivated by curiosity and are known for causing disruptions ranging from simple mischief to inconvenience to malicious damage. The ease of use of malware toolkits makes attacks more likely to happen, although historically software patches and detection signatures have also tended to be readily available to counteract the effects.

Diana J. Ringquist

See also: Hacktivism; Malware

Further Reading

Furnell, Steven. *Cybercrime: Vandalizing the Information Society.* Boston: Addison-Wesley, 2002.

Vacca, John R. *Computer and Information Security Handbook.* Cambridge, MA: Elsevier, 2017.

Wori, Okechukwu. "Computer Crimes: Factors of Cybercriminal Activities." *International Journal of Advanced Computer Science and Information Technology* 3, no. 1 (2014), 51–67.

Security Information and Event Management (SIEM) Tools

Tools providing real-time analysis of both security information, such as login and access patterns, with security event management, like records of attacks on a firewall or behavior by a computer on the network that may be infected by a virus. These data aggregations allow a network administrator to collect and normalize acceptable everyday usage, while finding anomalies and patterns that are clues to attacks on the system. Doing this effectively requires correlations from multiple sensors and the ability to store records over time and generate analytics reports.

Although important to the functioning of a secure network, SIEM tools are an expensive and labor-intensive investment available to large-scale and large-budget enterprises. Among the commercially available SIEM packages are IBM Security QRadar, SolarWinds Log and Event Manager, and McAffee Enterprise Security Manager, all of which represent expensive security measures, which must be managed by dedicated cybersecurity professionals to be fully effective.

Margaret Sankey

See also: Computer Network Defense (CND)

Further Reading

Bhaat, Sandeep, Pratyusa K. Manadhata, and Loai Zomlot. "The Operational Role of Security Information and Event Management Systems." *IEEE Security & Privacy* (October 2014).

Lane, Adrian. "Understanding and Selecting SIEM/LM: Use Cases, Part 1." *Securosis* (April 30, 2010).

Tankard, Colin. "Big Data Security." *Network Security* 7 (July 2012), 5–8.

Shamoon

A cyberattack independently identified in August 2012 by analysts at Kaspersky Lab, Symantec Corporation, and Seculert. Dubbed "Shamoon," the attack consisted of a self-replicating virus that primarily served as a means of cyber sabotage. The Shamoon virus has a modular design, allowing it to be quickly modified to tailor attacks for specific targets. In the 2012 case, it contained three primary components, namely a dropper, a wiper, and a reporter.

The dropper function of Shamoon served to spread the malware to as many computer workstations as possible within a network. It copied itself to uninfected systems and in the process activated the other program components. Shamoon proved capable of infecting both Microsoft Windows client and server-based machines, making it able to target the vast majority of computers in the world, particularly if they do not have the latest security patches. The dropper was equipped with multiple subroutines that could be activated depending on the operating system on the infected computer.

Once activated, the wiper component sought to delete files and overwrite them with deliberately corrupted JPEG images. Specifically, the 2012 version of Shamoon used an image of a burning U.S. flag, possibly as an additional form of propaganda. For all intents and purposes, this served to completely eliminate all of the data and programs on a computer, and at the same time prevent it from being repaired. Upon completion of this activity, the reporter component transmitted information back to the attacker, allowing tabulation of the amount of destruction caused.

Shamoon appears designed to target energy companies, although its modularity indicates that it could be easily adapted for other attacks. On August 15,

2012, it infected approximately 30,000 computers belonging to Saudi Aramco. On August 27, RasGas, a natural gas company in Qatar, experienced a similar attack. In each case, the affected workstations had to be completely replaced, as the master boot file of the computers was one of the systems affected by the attack. Both Aramco and RasGas lost more than a week of energy production due to the attacks.

A group calling itself Cutting Sword of Justice claimed responsibility for the attacks, ostensibly in retaliation for earlier cyberattacks against Iranian energy companies. An insider may have initiated the attack in both instances, possibly in return for financial compensation from an Iranian source. Although not particularly sophisticated, Shamoon's brute-force and scorched-earth approach make it one of the most costly viruses ever unleashed.

Paul J. Springer

See also: Cyberweapons; Malware; Stuxnet

Further Reading

Libicki, Martin. *Cyberspace in Peace and War.* Annapolis, MD: Naval Institute Press, 2016.

Zetter, Kim. "Qatari Gas Company Hit with Virus in Wave of Attacks on Energy Companies." *Wired,* August 30, 2012.

SIPRNet

An acronym that stands for SECRET Internet Protocol Router Network. This system has a different purpose than the NIPRNet, which means Non-Classified Internet Protocol Router Network. As seen from the acronyms, the main function of SIPRNet is to facilitate the transfer of classified information through a network of authorized users. The Department of Defense established and maintains SIPRNet for its operations, but the Department of State used the system as well to handle SECRET information until 2010. The Director of National Intelligence, the Department of Homeland Security, and the Federal Bureau of Investigation (FBI) have access to SIPRNet as well.

SECRET information is that information, which if it became publicly available, would cause serious damage to national security. This stands in contrast to TOP SECRET information, which, if publicly available, would cause exceptionally grave damage to national security. TOP SECRET information is transferred via a different system, the Joint Worldwide Intelligence Communication System (JWICS), which differs in construction from SIPRNet and NIPRNet. NIPRNet and SIPRNet commonly share the same communication lines with SIPRNet being the encrypted signal that runs alongside the unclassified data. JWICS is a completely separate system that is isolated from all others because of the sensitivity of the information carried on the network.

SIPRNet transmits classified information through various forms. Department of Defense users can send e-mails, files, and host websites on SIPRNet just like an unclassified system, but the information is only available to authorized users. SECRET information on the system can be available to U.S. personnel and members of FIVE EYES (a term referring to an intelligence-sharing alliance between the United States, Great Britain, Australia, New Zealand, and Canada).

Although SIPRNet information is only available to authorized users, controversy arose following the attacks on September 11, 2001. In light of the attacks and the coming "Global War on Terrorism" campaign, the number of users on SIPRNet increased rapidly. It was the increase in authorized users that many blame for the leak of classified information to the press. In 2010, WikiLeaks published more than 250,000 Department of State classified cables that came from the SIPRNet. As a remedy, the Department of State took itself off SIPRNet and sought to establish a classified system of its own to handle classified information and cables within the department.

Today, SIPRNet still remains the preferred means of transferring SECRET information among Department of Defense authorized users and allied nations.

Melvin G. Deaile

See also: NIPRNet; WikiLeaks, Impact of

Further Reading

Bonds, Timothy M., et al. *Army Network Enabled Operations: Expectations, Performance, and Opportunities for Future Improvements*. Arroyo, CA: RAND, 2012.

Libicki, Martin C. *Cyberspace in Peace and War*. Annapolis, MD: Naval Institute Press, 2016.

Singer, P. W., and Allan Friedman. *Cybersecurity and Cyberwar: What Everyone Needs to Know*. Oxford: Oxford University Press, 2014.

Skygrabber

A software used to hack into the video feeds of U.S. military unmanned aerial vehicles (UAVs) during the Iraq War. The software, which could be downloaded at a price of $26, is thought to have been used as recently as 2009 by anticoalition insurgents to eavesdrop unencrypted feed as it was electronically communicated from MQ-1 Predator systems to satellites to controller pilots in the United States. The software reportedly exploited the fact that UAVs in use during the Iraq conflict and occupation phases from 2003 to 2010 had not yet been provided with encryption for data links.

Skygrabber, and similar software, allowed insurgents to monitor video feeds and derive relevant data on U.S. surveillance information, but the software did not include any ability to control the UAV itself, either in terms of its flight or its weapon deployment. RAND analysts have characterized such activities as "the least-threatening and most-prolific form of signals exploitation" (Weinbaum, 2017, 5), although awareness of surveillance imagery can provide implicit information about surveillance patterns and priorities. The software itself is reportedly Russian, although analysts believe that groups connected to Iran were involved in the deployment of Skygrabber targeting U.S. UAV signals.

Nicholas Michael Sambaluk

See also: Encryption; MQ-1 Predator; Unmanned Aerial Vehicle (UAV)

Further Reading

Cabello, Carlos S. *Droning On: American Strategic Myopia toward Unmanned Aerial Systems.* Monterey, CA: Naval Postgraduate School, 2013.

Weinbaum, Cortney, et al. *SIGNINT for Anyone: The Growing Availability of Signals Intelligence in the Public Domain.* Santa Monica, CA: RAND, 2017.

Snake, Cyber Actions against Ukraine, 2014

A series of deliberate network compromises occurring between May and August 2014 and harbinger to subsequent and well-publicized nationwide cyberattacks conducted against Ukraine during 2015. As such, the 2014 attack is believed to have served as a test for the subsequent intrusions. The efforts, which experts have attributed to Russia, coincide with tension and conflict as Russia annexed the Crimea (formerly part of Ukraine) and as pro-Russia satellite republics were carved out of Eastern Ukraine.

The first attack in 2014 hit localized power stations and disrupted electricity to both commercial and private consumers. The second and third, later in the summer, assailed the national railway and five regional-level government administration systems. Both sets of cyberattacks originated from deceptive e-mails, intentionally designed to mimic normal e-mail traffic and mask destructive intent. As with the majority of cyberattacks in former Soviet states, especially ones with a favorable relationship with the West, blame is assigned to the Russian state. Typical of these attacks is an inability to attribute responsibility to a degree that merits international sanction. The ungoverned space of the Internet provides sophisticated hackers and belligerent state actors cover to behave without ascription.

Three target categories suffered attacks in the summer of 2014, each constituting key infrastructure and a logical choice for testing defensive networks in preparation for a more comprehensive attack. The electrical grid of specific regions of Ukraine became the first objective of hackers in May. In August, the railroad transportation systems and a handful of regional governments' websites suffered a directed network attack. Although each attack that summer appeared discrete, the combined effect of having no power, transportation, or public administration, especially if executed simultaneously, would be highly disruptive and even destabilizing to local populations. These attacks provided hackers with the opportunity to penetrate the network safeguards of three key systems.

In order to gain access to the electrical, railroad, and governmental systems, the hackers behind the breach utilized a combination of two techniques: e-mail phishing and malicious documents. Phishing scams generate e-mails to users within a specific web domain and emulate legitimate messages. An unknowing computer user can open an e-mail and open a hyperlink believing it to be real, only to then infect that computer with a number of possible viruses. Similarly, malicious documents appear in e-mails as normal file attachments. An unwitting user opens the document, thereby infecting the computer, and possibly the network, with dangerous software. Hackers successfully created e-mails that mimicked genuine organizational communications and counterfeited Microsoft Office documents to gain access to those networks.

The sophistication with which the 2014 attack occurred and the specific targets as well as the 2015 and 2016 cyberattacks in Ukraine all support the theory that the Russian state was behind the breaches. These attacks occurred after Russia's annexation of the Crimean Peninsula but while pro-Russian separatists in the eastern sections of Ukraine formed and prepared for an insurgent campaign. In 2015, a massive cyberattack struck the power grid at a nationwide level in Ukraine and forced the suspension of some government services as well as the shutdown of public utilities and transportation. Obviously, an attack on smaller-scale versions of the power grid, railway transport, and governmental administration provide an opportunity for experimentation, and evidence, ex post facto, of Russian culpability.

The year 2014 became a bellwether year for Russian foreign policy. After successfully annexing the Crimean Peninsula and inciting insurgent movements in Eastern Ukraine, multiple waves of cyberattacks ripped through Ukraine. That summer witnessed an initial cyberattack foray into strategic networks regulating utilities, transportation, and administration. The attacks were brief but intense and demonstrated a dynamic propensity to change based on new defensive measures. Hackers gained access by duping computer users into opening e-mails and documents infected with malware that experts attribute to Russian state security services. The timing, targeting, and methods of the attack reveal a high degree of technical knowledge and support. Further, Ukraine would experience a substantial attack on its power grid the following year, also blamed on Russia, but without clear attribution. These attacks illustrate a belligerent state experimenting with a new set of weapons. The Russians' experiment with the Near Abroad demonstrated the potential success for further operations, which the world saw during the 2016 U.S. presidential elections.

Andrew Akin

See also: Malware; Ukraine Power Grid Cyberattack, December 2015

Further Reading

Connell, Michael, and Sarah Vogler. *Russia's Approach to Cyber Warfare.* Arlington, VA: CNA, 2017.

Libicki, Martin C. *Cyberspace in Peace and War.* Annapolis, MD: Naval Institute, 2016.

Polyakova, Alina, and Spencer P. Boyer. *The Future of Political Warfare: Russia, the West, and the Coming Age of Global Digital Competition.* Washington, D.C.: Brookings Institution, 2018.

Weeden, Jen. "Beyond 'Cyber War': Russia's Use of Strategic Cyber Espionage and Information Operations in Ukraine." In *Cyber War in Perspective: Russian Aggression against Ukraine,* ed. Kenneth Geers. Tallinn, Estonia: NATO Cooperative Cyber Defence Centre of Excellence, 2015.

Sony Hack, 2014

Cyberattack typically attributed to the North Korean regime against Sony Pictures Entertainment as retribution for producing a movie that depicted the assassination of North Korean dictator Kim Jong Un. The U.S. Federal Bureau of Investigation (FBI) and Directorate of National Intelligence (DNI) publicly attributed the Sony attack to the government of North Korea. Some cyber experts question the strength of supporting evidence and note the attack could have been carried out by hackers who have previously threatened Sony or by a disgruntled Sony employee.

The 2014 Sony attacks preceded the Christmas Day release of the action comedy *The Interview*, which portrayed actors Seth Rogan and James Franco as journalists who secure an interview with North Korean dictator Kim Jong Un. In the satirical film, the Central Intelligence Agency recruits the journalists in a fictional plot to assassinate Kim Jong Un. The North Korean Foreign Ministry referred to the film as an act of terrorism and war and threatened a harsh response.

Sony became aware their computer systems were hacked when malware installed on employee computers made them inoperable on November 24, 2014. Computers displayed neon skeletons accompanied by a threat to release secret company information. Employee information, e-mails, and unreleased Sony films were made public throughout late November 2014. The attack went beyond disruption of Sony's plans to distribute the film and aimed to destroy network infrastructure. Some computers were damaged beyond repair. Although the duration of the hack is unknown, it is thought to have lasted several months due to the amount of data stolen.

The North Korean regime publicly denied a role in the hack and offered to conduct a joint investigation with the United States, but the regime also signaled approval of the attack through statements disparaging Sony. A group called Guardians for Peace claimed responsibility for the hack and demanded that Sony Pictures cancel distribution, otherwise threatening attacks on outlets affiliated with the distribution of the film. On December 16, 2014, the hackers threatened the New York City premiere of the movie and any theaters screening the film with physical violence. Despite the lack of evidence to support the existence of a planned plot, Sony announced it would suspend distribution of the film, and numerous theater chains canceled film showings.

During a press conference on December 19, 2014, President Barack Obama blamed the North Korean regime for the hack, expressed disappointment in Sony's decision to capitulate to hacker demands, and warned of a future proportional response against North Korea. The FBI also released circumstantial and technical evidence pointing to North Korean involvement. The following day the media reported a 10-hour-long Internet outage in North Korea. Analysts were uncertain whether it was a result of an outside attack on network infrastructure or a preventative shutdown. American officials would not confirm if the Internet outage was the proportional response but added only that some aspects of the response would be visible.

Skepticism by cyber analysts of the publicly divulged evidence used to attribute the Sony attack to North Korea prompted the FBI to add that IP addresses utilized by the Guardians for Peace matched those in other attacks conducted by North Korea. Due to the rapid attribution and response by the Obama administration, many assume additional evidence exists but likely remains classified.

In late December 2014, Sony distributed *The Interview* to a small number of independent theaters and online venues. Press related to the hack contributed to the film's popularity. Approximately 3 million people legally viewed the film, but many others watched it illegally, which resulted in an estimated $30 million in revenue losses. According to an April 2015 Sony report, the company incurred $41 million in costs associated with the investigation and immediate effects of the attack. The company paid employees an additional $8 million in damages associated with leaked personal information in an October 2015 lawsuit settlement. Damage to Sony's reputation is also estimated to have resulted in unknown financial losses.

The Obama administration imposed sanctions in January 2015 on a number of North Korean organizations and individuals in response to the hack. Although the sanctions did not target individuals who perpetuated the attack, they were meant to deter future cyberattacks. North Korea has increasingly bolstered cyberattack capabilities for financial gain and retribution against state and nonstate actors through Bureau 121 of the Reconnaissance General Bureau (RGB). The U.S. response was unique in that it is considered the first major penalty imposed on a state for its role in a cyberattack against a corporation, particularly because Sony's corporate headquarters is in Japan. The United States justified retaliation against North Korea on the grounds that the Sony attack was meant to have a chilling effect on freedom of speech.

Melia T. Pfannenstiel

See also: Cyberweapons; WannaCry Virus

Further Reading

DeSimone, Antonio, and Nicholas Horton. *Sony's Nightmare before Christmas: The 2014 North Korean Cyber Attack on Sony and Lessons for US Government Actions in Cyberspace*. Baltimore: Johns Hopkins, 2015.

Gause, E. Ken. *North Korea's Provocation and Escalation Calculus: Dealing with the Kim Jong-un Regime*. Arlington, VA: CNA, 2015.

Libicki, Martin C. *Cyberspace in Peace and War*. Annapolis, MD: Naval Institute Press, 2016.

Sharp, Travis. "Theorizing Cyber Coercion: The 2014 North Korean Operation against Sony." *Journal of Strategic Studies* 40, no. 7 (2017), 898–926.

U.S. Library of Congress, Congressional Research Service. *North Korean Cyber Capabilities: In Brief*, by Emma Chanlett-Avery, Liana Rosen, John Rollins, and Catherine A. Theohary, R44912 (August 3, 2007), 1–13.

SQL Slammer

Also known as "Sapphire," a computer worm whose small size facilitated its infecting an estimated 75,000 computers worldwide in early 2003 and precipitated a temporary slowdown in Internet traffic. The event also pointed to other computer security factors, such as the relationship between the pace of updating computers with patches and global vulnerability to hacking events.

A notable characteristic of SQL Slammer was its small size, under 400 bytes, allowing it to fit within a single packet. Once infected, computers projected more copies of the worm to random IP addresses, and the compactness of the code has led some to liken it to a "fire and forget" weapon in the physical combat realm, which is generally deemed more versatile because it allows the user to rapidly reposition or to repeat further attacks.

The ease with which the worm spread raised the level of Internet traffic enough to overburden some routers into crashing, a status communicated to other routers and thus leading to redirection of traffic onto remaining routers, some of which in turn also crashed. The cascading effect arguably resembled the spread of a major power outage, as routers failed and 90 percent of the infected computers were impacted within a space of minutes. At the time this made SQL Slammer the fastest-spreading worm in history. The collapse of Internet routers was eventually reversed through manual action of managers restarting them, although Internet traffic was briefly impacted by the volume of signals between newly restarted routers identifying themselves to peer systems as being again part of the global network.

Rather than being a zero-day attack, SQL Slammer leveraged a vulnerability that was already known in the computer security community. The vulnerability existed on specific kinds of Microsoft software and was identified in late July 2002, accompanied by a patch addressing the vulnerability. Personal computers without that particular software proved not to be vulnerable to the worm, because they lacked the software and thus also the vulnerability that the software carried. However, during the six months between the release of the patch and the launch of the worm, many operators neglected to perform the update that installed the patch, and thus many computers remained vulnerable. This reportedly included a number of machines operated by Microsoft itself. Research a dozen years after the major outbreak from 2003 suggested that the worm might have reappeared, inviting analysts to consider parallels between the worm's proliferation and patterns in epidemic disease.

Nicholas Michael Sambaluk

See also: Code Red Worm; Malware; Vulnerability

Further Reading

Chindipha, Stones Dalitso, et al. *An Analysis on the Re-Emergence of SQL Slammer Worm Using Network Telescope Data.* Southern Africa Telecommunications Networks and Applications Conference, 2017.

Piker, Joanne. *MS SQL Slammer/Sapphire Worm.* Bethesda, MD: SANS Institute, 2003.

Zou, Cliff D., et al. "Monitoring and Early Detection for Internet Worms." *Computer Science Department Faculty Publication Series University of Massachusetts Amherst,* 2004.

Stiennon, Richard

U.S. author on cybersecurity issues and founder of IT-Harvest. He has vocally asserted that cyberwar is both possible and already in existence. This perspective sets Stiennon in opposition to authors such as Thomas Rid as to the likelihood and character of cyberwar. His career as a cybersecurity executive included directing of a research effort meant to detect advanced persistent threats while he worked for Kaiser Permanente.

Stiennon's writing frequently points to dangers and threats in the cyber domain, prominently including organized criminal activities and especially the persistent hacking activities attributed to the People's Republic of China and their strategic implications. With respect to cyberwar, Stiennon has taken issue with statements of then Defense Secretary Leon Panetta about the possibility of a "Cyber Pearl Harbor." Although agreeing that such an attack is possible, Stiennon has emphasized that an adversary's targeting of a victim's infrastructure would fail to meet the appropriate definition, which entails "a crippling military defeat thanks to overwhelming control of the cyber domain" (Stiennon, 2015, 19). Citing the complexity of anticipating which entities might become targets of a particular adversary, he has also proven critical of attempts to apply traditional risk management to cybersecurity issues.

Nicholas Michael Sambaluk

See also: Advanced Persistent Threat; Cyber Pearl Harbor; Rid, Thomas

Further Reading

Stiennon, Richard. *Surviving Cyber War.* Plymouth, UK: Government Institutes, 2010.

Stiennon, Richard. *Up and to the Right: Strategy and Tactics of Analyst Influence: A Complete Guide to Analyst Influence.* Birmingham, MI: IT-Harvest, 2012.

Stiennon, Richard. *There Will Be Cyberwar: How the Move to Network-Centric War Fighting Has Set the Stage for Cyberwar.* Birmingham, MI: IT-Harvest, 2015.

Stuxnet

The malware engineered to sabotage the centrifuges at the Natanz Fuel Enrichment Plan in Iran. Experts believe the malware to have fit into the United States' anti-nuclear proliferation strategy. Stuxnet marks a clear turning point in the history of cybersecurity and military history, since it is the first known instance of a cyber-physical attack. The earliest version of Stuxnet was first "discovered" in 2007 when an unidentified person submitted a sample of code to Virustotal, a collaborative antivirus platform. However, at the time, the code could not be deciphered. It was not until 2010, when VirusBlokAda, a Belarusian antivirus company, was hired by an Iranian client to investigate why a computer continually kept rebooting itself, that the purpose and unique engineering of Stuxnet would come to light.

An analysis of the 2010 code discovered on the malfunctioning computer revealed that there were *two* attack routines embedded in Stuxnet. The first attack routine, also known as the first variant of Stuxnet (and most likely the 2007 code submitted to Virustotal), was probably carried out between 2006 to 2009 and contained a payload for severely interfering with the Cascade Protection System (CPS) at Natanz, which was a fault protection system designed to deal with centrifuge overpressure problems. This first attack variant was an extremely stealthy, sophisticated, and complex engineering routine. This first variant is also often "forgotten," since it was only discovered because it remained embedded (but nonfunctional) in the second variant's code. If the engineers had thought to remove the first routine from the second variant, we might never have known how sophisticated the first routine was designed to interfere with the CPS. This failure to remove the first attack routine from the code from the second variant of Stuxnet still remains somewhat of an operational security mystery.

The second attack code, also known as the second variant of Stuxnet, which was probably first implemented in 2009, was simpler and less stealthy than its original variant. This second variant was designed to interfere with the Centrifuge Drive System (CDS) that controls rotor speeds. It is this second attack routine that is usually referred to as "Stuxnet." However, understanding the design and purpose of the first variant is also important for an overall grasp of "Stuxnet" as a whole. Further, in order to understand the unique character of Stuxnet as a cyber-physical attack, not only must the malware engineering be understood, but the design features and processes of the physical plant that was attacked must be understood as well.

Different from "cyberattacks," which usually only target an information technology (IT) layer, a cyber-physical attack must exploit vulnerabilities in three specific layers: (1) the IT layer, which spreads the malware; (2) the industrial control system (ICS) layer, which manipulates (but does not disrupt) process control; and (3) the physical layer, for example, valves and electrical drives. The Stuxnet attack is a textbook example of how interaction of these layers can be leveraged to create physical destruction by a cyberattack.

In the technical paper "To Kill a Centrifuge," analyst Ralph Langer points out that both attack routines manipulate the ICS layer to achieve physical damage by exploiting physical vulnerabilities in the centrifuge rotors. As previously mentioned, most of the literature on Stuxnet focuses only on the second attack routine designed to *overspeed* the centrifuge rotors. However, Langer notes that Stuxnet's original "forgotten" attack routine, probably first executed in 2006, was designed to *overpressurize* the centrifuge, which also affects rotor speed. This original attack code, he claims, is of an order of magnitude more complex than the later rotor speed routine and sheds more light on the reasoning behind why the attack was designed the way it was.

Iran took a low-tech approach to its uranium enrichment program. They used the IR-1 centrifuge, an obsolete design of the late 1960s and early 1970s, which they obtained from the Pakistani nuclear trafficker A.Q. Khan. The IR-1s always had difficulty maintaining a constant rotor speed, and the Iranians could never get the IR-1 centrifuges to operate efficiently. In order to address this problem, Iran reduced the operating pressure of the centrifuge in order to lower rotor wall pressure. However, despite its flaws, it is hypothesized that Iran proceeded with the use of the IR-1 centrifuges because they could produce the antiquated design cheaply and on an industrial scale. So, if one of the centrifuges broke or was damaged, they could be easily replaced. As a matter of fact, the centrifuges could be manufactured faster than they crashed.

In order to achieve stability in the fuel enrichment process and maintain effective control tolerances with the inefficient IR-1s, Iran used a unique Cascade Protection System (CPS) that was designed to cope with the ongoing centrifuge pressure troubles. The CPS consists of two layers. The lower layer was at the centrifuge level, which had three fast-acting shutoff values, if a centrifuge should run into trouble. Once a centrifuge was shut down, it was isolated from the rest of the system and could be replaced while the rest of the centrifuges remained running. However, if multiple centrifuges are isolated and shut down within the same cascade before they can be replaced, then UF6 gas pressure will increase, which could then cause it to eventually solidify. Iran decided to work around this problem by adding an exhaust valve at every enrichment stage to compensate for the overpressure. When the exhaust valve is opened, the overpressure is relieved into the dump system. The stage exhaust valves were operated by a Siemens S7-417 controller.

The cyberattack against the CPS was designed to infect the S7-417 controllers with a matching configuration. This allowed the malware to take over control completely. Legitimate control logic was executed only as long as the malicious code permitted it do so. When the attack sequence was activated, pressure sensor values suggesting normal operation were recorded for 21 seconds and then replayed in a constant loop while an attack was being executed. These normal operational values appeared on the supervisory control and data acquisitions (SCADA) screen in the control room, where the human monitors were deceived into thinking that everything was fine. In reality, the malicious code was sending signals to various controllers to keep the various exhaust valves closed, so that pressure would build up in the centrifuge and cause them to fail. Interestingly, the goal of the attackers did not seem to be to destroy all the centrifuges, since by doing so the attack would have been obviously detected. Rather, their goal seemed to be to cause them to fail for reasons undetectable to the controllers, causing a "demon in the machine" phenomena that would have kept the Natanz engineers entirely perplexed by the problem.

For reasons still unknown, the attackers decided to switch from the overpressure attack to a more direct rotor speed attack in 2009. This second attack routine was designed to overspeed the rotors, which also affects centrifuge pressure. This attack was not directed at the CPS but rather at a different control system, the Centrifuge Drive System (CDS), which was controlled by a much smaller S7-315. Like the first attack routine, this second routine operated about once a month. However, unlike in the overpressure attack, which recorded and played back legitimate control code to deceive the SCADA application, the overspeed attack did not rely on fake code being sent to the SCADA. In the overspeed attack, all legitimate control code was simply suspended during the attack sequence. Since the SCADA application monitoring of rotor speeds got its information directly from memory in the controller, and since the control logic was suspended during the attack, memory updates were no longer taking place. As a result, the SCADA application displayed the static rotor speed values in the controller memory that matched normal operation before the attack to increase rotor speed took place.

Stuxnet revealed to the international community the difficulty for defenders to prevent cyber-physical attacks. Although antivirus software, network segregation defenses, intrusion detection and prevention systems, and security patches are standard forms of defense against cyberattacks aimed at traditional IT layers, these defenses are not, in and of themselves, sufficient to protect a physical infrastructure against attacks aimed at the industrial control system layer. Difficultly, a nation's critical physical infrastructures are mostly owned, operated, and maintained by civilian agencies and personnel. Consequently, the burden lies mainly on the private sector to procure the resources to defend against cyber-physical attacks. This reality will most likely result in increasing cooperation among private, public, and military sectors to figure out how to defend a nation's critical infrastructure against offensive cyber-physical attacks. It may also lead to more complicated and expensive regulatory requirements as well. Currently, approximately thirty nations employ offensive cyber programs, including North Korea, Iran, Syria, and Tunisia. Iran's Oghab 2 has been connected to several attacks following Stuxnet, including one dubbed Shamoon that targeted Saudi Aramco computer systems. However, the countries believed to have the monetary assets and expertise to develop a Stuxnet-like cyberweapon are the United States, Israel, the United Kingdom, China, and France.

Deonna D. Neal

See also: Computer Network Attack (CNA); Cyberweapons; Gas Centrifuge; Malware; Oghab 2; Operation Olympic Games; Shamoon; Supervisory Control and Data Acquisition (SCADA)

Further Reading

Collins, Sean, and Stephanie McCombie. "Stuxnet: The Emergence of a New Cyber Weapon and Its Implications." *Journal of Policing, Intelligence, and Counter Terrorism* 7, no. 1 (2012), 80–91.

Jenkins, Ryan. "Is Stuxnet Physical? Does it Matter?" *Journal of Military Ethics* 12, no. 1 (2013), 68–79.

Langer, Ralph. "To Kill a Centrifuge: A Technical Analysis of What Stuxnet's Creators Tried to Achieve." Munich, Germany, November 2013.

Lindsay, Jon R. "Stuxnet and the Limits of Cyber Warfare." *Security Studies* 22, no. 2 (2013), 365–404.

Zetter, Kim. *Countdown to Zero Day: Stuxnet and the Launch of the World's First Digital Weapon.* New York: Crown, 2014.

Supervisory Control and Data Acquisition (SCADA)

Instruments used in the operation and regulation of systems involving critical infrastructure and industrial sectors. Although varying definitions can exist, national critical infrastructure can be thought of as embracing nine areas: energy, food, water, transportation, communications, emergency services, health care, financial services, and government. Diverse and complex, national critical infrastructure is composed of widely distributed networks, and interdependent functions and systems operate in both the physical space and cyberspace. These systems must be regularly monitored, maintained, and upgraded, making remote supervision and management necessary for the efficient operation and, in some cases, personnel security.

SCADA systems are used in industry as well as for critical infrastructure: power generation and distribution, water supplies, manufacturing of steel, and in experimental facilities (i.e., nuclear fusion). Rather than a full control system, SCADA is a software package that operates on the supervisory level allowing for data collection from one or more facilities or systems distributed over significant distances, in remote areas, or in highly dangerous environment while allowing users to send limited control instructions to those facilities and systems. Through the use of SCADA, it becomes unnecessary for operators to be assigned to or regularly visit normally operating remote facilities.

A SCADA system's basic functions include data acquisition and transmission, historical data analysis, remote control, and report writing that are common to generation, transmission, and distribution systems. Data is acquired through the use of sensors, transducers, and status point information in the operational location. This enables greater automation, so that a user in a control room can remotely monitor processes in an array of formats, including digital and pulse formats. The data provided to the user is displayed and updated on the user's operator console at appropriate time intervals. In return, the control commands issued by the user are received by the remote system, and appropriate action occurs. The analysis of historical data and production of reports can be an important function performed by the SCADA system. In a postevent analysis, using the data available after the event the data provided by the SCADA system can provide insights into the sequence of events during the event, if there were any malfunctioning devices within the system, and the actions taken by the user. This information could be used to guide future planning and implementations.

For SCADA systems fast, accurate, and secure communication is of utmost importance, particularly when operating across vast geographical areas. Data transmission is time bound and measured in milliseconds. Originally, SCADA systems were closed, standalone systems used to control the operations of power transmission, water distribution, and gas pipelines. As SCADA systems have become widely used in industry, the need for real-time communication is accessed using cyber networks, bringing the need for cybersecurity of these systems into focus as cyberattacks on SCADA systems are a reality.

SCADA systems are vulnerable to a wide range of cyber threats. These include various forms of malicious code, some of which deploy exploits targeting zero-day vulnerabilities, as was the case with Stuxnet as discovered in 2010. Advanced persistent threats (APTs) include SCADA among their targets. The standardization of hardware components, communication protocols, and the growing interconnectivity of these systems have led to increased cyberattacks since the 1982 cyberattack on the Trans-Siberian pipeline. Although the likelihood of catastrophic cyberattacks on SCADA systems is relatively low, such an attack, if successful, could cause long-term, widespread outages of critical infrastructure systems. The detrimental impact of such outages would range from financial losses, environmental damage, or even loss of human life. Although the National Institute of Standards and Technology (NIST) has been publishing cybersecurity analyses and recommendations for SCADA systems since 2004, SCADA systems continue to be vulnerable targets.

Diana J. Ringquist

See also: Advanced Persistent Threat; Malware; Operation Olympic Games; Stuxnet; Zero-Day Vulnerability

Further Reading

Boyer, Stuart A. *SCADA Supervisory Control and Data Acquisition.* Pittsburgh: The Instrumentation, Systems, and Automation Society, 2018.

Cherdantseva, Yulia, Pete Burnap, Andrew Blyth, Peter Eden, Kevin Jones, Hugh Soulsby, and Kristan Stoddart. "A Review of Cyber Security Risk Assessment Methods for SCADA Systems." *Computers & Security* 56 (2016), 1–27.

Knapp, Eric D., and Joel Thomas Langill. *Industrial Network Security.* Waltham, MA: Syngress, 2015.

Macaulay, Tyson, and Bryan Singer. *Cybersecurity for Industrial Control Systems: SCADA, DCS, PLC, HMI, and SIS.* Boca Raton, FL: CRC Press, 2011.

Thomas, Mini S., and John D. McDonald. *Power Systems SCADA and Smart Grids.* Boca Raton, FL: CRC Press, 2015.

Syrian Electronic Army (SEA)

A hacking entity sympathetic to the Syrian regime of Bashar al-Assad. The organization's activities have led to Syria being called "the first Arab country to have a public Internet Army hosted on its national networks to openly launch cyberattacks on its enemies" (Noman).

The group emerged in April 2011, swiftly following a shift in regime policy that ended years of strict Internet censorship. This policy change had come as a concession to prodemocracy protests occurring in Syria within the context of a wider "Arab Spring" movement across northern Africa and the Middle East. As protests gained momentum the demands moved increasingly to become calls for al-Assad's removal from power, and the regime replied with violence and arrests. A variety of antiregime entities, with widely disparate political objectives beyond opposition to the al-Assad regime, mobilized. A multiside civil war erupted by June 2011, and each of the eventually four sides (the Syrian Arab Republic, the Syrian Opposition, the Islamic State of Iraq and the Levant, and the Democratic Federation of Northern Syria) sought to raise both formal and informal fighters and assets to advance their respective political goals. The Syrian Electronic Army is aligned with the al-Assad regime in this struggle.

The SEA has used different methods and tools to promote its cause in the cyber domain. On the more highly sophisticated side of the spectrum, some of its efforts have included the use of malware. Other activities have involved phishing, spamming, and launching distributed denial-of-service (DDoS) attacks. These efforts require much less technical skill, but depending on the objective they can nonetheless be impactful. For example, the SEA has frequently defaced adversarial and neutral websites by illicitly replacing content with messages supportive of the al-Assad regime.

A larger portion of the SEA's DDoS actions have been directed against outside sources of professional media, such as BBC News and al-Jazeera. This, coupled with the Syrian regime's prompt eviction of foreign journalists from the country at the start of the civil war, points to the regime's interest in controlling messaging about the war both within the country and to some extent beyond it. The SEA over time diversified from a messaging focus at the start of the conflict to include espionage and cyberattack.

The group has used Facebook as a tool to transmit software to sympathizers interested in helping the SEA launch DDoS attacks. Analysts have pointed to conflicting aspects of the SEA's use of Facebook, however, as also indicating a preference within the SEA to avoid attracting active scrutiny and adverse action by the social media giant. Although the group has targeted many websites that are associated with the Syrian Opposition and the websites of some public figures who had expressed sympathy for the establishment of a democratic postwar regime, other sites targeted by the SEA bear little or no relation to the conflict within Syria. This has prompted speculation about the degree to which the organization is open simply to defacing websites that are soft targets and therefore less challenging to populate with their own messages.

The SEA's relationship with the al-Assad regime in Syria is unconfirmed. When it was first launched, the group's homepage had asserted that it was a collection of young and tech-savvy sympathizers of the regime who were not officially connected with the country's government. In subsequent weeks, however, this claim was removed from the group's website. Circumstantial evidence suggests that the organization includes at least some number of expatriate ethnic Syrians living outside the country. Colloquial English has appeared within some SEA messages, supporting this conclusion. Information also points to a potentially close, rather than merely sympathetic, relationship between the SEA and the regime, including the fact that the SEA's cyber domain name belongs to an organization Computer Science Club (CSC) that al-Assad's brother Bassel had founded in 1989. Bashar, the current regime ruler and at the time the country's heir apparent, inherited the CSC leadership role upon Bassel's accidental death in 1994.

Nicholas Michael Sambaluk

See also: Distributed Denial-of-Service (DDoS) Attack; Facebook, Use of; Hacktivism

Further Reading

Cyber Defense Project. *The Use of Cybertools in an Internationalized Civil War Context: Cyber Activities in the Syrian Conflict.* Zurich, Switzerland: CSS Cyber Defense Hotspot Analysis, 2017.

Grohe, Edwin. *The Cyber Dimensions of the Syrian Civil War: Implications for the Future.* Applied Physics Laboratory: Johns Hopkins, 2015.

Kaplan, Fred. *Dark Territory: The Secret History of Cyber War.* New York: Simon & Schuster, 2017.

Lee, Bryan. "The Impact of Cyber Capabilities in the Syrian Civil War." *Small Wars Journal,* April 26, 2016, http://smallwarsjournal.com/jrnl/art/the-impact-of-cyber-capabilities-in-the-syrian-civil-war.

Noman, Helmi. "The Emergence of Open and Organized Pro-Government Cyber Attacks in the Middle East: The Case of the Syrian Electronic Army." https://opennet.net/emergence-open-and-organized-pro-government-cyber-attacks-middle-east-case-syrian-electronic-army.

Tallinn Manual

A nonbinding but influential resource for legal advisers regarding cyberwarfare. Originally titled *Tallinn Manual on the International Law Applicable to Cyberwarfare,* the *Tallinn Manual* was produced in 2013 by a group of legal scholars known as the International Group of Experts (IGE) at the invitation of the NATO Cooperative Cyber Defense Center of Excellence (CCDCOE) located in Tallinn-Croatia. During this process, the IGE unanimously agreed that both the *ius ad bellum* and the *ius in bello* just war principles apply to cyber operations.

The *Tallinn Manual* enumerates ninety-five "rules," and it is divided into two parts. Part 1, "International Cyber Security Law," outlines the *ius ad bellum* for engaging in justified cyber conflict. Part 2, "The Law of Cyber Armed Conflict," describes how cyberattacks that are determined to have risen to a level equivalent to the use of conventional force must be conducted.

The original manual focused on the use of cyber operations that violate the use of force in international relations and entitle states to exercise the right to self-defense as during armed conflict. The original manual was significantly expanded and updated in 2017 and the second edition was released as *The Tallinn Manual 2.0*. This second edition adds a legal analysis of the more common cyber incidents that states encounter on a day-to-day basis, which would fall below the thresholds of the use of force or armed conflict.

Deonna D. Neal

See also: Computer Network Attack (CNA); Computer Network Defense (CND); Estonia, 2007 Cyber Assault on

Further Reading

Beard, Matthew. "Beyond Tallinn: The Code of the Cyberwarrior?" In *Binary Bullets: The Ethics of Cyberwarfare,* ed. Fritz Allhoff, Adam Henschke, and Bradley Jay Strawser. Oxford: Oxford University Press, 2016, 139–156.

Lucas, George. *Ethics and Cyberwarfare: The Quest for Responsible Security in the Age of Digital Warfare*. Oxford: Oxford University Press, 2017, 57–84.

Schmitt, Michael N., ed. *Tallinn Manual on the International Law Applicable to Cyberwarfare*. Cambridge: Cambridge University Press, 2013.

TAXII

One of the protocols used by cyber defenders for sharing information about hacker adversaries. The acronym stands for Trusted Automated eXchange of Indicator Information. Although not an information-sharing application per se, the tool does establish definitions for protocols and message exchanges to support efforts that share information about the Indicators of Compromise (IoC) such as the IP addresses and other digital evidence about malicious files. As such, it is considered to be a tool to facilitate greater automation of computer network defense (CND). TAXII was established by the U.S. Department of Homeland Security and has been operated by the nonprofit MITRE Corporation, due to a recognition that earlier largely nonautomated efforts for sharing cyber threat data caused response efforts to lag.

TAXII supports several different push-and-pull information-sharing models, including a source subscriber in which one source distributes information outward, peer-to-peer in which the senders and recipients of data share directly, and hub-and-spoke sharing in which a data hub stands essentially as a clearing house for data coming in and out among sharers.

Nicholas Michael Sambaluk

See also: Computer Network Defense (CND); Exploit; Vulnerability

Further Reading

Connolly, Julie, et al. "The Trusted Automated eXchange of Indicator Information (TAXII)." McLean, VA: MITRE, 2012.

Lock, Hun-Ya. "Using IOC (Indicators of Compromise) in Malware Forensics." *SANS Institute InfoSec Reading Room,* 2013, https://www.sans.org/reading-room/whitepapers/forensics/ioc-indicators-compromise-malware-forensics-34200.

Stiennon, Richard. *There Will Be Cyberwar: How the Move to Network-Centric War Fighting Has Set the Stage for Cyberwar.* Birmingham, MI: IT-Harvest Press, 2015.

10th Fleet

The operational unit for U.S. Fleet Cyber Command and is the Navy component of the joint U.S. Cyber Command (USCYBERCOM), coequal in that command with Army Cyber Command, the Marine Corps Cyberspace Command, and Air Forces Cyber/24th Air Force. The 10th Fleet's mission is to conduct operations in and through cyberspace, space, and the electromagnetic spectrum.

The 10th Fleet has a rich history dating back to World War II. Originally created in May 1943, the 10th Fleet was created under Fleet Admiral Ernest King and led the effort against German U-boats in the Battle of the Atlantic, primarily through intelligence operations and dissemination of information. In that role, the 10th Fleet assembled and coordinated shipping convoys, provided antisubmarine intelligence and training, and directed submarine hunter-killer formations. Before the establishment of the 10th Fleet, Allied forces sunk an average of only four enemy submarines per month. Once the fleet was created, Allied forces averaged twenty-three U-boats sunk per month for the remainder of the war. After the surrender of Germany, the 10th Fleet was disbanded in June 1945.

In 2009, as a reaction to a 2008 cyber threat and the subsequent American response in Operation Buckshot Yankee, U.S. Strategic Command (USSTRATCOM) created the subordinate U.S. Cyber Command. To form the Navy component of that command, the 10th Fleet was recommissioned on January 29, 2010, by Chief of Naval Operations Admiral Gary Roughead. During the recommissioning ceremony, Roughead argued that the emergence of cyberspace threats posed a danger similar the historic threat of U-boats to Allied shipping during World War II. He asserted that intelligence and information were more crucial elements of victory historically and would be even more so the case in the realm of cyber warfare.

Fleet Cyber Command and the 10th Fleet were headquartered at Fort George Meade, Maryland, with Vice Admiral Bernard J. McCullough III as the first commander of both organizations. The 10th Fleet's standing forces include ten command task forces and twenty-seven command task groups spread across the globe. These forces and groups are organized into four main sections: Network Operations and Defense, Information Operations, Service Cryptologic Component Operations, Fleet and Theater Operations, plus one task force dedicated to Research and Development. More than 16,000 active and reserve sailors serve as part of the 10th Fleet.

The 10th Fleet is responsible for Navy information network operations, cyber operations (both offensive and defensive), signals intelligence, and space operations. In addition to serving as the Navy component of USCYBERCOM, the fleet is the

Navy space component to USSTRATCOM, and the Navy Cryptologic Component Commander for the National Security Agency (NSA) and Central Security Service (CSS). The fleet reports directly to the Chief of Naval Operations (CNO).

Naval cyber operations became much more intensive after September 2013, when Iranian hackers infiltrated the Navy and Marine Corps Intranet. This network was unclassified, and no sensitive information was taken. The hackers' intent was primarily reconnaissance, not destruction. The effort to defeat the Iranian intrusion was known as Operation Rolling Tide. Led by 10th Fleet, then under the command of Vice Admiral Michael S. Rogers, that effort was the first-named operation specifically undertaken to counter cyberactivity. Rogers, who later became Director of the NSA and Commander of USCYBERCOM, was determined not just to expel the Iranian hackers but to use the operation as a galvanizing incident to introduce widespread change throughout the Navy in the realm of cybersecurity. For the 10th Fleet's successful effort in revolutionizing Navy cyber defense during that operation, Secretary of the Navy Ray Mabus bestowed the fleet with the Navy Unit Commendation. The hacking incident caused significant alarm and led to the creation of Task Force Cyber Awakening. This task force proceeded to establish the Navy Cybersecurity Safety Program Office (CYBERSAFE), dedicated to protecting Navy cyber systems through the creation of new programs and requirements as well as increasing training for and awareness of cyber threats throughout the service.

In 2015, the then-commander of 10th Fleet, Vice Admiral Jan Tighe, the first woman commander of a numbered fleet, reported that since the 2013 hacking incident, no enemy had conducted a successful cyberattack against the Navy or the Marines. In July 2016, Tighe became the Deputy Chief of Naval Operations for Information Warfare and the Director of Naval Intelligence. She was replaced as commander of Fleet Combat Command and 10th Fleet by Vice Admiral Michael Gilday. The fact that Gilday's background was more heavily involved with warships than with information warfare was an attempt to develop further integration of cyber operations with the rest of the Navy's conventional warfighting capability. As a further part of the Navy's shift toward a tighter integration of information defense and cyber warfare, in 2017, the Navy launched a training program for cyber specialists called the Information Warfighting Development Center, described as the cyber equivalent of the famous "Top Gun" school for fighter pilots. There, information warfare practitioners would receive the highest-level training available, then be sent to their regular fleet assignments to teach others.

Also in 2017, after a series of seemingly unlikely accidents, the 10th Fleet made it standard procedure to conduct cyber investigations after any accidental collisions at sea. This came in response to the 2017 incident in which the merchant ship *Alnic MC* unexpectedly swerved into the destroyer USS *McCain*. Although the Navy insisted there was no evidence of a cyberattack in that case, the inherent vulnerability of merchant vessels to such attacks led the Navy to use the event as a test case for the new procedure of regular cyber investigations.

Michael Hankins

See also: Operation Rolling Tide; Rogers, Michael S.; 24th Air Force; USCYBERCOM

Further Reading

Brantly, Aaron Franklin. *The Decision to Attack: Military and Intelligence Cyber Decision-Making.* Athens: University of Georgia Press, 2016.

Conti, Gregory, and David Raymond. *On Cyber: Towards an Operational Art for Cyber Conflict.* New York: Kopidion Press, 2017.

Kaplan, Fred. *Dark Territory: The Secret History of Cyber War.* New York: Simon & Schuster, 2016.

Libicki, Martin. *Cyberspace in Peace and War.* Annapolis, MD: Naval Institute Press, 2016.

Valeriano, Brandon, et al. *Cyber Strategy: The Evolving Character of Power and Coercion.* Oxford: Oxford University Press, 2018.

Trusted Integrated Circuits (TRUST)

A program led by the Defense Advanced Research Projects Agency (DARPA) since its creation in 2008, addressing the reliance of U.S. defense industries on integrated circuits that are designed and manufactured by companies outside of the United States. Frequently, sourcing is from competitors or opponents. This access to the many vital components of the military's hardware raises serious concern about "backdoors" and flaws built into the large volume of high-performance chips required for manufacture of products like the F-35. Although the Department of Defense operates the "Trusted Foundry" program to certify domestic production, it is not financially feasible to alter the supply chain to require U.S. made and designed hardware, so TRUST exists to screen existing products for malicious components in a time-sensitive and high-reliability process that does not destroy the chips in the process of testing them. Currently, TRUST is a cycle of three, one-year phases, each with escalating metrics designed to identify additional and possibly malicious components of sample chips intended for use in military hardware systems.

Margaret Sankey

See also: Computer Network Attack (CNA); Computer Network Defense (CND); Cyberweapons; Defense Advanced Research Projects Agency (DARPA); Malware

Further Reading

Adee, Sally. "The Hunt for the Kill Switch." *IEEE Spectrum*, May 1, 2008, https://spectrum.ieee.org/semiconductors/design/the-hunt-for-the-kill-switch.

Alkabani, Y., F. Koushanfar, and M. Potkonjak. "Trusted Integrated Circuits: A Nondestructive Hidden Characteristics Extraction Approach." In *Information Hiding*, ed. K. Solanki, K. Sullivan, and U. Madhow. Berlin: Springer, 2008, 102–117.

Collins, Dean. "TRUST, a Proposed Plan for Trusted Integrated Circuits, Presented to DARPA," March 2006, http://www.dtic.mil/dtic/tr/fulltext/u2/a456459.pdf.

24th Air Force

Established in August 2009 and headquartered at Lackland Air Force Base (AFB) in San Antonio, Texas, the entity tasked with overseeing cyberspace operations of the U.S. Air Force. The establishment of a numbered air force (NAF) served as an expression not only of the service's recognition of the importance of cyber but also of the service's interest in playing an important part in securing the new domain. The 24th Air Force serves as that service's component of the U.S. Cyber Command (USCYBERCOM).

The organization has a strength of approximately 7,000. Nearly half of this number are uniformed active-duty military personnel, working with 900 federal civilian employees and assisted by 1,400 contractors. Reservists account for more than 1,000 further personnel. The 24th Air Force's mission involves the establishment, maintenance, and defense of Air Force networks as well as "directed mission critical cyber terrain," providing friendly forces with appropriate cyber capabilities, and "engaging the adversary in support of combatant and air component commanders" (United States Air Force, 2017). Part of this force includes the 39 cyber teams that reportedly include 1,700 airmen able to serve the Joint Cyber Mission Force. These cyber teams are divided into three types, with a dozen designated as National Mission Teams for the Cyber National Mission Force, thirteen as Combat Mission Teams for the U.S. Strategic Command and U.S. European Command, and fourteen as Cyber Protection Teams serving various military organizations. The Mission Defense Teams have been described as small groups numbering between four and six cybersecurity experts each, existing as a pilot program to provide cyber expertise and security across all Major Commands.

In addition to the 624th Operations Center, three wings originally comprised the 24th Air Force. These were the 67th Network Warfare Wing, the 688th Information Operations Wing, and the 689th Combat Communications Wing. Although the latter of these organizations was inactivated in June 2013, these units point to the array of responsibilities that are connected to the security of Air Force networks. The operations center plans and coordinates activities, while the network warfare wing focuses on preparation of defensive and offensive computer network operations. The information operations component adds information operations capabilities because these dovetail with, but are not necessarily synonymous with, cyber operations. The combat communications wing trained and deployed communications in support of these activities. Outgoing 24th Air Force commander Major General Burke "Ed" Wilson emphasized in 2016 that the service's cyber professionals were organized "either [as] cross-functional team of teams or [as] specialized units" (Wilson, 2016, 29).

Prior to the solidification of a NAF focused on cyber, analysts had noted a "disparity between the Air Force's publicly perceived role in and its actual more-modest investment in cyberspace" (Mesic, 2010, 3), something that could be addressed through the establishment of dedicated organizations institutionalizing more robust capabilities. The 24th Air Force was first envisioned as a main component of an Air Force Cyber Command (AFCYBER), but by October 2008 this had been decided against by Air Force leaders. Instead, the 24th Air Force would answer the cyber warfare mission needs of the Air Force Space Command (AFSPC), arguably in keeping with the service's perception of space and cyber as aspects related to and implicitly supportive of the air-breathing mission set of the Air Force. In a potentially significant realignment within the service, the 24th Air Force was shifted to the authority of the Air Combat Command in July 2018. This may imply an impending emphasis on the operationalization of cyber capabilities and an increased awareness that these capabilities are used in the present instead of their existing as a future potential.

The 24th Air Force has undertaken a wide range of operations in the cyber domain, including not only defensive network operations but also reportedly offensive activity. Its leaders have continued to identify defense as the first priority mission, citing projects such as the use of Automated Remediation and Asset Discovery software, which "perform[s] vulnerability management, incident response, system health diagnostics," and other functions for Air Force networks (Weggeman Presentation, 2018, 3). The 24th Air Force briefings have underscored the point that cyber activities are "synchronized" as part of "all-domain integrated effects" that include cyber squarely among the other relevant domains of "air, space, land [and] sea." This is presented as demonstrating "cyber IN war . . . not cyber war" (Weggeman Slides, 2018, 4). Nonetheless, several ongoing challenges have been acknowledged, including the need for situational awareness in the domain, the need for agility in the training of cyber personnel and in the acquisition of requisite technologies, and the demands involved in maintaining a constant state of readiness.

The first months of the 24th Air Force's history witnessed extensive efforts to build toward a full operational capability, which was reached at the start of fiscal year 2010. Leaders aimed to nest the organization's activities and roles within the larger framework of the Air Force's missions, and this meant integrating actions in the cyber domain into the existing philosophy known as air-mindedness. The 24th Air Force first provided the cyber component to the service's Red Flag exercise in 2014.

Stephanie Marie Van Sant and Nicholas Michael Sambaluk

See also: Computer Network Attack (CNA); Computer Network Defense (CND); 10th Fleet; USCYBERCOM

Further Reading

Mesic, Richard, et al. *Air Force Cyber Command (Provisional) Decision Support.* Santa Monica, CA: RAND, 2010.

Sovada, Jennifer P. *Postured to Support Air Force and USCYBERCOM Cyber Needs?* Carlisle, PA: Army War College, 2013.

United States Air Force. "24th Air Force (Air Forces Cyber)." *Fact Sheets,* February 8, 2017.

Weggeman, Chris. *24th Air Force/AFCYBER: Delivering Outcomes through Cyberspace.* Briefing slides, June 2018.

Weggeman, Chris. "Military Cyber Programs and Posture." *Presentation to the Senate Armed Services Committee, Subcommittee on Cybersecurity, United States Senate.* Washington, D.C.: Department of Defense, 2018.

Wilson, Burke "Ed." "Embedding Airmanship in the Cyberspace Domain: The First Few Steps of a Long Walk." *Cyber Defense Review* 1, no. 1 (Spring 2016), 27–32.

Ukraine Power Grid Cyberattack, December 2015

The first occasion in public record in which hackers successfully caused power outages. Hackers illegally entered the supervisory control and data acquisition

(SCADA) systems of three major energy companies in Ukraine on December 23, 2015, disrupting the supply of electricity to 200,000 customers across much of that country for 3 hours.

The hackers used a number of methods to carry out their attacks, including spear phishing e-mails, BlackEnergy 3 malware, and manipulation of Microsoft Office documents containing malware, in order to gain access into the information technology (IT) networks of the companies. They managed to harvest credentials to gain access to the Internet Connection Sharing (ICS) network through virtual private networks (VPNs). They also were able to manage the ICSs using the supervisory control system. They targeted field devices, rendering them inoperable with custom malicious malware. They even interfered with telephone systems to generate thousands of calls to customer call centers, denying customers the ability to report outages. A modified KillDisk erased the master boot record of impacted systems and targeted and deleted some information logs. These steps to erase records and logs of the attack made it difficult to determine who was responsible. Although the initial outages only lasted a few hours, critical devices at some substations are still unable to communicate with remote commands and have to be operated manually.

Experts noted the months of planning, including extended reconnaissance of targeted networks and the gathering of necessary credentials. Moreover, the action itself was carefully synchronized, leading to speculation of collaboration between cybercriminals and nation-state actors. Ukrainian officials were quick to assign blame to Russia.

Although Russia has not been officially named as responsible for the attack, circumstances surrounding the attack and tensions between Russia and Ukraine add credibility to the theory. When Russia annexed Crimea in 2014, they began taking over control of Ukrainian-owned energy companies there. Ukrainian activists retaliated by attacking substations distributing power to Crimea, cutting power to millions of Crimean residents. There is strong speculation that the Ukrainian power grid attack was retaliation for the attacks on the Crimean power substations. However, preparation for the power grid attack began before the attacks on the Crimean power substations. Evidence suggests that hackers may have sped up their power grid attack after the Crimean attacks, but that was not the initial catalyst. Others suspect that the power grid attack could have been orchestrated to cause frustration among Ukrainians with their privately owned power providers and garner favor for a bill being considered to nationalize those companies.

This unprecedented power grid cyberattack has raised awareness of the need for increased security precautions for power companies across the globe. It is difficult to determine definitively who is responsible for a cyberattack, making it an attractive strategy and increasingly viable threat.

Amy Baxter

See also: Malware; Snake, Cyber Actions against Ukraine, 2014

Further Reading

Connell, Michael, and Sarah Vogler. *Russia's Approach to Cyber Warfare*. Washington, D.C.: CNA, 2017.

Lee, Robert M., et al. *Analysis of the Cyber Attack on the Ukrainian Power Grid.* Washington, D.C.: SANS, 2016.

Risk and Resilience Team. *Hotspot Analysis: Cyber and Information Warfare in the Ukrainian Conflict.* Zurich, Switzerland: Center for Security Studies, 2017.

USCYBERCOM

The subunified command under U.S. Strategic Command (USSTRATCOM) focused on the protection of U.S. military networks and responses to cyberattacks. On June 29, 2009, Secretary of Defense Robert Gates directed the Commander of Strategic Command to establish U.S. Cyber Command (USCYBERCOM), which became operational on October 31, 2010.

USCYBERCOM in its current form resembles the first joint cyber warfighting organization, Joint Task Force for Computer Network Defense (JTF-CND), which was created on December 30, 1998, and later expanded and was renamed Joint Task Force for Computer Network Operations (JTF-CNO) in 2000. Prior to the establishment of USCYBERCOM under USSTRATCOM, cyber military capabilities were under Joint Task Force for Global Network Operations (JTF-GNO), primarily focused on defensive missions and the offensive-focused Joint Functional Component Command-Network Warfare (JFCC-NW).

Several cyber threats and intrusions between the late 1990s and early 2000s, including Solar Sunrise, Moonlight Maze, and a range of Chinese and Russian espionage activities, served as a catalyst to reorganize capabilities under USCYBERCOM. Most notable, the worm that resulted in Operation Buckshot Yankee was transferred through a USB stick and allowed intruders to gain access to the military's unclassified SIPRNet and classified networks. This highlighted the vulnerabilities within the U.S. Department of Defense security protocols and the need to centralize cyber defenses.

USCYBERCOM is the result of these reorganizations to centralize cyber offensive and defensive responsibilities to better provide support to combatant commanders in their execution of missions and continuously work to strengthen the ability of the United States to withstand and respond to cyberattacks. Upon the establishment of USCYBERCOM, the 2010 Quadrennial Defense Review and 2011 National Military Strategy directed joint forces to focus operations toward the mission of securing the ".mil" domain, building relationships between U.S. government agencies and with international partners, strengthening capabilities by recruiting top workforce talent, and employing operating concepts to detect, deter, and deny security threats.

According to the stated mission, USCYBERCOM "Plans, coordinates, integrates, synchronizes, and conducts activities to: direct the operations and defense of specified Department of Defense information networks and; prepare to, and when directed, conduct full-spectrum military cyberspace operations in order to enable actions in all domains, ensure US/Allied freedom of action in cyberspace and deny the same to our adversaries" (U.S. Department of Defense, 2010).

USCYBERCOM remains a subordinate command as of early 2018, under the command of the 4-star general leading Strategic Command (based at Offutt Air

Force Base, Nebraska). USCYBERCOM is bureaucratically closely linked to the National Security Agency (NSA) and is located with the NSA at Fort Meade, Maryland. The director of the National Security Agency also leads USCYBERCOM, and the two organizations share many resources, including personnel. Cyber Command's relationship with the NSA, the intelligence organization focused on signal and information protection, is intended to strengthen effectiveness in the cyber warfighting domain. However, the future of this arrangement is under review by Secretary of Defense James Mattis. In an official statement released on August 18, 2017, President Donald Trump announced the U.S. Cyber Command will be elevated to the status of a unified combatant command, focused on cyberspace operations. The elevation of USCYBERCOM is expected to streamline funding and command and control of cyberspace operations.

The military services have each developed and maintained respective cyber commands. The service cyber components originally fell under the joint task forces preceding the establishment of a dedicated U.S. Cyber Command. The force assigned to Cyber Command's Fort Meade, Maryland, headquarters is multiplied by the joint forces cyber commands. The 24th Air Force is headquartered at Lackland AFB, Texas. It was originally part of the planned Air Force Cyber Command (AFCOM), which existed in provisional status until the 24th Air Force was designated in 2009 with the cyber warfighting focus (AFCYBER). The 2nd Army/Army Cyber Command (ARCYBER) based at Fort Gordon, Georgia, the Navy's 10th Fleet/Fleet Cyber Command (FLTCYBER), and Marine Forces Cyber Command (MARFORCYBER) based at Fort Meade, Maryland, collectively provide specialized cyber warfighting capabilities and report to USCYBERCOM. Coast Guard Cyber Command (CGCYBER) is subordinate to the Department of Homeland Security but maintains a support relationship with Cyber Command.

To better unify the joint forces cyber commands under USCYBERCOM, in 2012 the Joint Staff and U.S. Cyber Command directed the services to build a Cyber Mission Force consisting of 133 Cyber Mission Teams. According to the 2015 Department of Defense Cyber Strategy, the 133 Cyber Mission Teams expected to be operational in 2018 are divided among: 13 National Mission Teams to "defend the United States and its interests against cyberattacks of significant consequence"; 68 Cyber Protection Teams to "defend priority DoD networks and systems against priority threats"; 27 Combat Mission Teams to "provide support to Combatant Commands by generating integrated cyberspace effects in support of operational plans and contingency operations"; and 25 Support Teams to "provide analytic and planning support to the National Mission and Combat Mission teams."

The establishment of a unified combatant command and the future association with the National Security Agency are among the most consequential issues facing U.S. Cyber Command. Other current issues include mission area and responsibilities, specifically the expected role of the Department of Defense in the protection of and responses to attacks on civilian networks. Because the Department of Homeland Security also provides cyber defense of civilian networks, the role of USCYBERCOM is not clearly defined. This discussion is expected to continue alongside that of the future organizational structure of the unified U.S. Cyber Command.

Melia T. Pfannenstiel

See also: National Security Agency (NSA); Operation Buckshot Yankee; Rogers, Michael S.; SIPRNet; 10th Fleet; 24th Air Force

Further Reading

Healy, Jason. "A Brief History of US Cyber Conflict." In *A Fierce Domain: Conflict in Cyberspace, 1986–2012,* ed. Jason Healy. Alexandria, VA: CCSA/Atlantic Council, 2013, 14–87.

Libicki, Martin C. *Cyberspace in Peace and War.* Annapolis, MD: Naval Institute Press, 2016.

President Donald J. Trump. "Statement on the Elevation of Cyber Command" (White House, Washington, D.C., August 18, 2017).

Reveron, Derek. "An Introduction to National Security and Cyberspace." In *Cyberspace and National Security: Threats, Opportunities, and Power in a Virtual World*, ed. Derek S. Reveron. Washington, D.C.: Georgetown University Press, 2012, 3–19.

Secretary of Defense Ash Carter. "Department of Defense Cyber Strategy" (U.S. Department of Defense, Washington D.C., April 17, 2015), https://www.defense.gov/News/Special-Reports/0415_Cyber-Strategy.

U.S. Department of Defense. "U.S. Cyber Command Fact Sheet." May 25, 2010, https://nsarchive2.gwu.edu/NSAEBB/NSAEBB424/docs/Cyber-038.pdf.

Virtual Control System Environment (VCSE)

Technology created by Sandia National Laboratories and designed as a modeling and simulation environment to investigate cybersecurity vulnerabilities associated with control systems on a power grid. Modeling these systems allows researchers to analyze and develop an understanding of cyberattacks on these systems.

VCSE uses a hybrid system of modeling components that supports Simulated, Emulated, and Physical components for Investigative Analysis (SEPIA). Through incorporating simulation software and hardware, system-in-the-loop (SITL) analysis can be performed to determine the failure modes of the ensemble characteristics (hardware, software, configuration, and application) of the tested system. By providing a testing environment where current systems as well as new software and hardware upgrades can be evaluated, researchers can determine and address cybersecurity vulnerabilities to real and potential threats prior to installation.

Diana J. Ringquist

See also: Supervisory Control and Data Acquisition (SCADA)

Further Reading

Bergman, D. C. 2009. "The Virtual Power System Testbed and Inter-Testbed Integration." Proc. 2nd Workshop Cyber Security Exp. Test. August.

Coerbell, Marlon C.D. *Creating a Network Model for the Integration of a Dynamic and Static Supervisory Control and Data Acquisition (SCADA) Test Environment.* Thesis. Wright-Patterson Air Force Base, OH: Air Force Institute of Technology, 2011.

McDonald, Michael J., John Mulder, Bryan T. Richardson, Regis H. Cassidy, Adrian Chavez, Nicholas D. Pattengale, Guylain M. Pollock, et al. "Modeling and

Simulation for Cyber-Physical System Security Research, Development and Applications." University of North Texas. Sandia National Laboratories, February 1, 2010, https://digital.library.unt.edu/ark:/67531/metadc837446/m2/1/high_res_d/1028942.pdf.

Vulnerability

Also known as an attack surface, the term for a weakness that allows for illicit intrusion into a computer or computer network. Although different kinds of vulnerabilities exist, software vulnerabilities often garner the most overt concern. A software vulnerability is leveraged by an exploit, which enables a tool or technique to be used to gain access to a system.

Very large software packages often include weak points containing vulnerabilities unrecognized by software authors at the time of coding. When these are discovered, either by employees of the company releasing the software or by users, the originating software company evaluates the risks that the vulnerability represents and paths toward a solution or patch that closes the vulnerability. The process of identifying, analyzing, and addressing vulnerabilities is called vulnerability management. Some analysts have recommended that software companies offer "bug bounties," or cash rewards to individuals who can identify and report vulnerabilities in new software. Once discovered, these can then be remediated. Even among companies that offer bug bounties, the monetary awards for reporting to the company pale in comparison to the substantially larger sums that can be obtained by selling evidence of a discovered vulnerability on the black market (to entities interested in using a vulnerability to create malware or ransomware, in order to steal or extort money from victims) or on the gray market to intelligence agencies worldwide. The monetary value of a vulnerability also tends to correlate to the scale on which the software is used, as for example a vulnerability in the nearly ubiquitous Microsoft Windows software will likely fetch a higher price than a vulnerability in software that is used by far fewer people.

The majority of leveraged vulnerabilities are, perhaps surprisingly, types already known and identified by computer experts. Throughout the late 20th century and early 21st century, patches were included as part of the software updates, which were themselves presented to computer users in ways that allowed the user to select the optimum timing for an update to proceed or to dismiss an update entirely. Among many users of personal computers, frequent software updates brought inconvenience in the short term, turning many users off from the idea of updating software promptly or at all. Other systems, particularly those which are part of industrial controls such as supervisory control and data acquisition (SCADA) systems, were inhibited from updating systems for other reasons. Updating a system takes it offline for a period, and systems like water and electrical utilities are not in a position to shut off service to their users.

Still another reason contributing to the widespread global lethargy in updating software with new patches is that a large portion of the world's software is pirated. The exact proportion is impossible to estimate with confidence, but pirated software is expected to be rampant in much of Asia and elsewhere. Pirated software provides the general functionality of the original, but it lacks the certificates that link it to the creator of the legitimate software, and this means that pirated software may look and feel similar to the original but it will not receive word of software updates and thus represents a fairly close likeness to the original unpatched version of the legitimate software.

Cumulatively, these factors mean that vast numbers of computers lack the protection of patches for vulnerabilities that are already known to exist and for which solutions have been found. This means that, somewhat contrary to some arguments in the early 21st century, a piece of malware is not necessarily a "single-shot" weapon that cannot be used after its first introduction. Given that the patterns of patching machines are difficult to know with accuracy, the opportunity for repeated use is disproportionately of value to actors who care more about impacting some computers than on impacting specific computers. Thus repeated usability of malware is good news for criminals interested in theft and extortion.

Analysts have also noted that the challenges and expense involved in updating SCADA computers means that industrial and utility facilities may be especially at risk. Circumstantially, this concept was reinforced by the December 2015 targeting of Ukraine's electrical grid, and much of the eastern part of the country's electricity was interrupted for several hours. At the time of the attack, the vulnerability that the malware leveraged had been known by computer security experts for fifteen years, but it remained unpatched on these computers.

Even operators who meticulously update their systems remain at risk of attack via malware exploits that leverage vulnerabilities that are known only by their original discoverer, or by the discoverer and a purchaser of vulnerabilities. When the purchaser is the software company, a patch can be developed, and this allows the window of vulnerability between discovery of a vulnerability and its solution through the development and deployment of a patch to be narrowed. However, black market and gray market purchasers buy vulnerabilities for entirely different reasons. Because there is no time between a victim's discovery and its victimization, these are known as zero-day vulnerabilities, and attacks using exploits based on them are called zero-day attacks.

Black market buyers aim to use vulnerabilities to develop exploits that can empower new forms of malware, usually for theft. Gray market buyers reportedly stockpile vulnerabilities so that they can construct malware that is impervious to the defenses of carefully updated computers carrying antivirus packages, and the purpose in these cases is typically related to espionage.

The alleged stockpiling of vulnerabilities by intelligence agencies is highly controversial. Computer privacy advocates voice extreme concern about the idea that intelligence agencies can gain completely unfettered access to systems, either for espionage or for sabotage or a combination of both. Many experts note that the exploits that are built and deployed by intelligence agencies or other similar entities will ultimately be sold to parallel organizations in other countries or alternatively to criminal entities. Security experts have suggested that sophisticated surveillance software has been purchased and used by the law enforcement agencies of several countries in the Middle East and beyond, and that the targets of surveillance efforts are quite frequently journalists, democracy advocates, and human rights workers, whose activities are then curtailed by regimes in autocratic countries.

Concern arose in 2010 with the discovery of Stuxnet, which despite being very carefully tailored to damage only the physical systems enriching uranium and only corresponding to the types used by Iran's nuclear program, employed exploits against four separate zero-day vulnerabilities in the German software used by Iran's nuclear program. Copycat attacks using the now discovered zero-day vulnerabilities Stuxnet publicized has to date not occurred, though this does not allay concern regarding a similar scenario in the future when the zero days relate to commonly used software or to software used by systems that could be conveniently held for ransom.

Beyond software vulnerabilities are other less obvious kinds of vulnerabilities that can create significant problems for access or preservation of information. Insecure network architecture, for example, can magnify a security problem. Many kinds of vulnerabilities involve circumstances rather than targeting by adversaries. Although computer hardware is considerably more robust than computers of the latter decades of the 20th century, extreme temperatures, humidity levels, and the presence of dust can still cause failure. Insecurity among computer users can compromise computer systems, so personal and organizational factors can constitute vulnerabilities also. Physical placement of computers in areas subject to events such as flooding forms another vulnerability, and it is these types of challenges to which cloud computing is actually expected to be the most beneficial.

Thus software vulnerabilities are arguably the most salient type of weakness for computers in this part of the 21st century. Although other kinds of vulnerabilities exist, software vulnerabilities pose a notable area of concern because of the continually growing range of software systems that are available and used (and therefore the surface area for vulnerabilities to exist and be discovered), as well as the fact that software vulnerabilities can be leveraged by sentient adversaries, and these cannot be addressed with the regularity and reliability of environmental or other problems that are less likely to adapt against a solution.

Nicholas Michael Sambaluk

See also: Computer Network Defense (CND); Exploit; Malware; Stuxnet; Supervisory Control and Data Acquisition (SCADA); Zero-Day Vulnerability

Further Reading

Clark, Saender Aren. *The Software Vulnerability Ecosystem: Software Development in the Context of Adversarial Behavior.* Dissertation. University of Pennsylvania, 2012.

Hildick-Smith, Andrew. *Security for Critical Infrastructure SCADA Systems.* Bethesda, MD: SANS Institute, 2005.

Matwyshyn, Andrea M. "Material Vulnerabilities: Data Privacy, Corporate Information Security, and Securities Regulation." *Berkley Business Law Journal* 3, no. 1 (December 2005), 129–204.

Palmaers, Tom. *Implementing a Vulnerability Management Process.* Bethesda, MD: SANS Institute, 2013.

Robles, Rosslin John, et al. "Vulnerabilities in SCADA and Critical Infrastructure Systems." *International Journal of Future Generation Communication and Networking* (2008), 99–104.

Singer, P. W., and Allan Friedman. *Cybersecurity and Cyberwar: What Everyone Needs to Know*. Oxford: Oxford University Press, 2014.

Zetter, Kim. *Countdown to Zero Day: Stuxnet and the Launch of the World's First Digital Weapon*. New York: Penguin Random House, 2014.

WannaCry Virus

A massive ransomware attack using a cryptoworm program. First detected in May 2017, WannaCry relied like most ransomware attacks on social engineering to infect initial victim computers. Upon infection, a target system initiates a program that seeks to spread WannaCry as broadly and quickly as possible. At the same time, the users of an infected machine find that it is unresponsive, and then are informed that if they do not provide a ransom, all of the data on the device will be permanently encrypted and hence unusable. The creators of WannaCry typically demanded that a victim pay the ransom via the purchase and transfer of bitcoins, a cryptocurrency that is much harder to track than electronic funds transfers. At the time of the attacks, most victims were ordered to provide bitcoins worth approximately $300 U.S. within three days. Alternatively, they could pay $600 within seven days to regain control of their computer.

At the height of the attacks, cyber experts advised users not to pay the ransom for the return of their computers' data. In part, this was because paying ransom would almost certainly encourage more such attacks. However, it was bolstered by the fact that there were few if any demonstrable instances of the ransom actually triggering the return of control and decryption of files. Although it is impossible to determine how much ransom money was delivered to the attackers, it almost certainly reached millions of dollars of cryptocurrency. Cyber researchers collaborated on a means to defeat WannaCry and restore users' data. They released several open source toolkits that have allowed decryption without ransom payments in at least some instances, although it has certainly not provided a panacea against similar attacks in the future.

WannaCry relied on the EternalBlue exploit, essentially allowing it to seize control over target machines through an unpatched security vulnerability in Microsoft Windows. The Microsoft Corporation, having discovered the vulnerability, released a series of emergency patches in April 2017, meaning that many victims of WannaCry would have been protected if they had updated their computers. Other victims were operating legacy software no longer receiving support from Microsoft, such as Windows NT. Perhaps the largest category of victims were guilty of using pirated copies of Microsoft Windows, which are incapable of automatically receiving and uploading security patches.

The EternalBlue exploit, along with the toolkits used to create WannaCry, were released by the online entity The Shadow Brokers and were almost certainly initially developed by the U.S. National Security Agency (NSA). One major piece of evidence supporting this argument was that the initial WannaCry software contained a digital "kill switch." Infected machines attempted to communicate with a specific domain name, and if unable to do so, proceeded with their attack. When cyber researcher Marcus Hutchins discovered that such a kill switch was encoded in the WannaCry program, he inadvertently discovered an extremely effective means of slowing its spread. All that was required was to register the required domain name—when an infected computer contacted it, that computer received a signal not to proceed with the encryption and ransom demands. Because most individual hackers and hacking organizations would not deliberately hamstring their own efforts, the inclusion of the kill switch is evidence of a Western state organization operating under specific limits. Later iterations of the WannaCry virus were released without the kill switch, suggesting that the original attackers had not noticed its presence. By the time they updated the malware, Microsoft's security patches had closed the vulnerability of most systems.

Cybersecurity firms such as Cisco, McAfee, and Symantec have estimated that approximately 300,000 computers were infected by WannaCry, most within the first few days of the attack. Infections were found in more than 150 countries, although the hardest hit areas were inside India, Russia, Taiwan, and Ukraine, all nations with high rates of illicit copies of Microsoft Windows. One of the most prominent infections from WannaCry was in the networks of the National Health Service (NHS) in the United Kingdom. Later cyber forensic examination showed that the NHS was guilty of using long-outdated versions of Windows, making them especially prone to cyberattacks.

The WannaCry attacks created a significant amount of backlash for the NSA. The agency had discovered the potential vulnerability in Windows, but did not disclose it to Microsoft, leaving billions of computers open to such an exploit. Instead, the NSA created its own toolkits to exploit the weakness, but then failed to prevent the release of its own software. By handing such a tool to an infinite number of potential attackers, this type of attack was almost inevitable.

The motivation for WannaCry took little effort to deduce, as it was a simple cash grab for the perpetrators. Attribution of responsibility required significantly more time and included linguistic analysis of the ransom notes as well as cyber forensic examination of the software released in later versions. In December 2017, the governments of the United States, United Kingdom, and Australia publicly accused the government of North Korea of initiating the attacks. Unsurprisingly, North Korea denied all of the accusations, despite the similarities between this attack and earlier ones definitively pinpointed to North Korea.

Paul J. Springer

See also: Cyberweapons; EternalBlue Exploit; Malware; National Security Agency (NSA)

Further Reading

Bossert, Thomas P. "It's Official: North Korea Is Behind WannaCry." *Wall Street Journal,* December 18, 2017, https://www.wsj.com/articles/its-official-north-korea-is-behind-wannacry-1513642537.

Cameron, Dell. "Today's Massive Ransomware Attack Was Mostly Preventable; Here's How to Avoid It." *Gizmodo,* May 12, 2017, https://gizmodo.com/today-s-massive-ransomware-attack-was-mostly-preventabl-1795179984.

Fox-Brewster, Thomas. "An NSA Cyber Weapon Might Be Behind a Massive Global Ransomware Outbreak." *Forbes,* May 12, 2017, https://www.forbes.com/sites/thomasbrewster/2017/05/12/nsa-exploit-used-by-wannacry-ransomware-in-global-explosion/#49608d7ce599.

Khomami, Nadia, and Olivia Solon. "'Accidental Hero' Halts Ransomware Attack and Warns: This Is Not Over." *The Guardian,* May 13, 2017, https://www.forbes.com/sites/thomasbrewster/2017/05/12/nsa-exploit-used-by-wannacry-ransomware-in-global-explosion/#49608d7ce599.

Nakashima, Ellen. "The NSA Has Linked the WannaCry Computer Worm to North Korea." *Washington Post,* June 14, 2017, https://www.washingtonpost.com/world/national-security/the-nsa-has-linked-the-wannacry-computer-worm-to-north-korea/2017/06/14/101395a2-508e-11e7-be25-3a519335381c_story.html?utm_term=.fccf33e7c876.

Newman, Lily Hay. "The Ransomware Meltdown Experts Warned About Is Here." *Wired,* May 12, 2017, https://www.wired.com/2017/05/ransomware-meltdown-experts-warned.

Zero-Day Vulnerability

A vulnerability found in software or hardware that was previously unknown to the developers of the software or hardware who now have "zero days" to fix the vulnerability. A vulnerability in this case is an error often called a "bug" in the software or hardware that a threat can exploit to use the software or hardware for purposes other than for which it was created. These purposes can include retrieving sensitive data from the computer or device running the software and using the computer or device to attack other computing devices that are on the same network.

The existence of a vulnerability does not necessarily mean that a threat often called an "actor" has exploited the vulnerability; these threat actors usually have some sort of human origin and may be software written by a human but now acting in an independent manner for the most part. Instead, the creators of the software or hardware themselves may discover the vulnerability. However, even in such cases, it is nearly impossible to know whether the vulnerability has been exploited by a threat; naturally, even more harmful are instances where the awareness of the vulnerability only happens due to a threat exploiting the vulnerability. Such instances are known as zero-day attacks and constitute the most disruptive examples of a zero-day vulnerability. Some well-known zero-day attacks include Stuxnet in 2010 and Heartbleed in 2016.

As a result, it is extremely important for the software or hardware developers to move quickly to address the error and push the corrected version of the software or hardware to their users; such versions are often called "patches" to signify that the "hole" represented by the vulnerability has been covered. The users in turn often must apply the patch to ensure that the vulnerability has been addressed.

Accordingly, both the developers and the users form the first line of defense against such vulnerabilities and must stay vigilant in their watch and then diligent in their resolution once a vulnerability is known. An additional measure is to place the proper controls on the environments—such as the computer or network—in which the software or hardware functions; in other words, one may be able to limit the impact of a vulnerability by limiting external access to the computer or device with the vulnerability.

John A. Springer

See also: Computer Network Attack (CNA); Exploit; Malware; National Security Agency (NSA); Stuxnet; Vulnerability

Further Reading

Bilge, Leyla, and Tudor Dumitras. "Before We Knew It." In *Proceedings of the 2012 ACM Conference on Computer and Communications Security—CCS '12, 833.* New York: ACM Press, 2012. doi:10.1145/2382196.2382284.

Bishop, Matt C. N. *Introduction to Computer Security.* Boston: Addison-Wesley.

Kim, David, and M. G. Solomon. *Fundamentals of Information Systems Security,* 3rd ed. Jones & Bartlett Learning Information Systems Security & Assurance Series. Burlington, MA: Jones & Bartlett Learning, 2016.

Millett, Lynette I., Baruch Fischhoff, Peter J. Weinberger, Computer Science, Telecommunications Board, and Physical Sciences. 2017. *Foundational Cybersecurity Research.* doi:10.17226/24676.

Cyber Warfare Bibliography

Abomhara, Mohamed, and Ceir M. Koien. "Cyber Security and the Internet of Things: Vulnerabilities, Threats, Intruders and Attacks." *Journal of Cyber Security* 4 (2015), 65–88.

Allhoff, Fritz, Adam Henschke, and Bradley Jay Strawser, eds. *Binary Bullets: The Ethics of Cyberwarfare*. Oxford: Oxford University Press, 2016.

APT1: Exposing One of China's Cyber Espionage Units. Alexandria, VA: Mandiant, 2014.

Arquilla, John, and David Ronfeldt. *Networks and Netwars*. Santa Monica, CA: RAND, 2001.

Betz, David J., and Tim Stevens. *Cyberspace and the State: Toward a Strategy for Cyber-Power*. New York: Routledge, 2002.

Blunden, Bill, and Violet Cheung. *Behold a Pale Farce: Cyberwar, Threat Inflation, & the Malware Industrial Complex*. Walterville, OR: Trine Day, 2014.

Bodmer, Sean, et al. *Reverse Deception: Organized Cyber Threat Counter-Exploitation*. New York: McGraw-Hill, 2012.

Bowden, Mark. *Worm: The First Digital World War*. New York: Atlantic Monthly, 2011.

Brantly, Aaron Franklin. *The Decision to Attack: Military and Intelligence Cyber Decision-Making*. Athens: University of Georgia Press, 2016.

Brenner, Joel. *America the Vulnerable: Inside the New Threat Matrix of Digital Espionage, Crime, and Warfare*. New York: Penguin, 2010.

Brenner, Susan W. *Cyberthreats: The Emerging Fault Lines of the Nation State*. Oxford: Oxford University Press, 2009.

Buchanan, Ben. *The Cybersecurity Dilemma: Hacking, Trust, and Fear between Nations*. Oxford: Oxford University Press, 2016.

Campen, Alan D., and Douglas H. Dearth, eds. *Cyberwar 3.0: Human Factors in Information Operations and Future Conflict*. Fairfax, VA: AFCEA, 2000.

Carafano, James J. *Wiki at War: Conflict in a Socially Networked World*. College Station: Texas A&M University Press, 2012.

Carlin, John P. *Dawn of the Code War: America's Battle against Russia, China, and the Rising Global Cyber Threat*. New York: Public Affairs, 2018.

Chen, Dean. *Cyber Dragon: Inside China's Information Warfare and Cyber Operations*. Santa Barbara, CA: Praeger, 2017.

Clark, Saender Aren. *The Software Vulnerability Ecosystem: Software Development in the Context of Adversarial Behavior.* Dissertation. Philadelphia: University of Pennsylvania, 2017.

Clarke, Richard A. *Cyber War: The Next Threat to National Security and What to Do About It.* New York: HarperCollins, 2010.

Conti, Greg. *Googling Security: How Much Does Google Know About You?* New York: Addison-Wesley, 2009.

Conti, Greg, and David Raymond. *On Cyber: Towards an Operational Art for Cyber Conflict.* New York: Kopidion Press, 2017.

Cook, Sarah. "Chinese Governmental Influence on the US Media Landscape." *Hearing on China's Information Controls, Global Media Influence, and Cyber Warfare Strategy.* Washington, D.C.: Freedom House, 2017. New York: Kopodion Press, 2017.

Cordesman, Anthony H. *Cyber-Threats, Information Warfare, and Critical Infrastructure Protection, Defending the US Homeland.* Westport, CT: Praeger, 2002.

Corera, Gordon. *Cyber Spies: The Secret History of Surveillance, Hacking, and Digital Espionage.* New York: Pegasus, 2015.

Council of Economic Advisors. *The Cost of Malicious Cyber Activity to the US Economy.* Washington, D.C.: Government Printing Office, 2018.

Cummings, M. L. "Artificial Intelligence and the Future of Warfare." *International Security Department and US and the Americas Programme.* London: Chatham House, 2017.

Danchev, Dancho. "Study Finds the Average Price for Renting a Botnet." *ZDNet,* May 26, 2010, https://www.zdnet.com/article/study-finds-the-average-price-for-renting-a-botnet.

Demchak, Chris. *Wars of Disruption and Resilience.* Athens: University of Georgia Press, 2011.

DeNardis, Laura. *The Global War for Internet Governance.* New Haven, CT: Yale, 2014.

Direnzo, Joseph et al., eds. *Issues in Maritime Cyber Security.* Washington, D.C.: Westphalia, 2017.

Fraga-Lamas, Paula, et al. "A Review on Internet of Things for Defense and Public Safety." *Sensors* 16 (2016).

Gertz, Bill. *iWar: War and Peace in the Information Age.* New York: Threshold, 2017.

Giannetti, William. "A Duty to Warn: How to Help America Fight Back against Russian Disinformation." *ASPJ* (Fall 2017), 95–104.

Greenberg, Andy. *Sandworm: A New Era of Cyberwar and the Hunt for the Kremlin's Most Dangerous Hackers.* New York: Doubleday, 2019.

Han, Rai. *Data-Driven Malware Detection Based on Dynamic Behavioral Features.* Dissertation. Coral Gables, FL: University of Miami, 2017.

Hayden, Michael. *Playing to the Edge: American Intelligence in the Age of Terror.* New York: Penguin Press, 2016.

Healey, Jason, ed. *A Fierce Domain: Conflict in Cyberspace, 1986–2012.* Washington, D.C.: Cyber Conflict Studies Association, 2013.

Heckman, Kristin E., et al., eds. *Cyber Denial, Deception, and Counter Deception: A Framework for Supporting Active Cyber Defense.* New York: Springer, 2015.

Iran: Cyber Repression, How the IRGC Urges Cyberwarfare to Preserve the Theocracy. Washington, D.C.: National Council of Resistance of Iran, 2018.

Jasper, Scott. *Strategic Cyber Deterrence: The Active Cyber Defense Option.* Lanham, MD: Rowman & Littlefield, 2017.

Kaplan, Fred. *Dark Territory: The Secret History of Cyber War.* New York: Simon & Schuster, 2016.

Kello, Lucas. *The Virtual Weapon and International Order.* New Haven, CT: Yale University Press, 2017.

Klimburg, Alexander. *The Darkening Web: The War for Cyberspace.* New York: Penguin, 2017.

Kramer, Franklin D., Stuart H. Starr, and Larry K. Wentz, eds. *Cyberpower and National Security.* Dulles, VA: Potomac, 2009.

Libicki, Martin. *Conflict in Cyberspace: National Security and Information Warfare.* Cambridge: Cambridge University Press, 2007.

Libicki, Martin. *Cyberdeterrence and Cyberwar.* Santa Monica, CA: RAND, 2009.

Libicki, Martin. *Cyberspace in Peace and War.* Annapolis, MD: Naval Institute, 2016.

Mandel, Robert. *Optimizing Cyberdeterrence: A Comprehensive Strategy for Preventing Foreign Cyberattacks.* Washington, D.C.: Georgetown University Press, 2017.

Matwyshyn, Andrea M. "Material Vulnerabilities: Data Privacy, Corporate Information Security, and Securities Regulation." *Berkeley Business Law Journal* 3, no. 1 (2005), 129–204.

Maurer, Tim. *Cyber Mercenaries: The State, Hackers, and Power.* Cambridge: Cambridge University Press, 2018.

Mazanec, Brian M. *The Evolution of Cyber War: International Norms for Emerging-Technology Weapons.* Lincoln: University of Nebraska Press, 2015.

Mazanec, Brian M., and Bradley A. Thayer. *Deterring Cyber Warfare: Bolstering Strategic Stability in Cyberspace.* New York: Palgrave Macmillan, 2015.

Mitnick, Kevin D. *The Art of Deception: Controlling the Human Element of Security.* New York: Wiley, 2002.

Perkovich, George, and Ariel E. Levite, eds. *Understanding Cyber Conflict: Fourteen Analogies.* Washington, D.C.: Georgetown University Press, 2017.

Poindexter, Dennis F. *The New Cyberwar: Technology and the Redefinition of Warfare.* Jefferson, NC: McFarland, 2015.

Poindexter, Dennis F. *The Chinese Information War: Espionage, Cyberwar, Communications Control and Related Threats to United States Interests.* Jefferson, NC: McFarland, 2018.

Rabkin, Jeremy, and John Yoo. *Striking Power: How Cyber, Robots, and Space Weapons Change the Rules for War.* New York: Encounter, 2017.

Rattray, Gregory J. *Strategic Warfare in Cyberspace.* Cambridge, MA: MIT Press, 2001.

Reveron, Derek S., ed. *Cyberspace and National Security: Threats, Opportunities, and Power in a Virtual World.* Washington, D.C.: Georgetown University, 2012.

Rid, Thomas. *Cyber War Will Not Take Place.* Oxford: Oxford University Press, 2013.

Rosenzweig, Paul. *Cyber Warfare: How Conflicts in Cyberspace Are Challenging America and Changing the World.* Santa Barbara, CA: Praeger, 2013.

Sanger, David E. *The Perfect Weapon: War, Sabotage, and Fear in the Cyber Age.* New York: Crown, 2017.

Schmitt, Michael N., ed. *Tallinn Manual on the International Law Applicable to Cyber Warfare.* Cambridge: Cambridge University Press, 2013.

Schneier, Bruce. *Secrets and Lies: Digital Security in a Networked World.* New York: Wiley, 2000.

Schneier, Bruce. *Click Here to Kill Everybody: Security and Survival in a Hyper-Connected World.* New York: Norton, 2018.

Shehabat, Ahmad, Teodor Mitew, and Yahia Alzoubi. "Encrypted Jihad: Investigating the Role of Telegram App in Lone Wolf Attacks in the West." *Journal of Strategic Security* 10, no. 3 (2017), 27–53.

Singer, P. W., and Allan Friedman. *Cybersecurity and Cyberwar: What Everyone Needs to Know.* Oxford: Oxford University Press, 2014.

Soldatov, Andrei, and Irina Borogan. *The Red Web: The Kremlin's War on the Internet.* New York: Public Affairs, 2015.

Springer, Paul J. *Cyber Warfare: A Reference Handbook.* Santa Barbara, CA: ABC-CLIO, 2015.

Stiennon, Richard. *Surviving Cyberwar.* Lanham, MD: Scarecrow, 2010.

Ullah, Haroon K. *Digital World War: Islamists, Extremists, and the Fight for Cyber Supremacy.* New Haven, CT: Yale University Press, 2017.

Valeriano, Brandon, and Benjamin Jensen. *Cyber Strategy: The Evolving Character of Power and Coercion.* Oxford: Oxford University Press, 2018.

Valeriano, Brandon, and Ryan C. Maness. *Cyber War versus Cyber Realities: Cyber Conflict in the International System.* Oxford: Oxford University Press, 2015.

Weed, Scott A. *US Policy Response to Cyber Attack on SCADA Systems Supporting Critical National Infrastructure.* Maxwell AFB, AL: Air University Press, 2017.

Weimann, Gabriel, and Bruce Hoffman. *Terrorism in Cyberspace: The Next Generation.* New York: Columbia University Press, 2015.

Yannakogeorgos, Panayotis A., and Adam B. Lowther, eds. *Conflict and Cooperation in Cyberspace: The Challenge to National Security.* Boca Raton, FL: Taylor & Francis, 2014.

Zetter, Kim. *Countdown to Zero Day: Stuxnet and the Launch of the World's First Digital Weapon.* New York: Crown, 2016.

Section 2

Social Media

INTRODUCTION

Although the deployment of malware has already demonstrated an ability to compromise adversaries and in limited scenarios has been used to catalyze physical destruction, cyber warfare is not the only way in which the digital domain plays a role in modern warfare. As a communications environment, the cyber domain provides a rich landscape in which to use messaging to human beings and to leverage psychological reactions in support of war goals. As such, the rise of "Web 2.0" creates the space in which social media has begun to play a significant part in warfare.

Although the term "Web 2.0" dates to 1999, it did not achieve wide usage until about 2006, when *Time* magazine declared web users and their online presence to be "Person of the Year." In contrast to the presented content format of "Web 1.0," in which Internet users simply access and observe information provided by the creators of webpage content, Web 2.0 platforms are built for the purpose of active participation, forming "social" media. Users enjoy activity on free-to-use applications, and companies (particularly Facebook and Twitter) have risen to become commercial giants by monetizing personal information that users voluntarily provide, selling the information to third parties and especially to advertisers. Cyber privacy expert Bruce Schneider has observed that shortsighted bargains are rife in the information era and that data is the digital equivalent of what fossil fuel exhaust is in the physical realm, although data is also eminently prone to profitable exploitation (Schneider, 2015, 4–6, 17). The rampant activity on social media platforms suggests its value as a piece of digital terrain in conflict as well as in commerce.

Scholars who have argued against the notion of "cyberwar" point to the proliferation of subversiveness in the digital realm as evidence that electronic connectedness gives a credible and less destructive outlet as an alternative to war (Rid, 2013, 114–120). Another worthwhile interpretation is that diverse groups have already shown how the digital realm can be leveraged for subversive purposes that cooperate with, rather than compete against, the application of politically motivated violence. In this regard, social media has quickly become an adjunct overlapping with conflict. Other scholars have suggested that the rise of social media in warfare has coincided with and supported the popularity of "gray-zone" warfare and meant an eclipse of the modern state's monopoly on information and on violence (Patrikarakos, 2017, 3–11, 60, 196).

Much as how form follows function, the intent behind social media activity helps define which social media platforms are best suited to answer a recognized need. This is as true for applications related to warfare as it is for uses of technology in any field. Facebook and Twitter are each icons of the social media landscape, but they are importantly distinct from one another, and their differences tailor them more closely to some kinds of activities than to others.

Facebook is well suited to facilitating a one-to-many style of communication, and it is meant to maintain ties between individuals who already share relationships beyond the digital realm. Especially prior to the advent of Facebook Chat in April 2008, communications meant by one member to be read by one other specific member were nonetheless visible to third parties (Silvestri, 2015, 57, 69). Thus it is possible to engage with strangers through Facebook, but these encounters are peripheral compared to the connections between members who are already "friends," and typically interactions with strangers on Facebook happen because of an encounter catalyzed by two strangers reacting to the post made by (or about) a mutual acquaintance.

Combatants and sympathizers of various warring entities have already demonstrated innovative methods of using Facebook to support efforts in war. Concurrent with the latter phases of the U.S. occupation of Iraq, U.S. service personnel deployed between about 2007 and 2009 experienced a dramatic shift in the possibilities and implications of communicating with loved ones at home. Facebook enabled "a mediated form of talking" that boosted morale while also insinuating a more constant connectivity with families that research suggests may impose feelings of obligation to be ubiquitously and simultaneously engaged virtually with events at home *and* physically on the battlefield (Silvestri, 2015, 11, 46–47, 161).

Although pro-Russian paramilitary forces have used both physical and digital tools to detach parts of Ukraine since 2014, Ukraine's resistance has, perhaps surprisingly, been bolstered through a creative leverage of social media. Since Facebook excels in communalizing members who share affinities and join Facebook groups, the platform has been used in places including Ukraine to organize civic donations to support the nation's armed forces, and Facebook has also served as the platform by which photographic proof of donations reaching grateful military personnel has returned to the donors themselves (Patrikarakos, 2017, 107, 118–119).

Whereas Facebook is postured in a way that helps maintain links between friends, Twitter is designed to allow users to reach potentially much broader audiences, encompassing both the like-minded and the dissenting. Rather than facilitating communication between individuals who are already acquainted, the app's focus is on congregating different users interested in the same topic or idea. At the core of Twitter's model is the tweet, a brief message originally limited to 140 typed characters, and the hashtag, which identifies posts as sharing a theme or topic. Although Facebook includes a list of trending topics, the trending list is much more important to Twitter as a platform, given that app's focus on current events, broadly defined (Prier, 2017, 53). Twitter too has been used in potent ways.

One popular use has been to personalize the war experience, as has been done by Palestinians in resistance to Israeli military actions across Gaza and the West Bank. Subsequently, Israeli Defense Forces also began organizing social media effort to highlight the civilian suffering inflicted on Israel as a result of conflicts that it suggests Hamas and other militant groups precipitate (Patrikarakos, 2017, 30, 70–71). Scholarship has suggested that so far Palestinian accounts have succeeded better in projecting to the wider world, as demonstrated by the way in which a leading Palestinian hashtag #gazaunderattack outdid its leading Israeli counterpart, #israelunderfire, by almost tenfold during a clash between the Israeli Defense Forces and Hamas in 2014. The Palestinian handle appeared on Twitter more than 2.5 million times (Patrikarakos, 2017, 89). The attention paid to social media by both sides in that conflict suggests agreement with a point made in one U.S. Air Force analysis, that the group that "controls the trend will control the narrative—and, ultimately, the narrative controls the will of the people" (Prier, 2017, 81).

Dominating the narrative is a dynamic and potentially elusive goal, as has been demonstrated during war in Nigeria and across Syria and Iraq. In the former, the jihadist militant group Boko Haram determinedly leveraged a combination of social media messaging and physical violence to challenge the legitimacy of the nation's military and law enforcement capability. By launching and publicizing its attacks on communities sympathetic to the government, Boko Haram worked to undermine domestic faith in the authorities' ability to govern. Its most infamous and perhaps most brazen attack involved the taking of nearly 300 schoolgirls from the northeastern town of Chibok as hostages in mid-April 2014. A countercampaign on Twitter, #BringBackOurGirls, quickly emerged to spread awareness of the situation and support for the enslaved youths' release, although at the time of writing over four years nearly half of the women taken remain in bondage (Ringquist, 2018, 245–255). Somewhat in keeping with the #Kony2012 campaign that two years earlier had aimed to increase awareness of a militant cult leader in Africa wanted by the International Criminal Court, many people globally relied on Twitter to express outrage and solidarity and urge action. However, the social media platform has been equally of use to the jihadist militants who raided Chibok, and Boko Haram leader Abubakar Shekau has proven adept and propagandizing into the digital age.

The Islamic State of Iraq and Syria (ISIS) stridently used Twitter, and the group was characterized in April 2016 by then-Defense Secretary Ashton Carter as "the first social media-fueled terrorist group" (Carter, 2016). Although its origins as an Al Qaeda splinter group date to about 2005 and the galvanizing of the movement into an "Islamic State" occurred in 2013, it was the group's dramatic military successes in mid-2014 and its use of social media—especially Twitter—that grabbed global attention. Its "Dawn of Glad Tidings" social media campaign coincided with its capture of the northern Iraqi city of Mosul. The group advertised images and videos of its murderous rampages through Twitter. This, analysts noted, had the dual effect of impressing sympathizers with the power of the militant group and also of terrifying enemies through pictures of brutality (Prier, 2017, 63). ISIS operatives wove social media communications into the context of a larger recruitment strategy that often sought to play on personal and psychological needs such as for belonging and for a sense of larger purpose (Patrikarakos, 2017, 214–220). By enticing sympathizers to download their app that transformed smartphones into members of a botnet, and by invading other discourse through methods such as hashtag hijacking, ISIS projected a global message of power and inescapable ubiquity (Prier, 2017, 63–65).

Efforts to counteract jihadist propaganda, and especially its ability to attract recruits to travel to ISIS-controlled territory and become militants, had already prompted the U.S. State Department to establish its Center for Strategic Counterterrorism Communications (CSCC) in late 2011 (Patrikarakos, 2017, 234). However, CSCC's task of dissuading recruits was fundamentally a negative message rather than a positive one, making it both challenging to conduct and difficult to demonstrate success. The organization's "Think Again, Turn Away" social media campaign engendered both skepticism and ridicule when encountered by Western observers. ISIS's powerful voice in social media was mostly extinguished by 2016, but analysis suggests that this was not as much the result of counterpropaganda as it was the result of ongoing policies by social media app administrators of removing graphic and militant content and especially the fact that battlefield reverses provided ISIS pundits with less and less material depicting the victor or the viability of a jihadist quasi-nation-state (Prier, 2017, 65).

Beyond the sourcing, public relations, and recruitment uses to which Facebook and Twitter have notably been used, other Web 2.0 applications have also been utilized by various groups. As a video platform, YouTube stands as a ready venue for movie clips used in similar roles. Additionally, terrorist networks have videoed and uploaded how-to clips describing topics such as bomb making. Communication apps such as Telegram and Surespot have been used by groups including ISIS for tactical communications (Gertz, 2017, 19, 227). The apps provide real-time and encrypted communication opportunities, although with time and effort encrypted signals can eventually be compromised. Since tactical communications frequently deal with relatively immediate topics, the time required to break an encryption exceeds the opportunity to make timely use of the decrypted messages.

Although not yet reported as used on the battlefield, FireChat is an unencrypted app that has been utilized in political protest movements and implies some battlefield applicability as well. Users with compatible mobile phones can build a dynamic mesh network that functions in the absence of a formal cellular or WiFi network. Its range was reported to be under 200 feet, and the app's creator spurned the notion of advertising the app as permitting communication "hops" from device to device to device as a means of extending range (Petry, 2015, 41–43), but the potential usefulness of a robust mesh networking app in an austere environment is obvious. Such apps give nonstate actors, albeit in a limited sense, a relatively low-cost counterpart to state-of-the-art encrypted communication technologies possessed by the militaries of powerful nation-states.

The significant advantages and opportunities of using social media in warfare are matched with risks and dangers. Sloppy use of social media can have a negative, and potentially lethal, effect on operational security. Oversights in what images and information were posted online on personal accounts contributed in one case to forensic efforts that traced sophisticated hacking operations to the People's Liberation Army cyber units based in Shanghai, ultimately leading to the traced individuals being indicted (in absentia) in a U.S. court (Gertz, 2017, 120–122). Online imagery of Russian-supplied technology given to pro-Russian factions in Eastern Ukraine played a part in the crowdsourcing confirmation that Russia provided the surface-to-air missile that downed Malaysian airliner MH-17 in April 2014 (Patrikarakos, 2017, 177–182). Repeated recontouring of the same Russian Twitter account facilitated forensics research into a host of Russian propaganda campaigns (Prier, 2017, 67–75). It has also been reported that Jordanian original founder of ISIS Abu Musab al Zarqawi, propagandist and British national "Jihad John," and U.S. national and prelate Anwar al-Awlaki, were each said to have been traced and killed by coalition forces in large part because of occasions in which their social media activities inadvertently disclosed information that made geolocation possible (Atwan, 2015, 62). Such events demonstrate that social media cannot be imagined to solely advantage underdogs, nonstate actors, or aggressors generally.

Nation-states have long been alert to the shortcomings and operational security risks involved with social media usage. Certainly, the slow reactions of ponderous bureaucracy have been blamed and can be involved in institutional mistrust of decentralized communications that inherently resist control (Patrikarakos, 2017, 55). But realistic alertness to credible danger also matters, and it helps to explain examples of reluctance to embrace social media in an official capacity and also a hesitation regarding the toleration of social media use by service personnel, particularly when deployed overseas. Scholarship has confirmed the intuitive theme that military personnel, such as U.S. forces in Iraq and Afghanistan, enthusiastically reached for social media technologies to facilitate communication with loved ones at home and also to add a real-time character to that communication that is unparalleled in the history of warfare. However, the operational security risks associated with that usage prompted the U.S. government to prohibit its personnel from using social media (Silvestri, 2015, 2–6, 23).

Perhaps realizing that prohibition simply punished cooperative personnel while making offenders out of those who continued to use emerging communication technologies, the Defense Department unveiled some social media apps that were built with a degree of user security unmatched by popular commercial counterparts. Examples included HARMONIEWeb and TroopTube. U.S. government efforts to launch and maintain parallel social media platforms for safer use by personnel faced daunting challenges, however. Increased security tended to mean that the approved apps were notably less user-friendly than the already available alternatives. Additionally, security was achieved in part by narrowing the circle of users, meaning that personnel could not communicate with as many friends and family, even if all of them were willing to learn how to use the clunkier approved apps. As one Army National Guardsman reportedly remarked, "it's much easier to get a hold of a deployed soldier on Facebook than it is on A[rmy] K[knowledge] O[nline]," which is an officially created and sanctioned communication system (Silvestri, 2015, 3). During the 2010s, the approved apps gradually lost their technical support and their usership shrank, while use of the mainstream apps came to be accepted.

Largely grassroots efforts by military personnel enjoyed greater success. Two notable examples are CompanyCommand and PlatoonLeader, which are discussion forums for use by captains and lieutenants discussing challenges and socializing best practices with one another. The success of these forums speaks to the value of users contributing to the design and creation of their own communication tools, which is in keeping with the mythology if not always the reality regarding social media apps. The survival of the two forums also owes something to official recognition of their merit and insightful leaders lending key technical support to ensure the survival of a worthwhile idea (Carafano, 2012, 128–133, 144).

Experts have similarly noted an emerging relationship of mutual interdependency between social and traditional media, and this relationship has already shown political and strategic significance, including in military conflict. When social media movements emerge, as with Palestinian Twitter users caught in the crossfire of battles between Israeli Defense Forces and Hamas militants, journalists have already shown a tendency to identify a small number of archetypal individuals and project the voice of this small and selected subgroup to a wider world. In one case, this led to a more than thirtyfold increase in a Palestinian Twitter accounts followership in less than three weeks. The recipe for such success appears to be to "first seduce the gatekeepers and then, critically, to transcend them" (Patrikarakos, 2017, 33–39). Journalists' reliance on Twitter as a source for breaking news on events naturally leads to reports that frequently accept tweets at face value (Weimann and Hoffman, 2015, 138). One consequence of this tendency is to facilitate the impact of propaganda campaigns waged via social media by agents of nation-states who churn out messages that are subsequently retweeted by a throng of controlled bots in order to establish a semblance of consensus and factuality, when instead false information was posted in order to achieve a strategic purpose. The most obvious and controversial example of this dynamic involves the exploits of Russian trolls (Patrikarakos, 2017, 131–166).

Social media thus exerts important impact on warfare and strategic strife generally, and the examples from the first two decades of the 21st century will quickly be joined by additional ones, as participants and sympathizers involved in conflicts seek to use the tools at hand, digital as well as physical, to promote their objectives.

FURTHER READING

Atwan, Abdel Bari. *Islamic State: The Digital Caliphate*. Oakland: University of California Press, 2015.

Carafano, James J. *Wiki at War: Conflict in a Socially Networked World.* College Station: Texas A&M University Press, 2012.

Carter, Ashton. "News Transcript: Stennis Troop Talk." April 15, 2016, https://www.defense.gov/News/Transcripts/Transcript-View/Article/722859/stennis-troop-talk.

Gertz, Bill. *iWar: War and Peace in the Information Age.* New York: Threshold, 2017.

Patrikarakos, David. *War in 140 Characters*. New York: Basic Books, 2017.

Petry, Joshua W. *Let the Revolution Begin, 140 Characters at a Time: Social Media and Unconventional Warfare*. Maxwell AFB, AL: School of Advanced Air and Space Studies, 2015.

Prier, Jarred. "Commanding the Trend: Social Media as Information Warfare." *Strategic Studies Quarterly* 12, no. 4 (Winter 2017), 50–85.

Rid, Thomas. *Cyber War Will Not Take Place.* Oxford: Oxford University Press, 2013.

Ringquist, John P. "War by Tweet, Hashtag, and Media Messaging." In *Paths of Innovation*, ed. Nicholas Michael Sambaluk. Lanham, MD: Lexington, 2018, 245–255.

Schneider, Bruce. *Data and Goliath: The Hidden Battles to Collect Your Data and Control Your World.* New York: Norton, 2015.

Silvestri, Lisa Ellen. *Friended at the Front: Social Media in the American War Zone.* Lawrence: University Press of Kansas, 2015.

Weimann, Gabriel, and Bruce Hoffman. *Terrorism in Cyberspace: The Next Generation.* New York: Columbia University Press, 2015.

Al-Furqan

The principal organization established by the Islamic State of Iraq and Syria (ISIS) for the purpose of overseeing the spread of its propaganda message. Officially called the Al-Furqan Foundation for Media Production, it was established in November 2006 to oversee the print, multimedia, and social media propaganda efforts of ISIS.

Early propaganda materials reportedly prioritized posters and pamphlets, as well as CDs and DVDs that would be physically transported to sympathizers. At its foundation, al-Furqan propaganda represented a jihadi movement whose brazen violence had alienated even Al Qaeda, and the senior leader of the group had only months earlier died in a U.S. air strike. ISIS, and its messaging branch al-Furqan, survived in the convulsive environment of Syria and Iraq during the latter phases of the U.S. occupation of Iraq that ended in 2011 and the onset of the Syrian civil war the same year. In 2013, ISIS prepared for major territorial expansion, and its leaders apparently also recognized a need for a robust messaging capability to accompany battlefield offensives. Additional media apparatuses, the al-I'tisam Media Foundation and the Ajnad Foundation for Media Production, appeared that year. The family of media entities allowed respective offices to concentrate on particular target audiences or media types. For example, the Ajnad Foundation focused on creation of audio content, including nasheed chants touting themes of jihad.

Another entity was formed in 2014 to join the growing constellation of ISIS propaganda agencies. The Al-Hayat Media Center was formed in mid-2014 as ISIS forces overran some parts of Syria that had been in the hands of the Bashar al-Assad regime, other sections of the country that had earlier been controlled by anti-Assad rebels, and a large swath of northern Iraq, including that country's second largest city, Mosul. The capture of Mosul witnessed the unveiling of an ISIS media campaign in Twitter that leveraged the Dawn of Glad Tidings smartphone app and the tweeting of more than 40,000 announcements dealing with the capture of Mosul. This media campaign included graphic images and videos of the murder of Iraqi soldiers and prisoners at the hands of ISIS fighters. In the coming months, ISIS would come to dominate the eastern half of Syria and the northwestern fifth of Iraq. Having conquered enough terrain to credibly simulate a nation-state, ISIS also leveraged its new al-Hayat organization, which specialized in the creation and distribution of ideological materials in major Western languages. Its focus was on English, French, German, and Russian in particular, although the then Federal Bureau of Investigation director James Comey indicated at the end of 2014 that ISIS "propaganda is unusually slick" (Sullivan, 2014) in its technical professionalism and that it was being distributed in almost two dozen languages.

Almost as soon as al-Hayat had been established, it launched a visually professional digital magazine in several languages. The journal was dubbed *Dabiq*, in reference to references in the Islamic hadith that ISIS interprets to mean a cataclysmic battle and apocalyptic victory for jihadists will occur at a town by the same name, located in northwestern Syria. The journal, which was distributed online via the deep web about monthly as a part of a larger ISIS strategy for recruiting ISIS members among populations overseas, consciously mimics the style of professional advertising and entertainment in Western countries. Some analysts believe that *Dabiq* also aimed to copy the message of the older Al Qaeda journal *Inspire*, which promoted the idea of launching lone wolf attacks in Western countries. Analysts have noted a tendency in ISIS to plagiarize the style and sometimes the strategies of other movements that have already demonstrated success. ISIS efforts to reach a global audience are also evident in the activities of its radio outlet al-Bayan, which broadcasts in English and Russian as well as in Arabic.

Nicholas Michael Sambaluk

See also: #AllEyesOnISIS; *Dabiq*; Dawn of Glad Tidings Campaign; ISIS Recruitment and War Crimes

Further Reading

Atwan, Abdel Bari. *Islamic State: The Digital Caliphate.* Oakland: University of California Press, 2015.

Rogers, Paul. *Irregular War: ISIS and the New Threat from the Margins.* London: I. B. Tauris & Co., 2016.

Salzar, Philippe-Joseph. *Words Are Weapons: Inside ISIS's Rhetoric of Terror.* New Haven, CT: Yale University Press, 2017.

Sullivan, Kevin. "Three American Teens, Recruited Online, Are Caught Trying to Join the Islamic State." *The Washington Post,* December 8, 2014, https://www.washingtonpost.com/world/national-security/three-american-teens-recruited-online-are-caught-trying-to-join-the-islamic-state/2014/12/08/8022e6c4-7afb-11e4-84d4-7c896b90abdc_story.html?tid=hybrid_1.1_strip_1&utm_term=.800e156f5fe5.

Warrick, Joby. *Black Flags: The Rise of ISIS.* New York: Anchor, 2016.

#AllEyesOnISIS

Part of the coordinated social media propaganda campaign launched by Islamic State of Iraq and Syria (ISIS) operatives on June 19, 2014, in coordination with major ISIS offensive actions. Various social media users tweeted photos of martyred fighters and landmarks across the globe, including Rome and New York, while others tweeted photos of handwritten notes proclaiming their support for ISIS.

Researchers Shiarz Maher and Joseph Carter identified user @Ansaar999 as the first to announce the planned media campaign, in which the user identified

حملة_المليار_مسلم_لنصرة_الدولة_الإسلامية# as the unifying hashtag; this roughly translates to "one billion Muslims in support of the Islamic State." Maher and Carter also found that between June 19 and 21, 20,000 tweets were posted with the hashtag, with roughly two-thirds of all tweets posted within the first 7 hours of the timed rollout. Their research also suggested this was not an entirely organic operation, in that rather than relying on social media users to naturally spread the hashtag, a majority of the tweets originated from already previously affiliated ISIS accounts.

Regardless, the use of social media outlets marks a new age in terrorism as groups now pair their military operations with online campaigns of influence. Since the inception of ISIS in 2006, the group has developed a large social media following and arguably brought their "jihad" (struggle) onto several online platforms. Shortly after this Twitter campaign, ISIS leader Abu Bakr al-Baghdadi declared himself the caliph and proclaimed the caliphate, on June 29, 2014, in Mosul, Iraq.

Kate Tietzen

See also: Dawn of Glad Tidings Campaign; Facebook, Use of; ISIS Recruitment and War Crimes; Twitter, Use of

Further Reading

Gartenstein-Ross, Daveed, et al. *The Islamic State's Global Propaganda Strategy.* The Hague, Netherlands: International Center for Counter-Terrorism, 2016.

McCants, William. *The ISIS Apocalypse: The History, Strategy, and Doomsday Vision of the Islamic State.* New York: St. Martin's Press, 2015.

Stern, Jessica, and J. M. Berger. *ISIS: The State of Terror.* New York: HarperCollins, 2015.

Anonymous

A highly decentralized hacktivist movement dating to 2003. Adopting the Guy Fawkes mask borrowed from the dystopian story and film *V for Vendetta,* the movement engages in digital campaigns against a diverse array of targets. The decentralized nature of the group carries important implications about its impact in cyber conflict.

The lack of a leader and of formal organization means that Anonymous as a movement is unable to plan and launch actions in a manner resembling that of a nation-state possessing robust cyber capabilities. Instead, campaigns occur in emergent patterns as the numbers of self-identified members engaging a particular target happen to mount. The movement's leaderless model does, however, also mean that no arrest and prosecution of any "key leader" has yet brought the movement's activities to a halt. Dozens of suspected Anonymous hacktivists have been arrested in the United States, Australia, the Netherlands, India, Turkey, and elsewhere.

Since participation in the movement is so loosely defined, the levels of skill of various members of the movement are suspected to vary considerably. Analysts have suggested that a very large part of the group's population may lack sophisticated expertise and is instead drawn to the overarching irreverent zeitgeist of the movement and the infusion of self-defined righteousness when Anonymous intermittently conducts actions against state and nonstate actors that its members view negatively. Extreme decentralization means that attributing specific actions to Anonymous can prove extremely difficult. When combined with the ability of anyone to self-identify with the group, and with a reported tendency of Anonymous's announcements to fabricate claims, assessing the effectiveness and even the overall priorities of the movement is a challenge.

Broadly, Anonymous members mistrust state and corporate entities that the hacktivists associate with censorship, exploitation, or even simply the concentration of power. As a result, Anonymous embraced an eclectic range of causes. This cyber vigilantism sometimes places Anonymous's efforts on the side of law enforcement (although not in keeping with legal activities). An example was the coercive doxing of personal information about child pornography customers and about the makers of revenge pornography websites.

Opposition to censorship has frequently led Anonymous to be engaged in the cyber elements of physical intrastate conflicts. For example, the movement supported the Arab Spring democracy movements across North Africa and the Middle East in 2011. Typically, and including during the Arab Spring, Anonymous leveraged distributed denial-of-service (DDoS) attacks against autocratic regimes as those governments sought to stifle protests. However, Anonymous appears to interpret the constraints of "censorship" within very wide bands, since Anonymous has also targeted various U.S., Australian, and Spanish legal and government and entities that uphold copyright law.

Operation Payback is another Anonymous campaign that reflects the extremely broad way in which many Anonymous members appear to interpret "censorship." WikiLeaks, a web organization dedicated to divulging classified information, began releasing leaked U.S. documents in November 2010. The U.S. government responded to the release of hundreds of thousands of documents by pressing financial service companies such as PayPal, Visa, and Mastercard to halt payments by would-be donors to the WikiLeaks organization; many members of Anonymous retaliated against the financial institutions. An Anonymous DDoS action supported by botnets briefly took down PayPal's website on two separate occasions in December 2010. Cyber forensics efforts led to the arrest and prosecution of 14 of the people involved in that DDoS attack.

Intermittently, Anonymous has also involved itself in cyber efforts to combat Al Qaeda and related groups such as the Islamic State of Iraq and Syria (ISIS). Anonymous members claimed responsibility for the crash of some jihadist websites following the January 2015 ISIS-affiliated attack on the irreverent French media outlet Charlie Hebdo. Two further actions, #OpISIS and #OpParis, came later in the year and also targeted jihadi cyberactivity. Another Anonymous initiative was an ISIS Trolling Day on December 11, 2015, in which the terrorist organization that was seeking to achieve the qualities and power of a geopolitical state was openly ridiculed online.

The impact of these campaigns on jihadi movements has been disputed. On the one hand, ISIS is known to have invested attention and strategic hopes in recruiting sympathizers to become home-grown fighters across several Western countries including the United States. On the other hand, surveillance experts note that causing jihadi sites to crash has an indeterminate impact on recruiting while exerting an undeniable impact on the ability of counterterrorism experts to observe enemy recruitment activities. ISIS leaders publicly claimed the irrelevance of hacktivist activities, but these protestations cannot be accepted at face value. Meanwhile, Anonymous members seem to consider these actions against jihadist groups to be part of the Anonymous movement's overall philosophy opposed to hate-based political beliefs, and Anonymous has also targeted the websites of the Ku Klux Klan and other organizations that Anonymous members identified as racist.

In response, characterizations of the movement differ greatly, as some globally consider the organization to serve as digital freedom fighters and others who disapprove of the targets Anonymous selects deem the movement to be on the verge of cyberterrorism. In terms of organizational structure, some analysts have compared Anonymous with the distributed structure of the jihadi terrorist group Al Qaeda. This characterization notwithstanding, Anonymous appears to have even less formal structure and centralization than the latter terrorist group. Experts on the group identify occasional coordination in a given campaign but note that this does not apparently translate into an impetus to solidify a more permanent set of objectives or organizational structure. Journalists attempting to report on Anonymous struggle with the challenge of reporting on an amorphous entity that consciously avoids identification of personnel. To a degree, this has contributed to the mystique of the Anonymous movement.

Nicholas Michael Sambaluk

See also: Botnet; Distributed Denial-of-Service (DDoS) Attack; Doxing; Hacktivism; ISIS Recruitment and War Crimes; ISIS Trolling Day; Social Media and the Arab Spring; Twitter, Use of; War on Terrorism and Social Media; WikiLeaks, Impact of

Further Reading

Coleman, Gabriella. *Anonymous in Context: The Politics and Power behind the Mask.* Internet Governance Papers No. 3. Waterloo, Canada: The Centre for International Governance Innovation, 2013.

Coleman, Gabriella. *Hacker, Hoaxer, Whistleblower, Spy: The Many Faces of Anonymous.* London: Verso, 2014.

Fish, Adam, and Luca Follis. "Gagged and Doxed: Hacktivism's Self-Incrimination Complex." *International Journal of Communication* (2016), 3281–3300.

Olson, Parmy. *We Are Anonymous: Inside the Hacker World of LulzSec, Anonymous, and the Global Cyber Insurgency.* New York: Hachette, 2012.

Rid, Thomas. *Cyber War Will Not Take Place.* New York: Oxford University Press, 2013.

Stoehrel, Rodrigo Ferrada, and Simon Lindgren. "For the Lulz: Anonymous, Aesthetics, and Affect." *Triple* 12, no. 1 (2014), 238–264.

Xiang Li. "Hacktivism and the First Amendment: Drawing the Line between Cyber Protests and Crime." *Harvard Journal of Law & Technology* 27, no. 1 (Fall 2013), 301–330.

Army Knowledge Online (AKO)

A primary component of Army Knowledge Management (AKM). This platform serves as the single point of entry into a robust and scalable knowledge management system; as such, AKO strategically changes the way the Army does business.

By enabling greater knowledge sharing among Army communities, AKM fosters improved decision dominance by commanders and business stewards in the battle space, organizations, and Army's mission processes. In essence, it serves as a one-stop platform for Active Duty Army; Army Reserve; Army National Guard; Department of the Army civilians' Army retirees; USMA Cadets; and ROTC Cadets to access, gather, and maintain information for and about themselves.

Originally championed in 1992 by then Vice Chief of Staff of the Army General Dennis Reimer, the military envisioned AKO as part of its plans to use information technology to help transform the Army and bring it into the 21st century. In 1999, AKO originally opened as a limited channel for general officers to communicate with each other. As the system proved its validity, programmers added additional capabilities to the platform, mainly focusing on access to information such as after-action reviews of training events, a directory of addresses of Army personnel, and a standard e-mail address to be used by individuals for the duration of their Army career, all of which can be accessed by any computer through a secure login. By the end of the year, the platform opened up to the Army at large, with 70,000 users signing up for the service. The Army further capitalized on this initial success in August 2001 with the signing of "Army Knowledge Management Guidance Memorandum Number 1," which required all soldiers, Department of the Army civilians, and employees to obtain AKO addresses, thereby endorsing AKM as a vital part of the Army's transformation process and forever changing the way the Army manages and uses information concepts and technologies.

In its present state, AKO has moved well into the 21st century and is a powerful tool that offers its users a multitude of options that can be controlled with a few strokes of a keyboard and clicks of a computer mouse. Personnel have access to a variety of features to enable them to do their day-to-day job and manage their careers. Financial documents, military records, and medical readiness are but a few features that can be currently managed through AKO. In addition, the Army has launched the Army Learning Management System that allows users to conduct online training over a variety of topics required both annually and for specific jobs and professional development courses. In addition, individuals can gain information through a variety of links from subjects ranging from legal assistance to doctrinal publications and forms to traveling to different duty stations with pets. With the ability to customize the platform to issues directly relating to the specific user, AKO will continue to be utilized and improved for the indefinite future by the Army in order to allow its personnel to manage many of their requirements at their own level.

Christian Garner

See also: Battle Command Knowledge System (BCKS)

Further Reading

Lausin, Anthony, Kevin C. Desouza, and George D. Kraft. "Knowledge Management in the US Army." *Knowledge and Process Management* 10, no. 4 (December 2003), 218–230.

Oliver, Rebecca. "NCOA: Moving Forward with Knowledge Management." *Military Intelligence* 39, no. 1 (January 2013), 21–23.

Battle Command Knowledge System (BCKS)

The U.S. Army's formal knowledge management tool. Advertised through the rhetorical question, "What if a single warrior had the knowledge of thousands?" and touting the motto, "Share what you know, find what you need," BCKS reflects an institutional awareness of the need for using digital age information sharing to promote the sharing of knowledge relevant to U.S. military personnel.

BCKS was officially created in September 2004, when the U.S. military was simultaneously engaged in operations to stabilize Afghanistan and Iraq after hostile regimes there had been ousted, but the countries, particularly Iraq, were sliding into violent chaos that analysts compared to civil war. This context reinforced institutional interest in facilitating military professionals sharing knowledge and building relevant communities of practice. At establishment, staff numbering about fifty personnel monitor posts to ensure that the content does not violate either copyright laws or classification regulations. Although within the fold of official military platforms since 2004, BCKS and related communities of practice such as companycommander.mil and platoonleader.mil do not exist under the umbrella of military doctrinal authorities. Structurally, this might arguably facilitate more fluid dialog among military professionals.

The origins of the communities of practice approach evident in BCKS dates to the early 1990s, when a cadre of particularly tech-savvy noncommissioned officers began building and maintaining bulletin board systems that dealt with Army-relevant information such as tactics, techniques, and procedures (TTPs). These initial activities were informal, although by 2003 they gained implicit recognition from the Army's chief information officer. The turn of the millennium also witnessed greater technical support for these early efforts. Gradually, however, a larger proportion of the U.S. military and intelligence authorities noted that dramatic improvements in commercial information technology were not only outstripping U.S. military structures for disseminating information, but also that adversaries were using commercial technologies to cheaply compete with U.S. combat forces.

Notable elements of BCKS include training opportunities, as well as forums and threaded discussions whose format generally resembles those of university online courses and the military's own distance learning courses for professional military education. Disseminating pertinent information in a timely and digestible manner constitutes an ongoing challenge, and the BCKS staff have undertaken studies on the effectiveness of knowledge management efforts. These studies have pointed to opportunities for improvement in ensuring that information is easily findable and that users appreciate the purpose of knowledge management tools. Despite official knowledge management initiatives, attaining a shared operational perspective among different organizations and guaranteeing that corporate knowledge can be retained by the organization despite turnover in personnel suggest that the digital age makes possible new tools but also adjusts the challenges related to knowledge management.

Nicholas Michael Sambaluk

See also: Army Knowledge Online (AKO)

Further Reading

Lewis, James M. *Trust and Dialog in the Army Profession.* Ft. Leavenworth, KS: School of Advanced Military Studies, 2008.

McGurn, Linda. "Knowledge Management Assessment Trends." *Army Communicator* (Summer 2011), 43–45.

Robertson, Gregory D. *Applying Knowledge Management Theory to Army Doctrine Development: Case Study of a Web-Based Community of Practice.* MA thesis. Ft. Leavenworth, KS: Command and General Staff College, 2007.

Williams, Daniel S. *Using Innovative Knowledge Management Tools for Information Technology Development, Acquisition, and Integration in the United States Army.* Ft. Leavenworth, KS: Command and General Staff College, 2007.

Blogs

Also called weblogs, a form of hypertext online diaries expanding participation in an electronic space. Originating in 1999, blogs have changed how communication, interaction, and the proliferation of information are generated within the infosphere. Warblogs and milblogs, dealing specifically with ongoing military affairs, grew enormously in prominence with U.S. participation in the Iraq War from 2003 to 2011. The term "milblog" commonly implies that the blogger was or is an active member of the military. A prominent example is the satirical Duffel Blog, which focuses on military-related topics. Warblogs became a torrent of dispatched information where personalized social and political commentaries served as international digital conduits for war reports. Soldiers deployed in Afghanistan and Iraq often created their own milblogs, filling in the missing information excluded by both the media and the military.

Warblogs specifically enable significant access to military-related information. Staff sergeant C. J. Grisham, a military intelligence analyst for the 3rd Infantry Division, created his blog *A Soldier's Perspective* to document his experiences in Iraq, providing brief details on how his job involved searching dead bodies for "intelligence value" (Burden, 2006, 45). Some media consumers have come to prefer to rely increasingly on warblogs over traditional media and publication presses because such blogs are uncensored and therefore widely considered to be more frank and reliable. Studies have concluded that over three-quarters of users accessed warblogs as information sources.

Although many warblogs are authored by combatants, other war-related blogs are used by citizens living in the midst of conflict. In such cases, blogs typically report on citizens' personal experiences within war. For example, Salam Pax, a gay Iraqi citizen, authors the blog *Where Is Raed?* and described bombings in Baghdad that can be filtered out of Western blogs. Blogs are thus powerful digital tools used to challenge the exclusionary practices of traditional war reporting.

Military and war blogging initiated several opportunities, radically changing the production and dissemination of information online. The unfiltered character of warblogs brings attendant concerns about operational security (OPSEC), since uncensored writing can inadvertently provide valuable information in near real time to alert adversaries. Adjustments to U.S. Army regulations relating to OPSEC and blogging have impacted the activity of serving personnel in maintaining warblogs.

Kathy Nguyen

See also: Doctrine Man; Facebook, Use of; Social Media Intelligence (SOCMINT); War on Terrorism and Social Media; Web 2.0

Further Reading

Burden, Matthew C. *The War Blog: Front-Line Dispatches from Soldiers in Iraq and Afghanistan.* New York: Simon & Schuster, 2006.

Cammaerts, Bart, and Nico Carpentier. "Blogging the 2003 Iraq War: Challenging the Ideological Model of War and Mainstream Journalism." *Observatorio* 9 (2009), 1–23.

Kaye, Barbara K., and Thomas J. Johnson. "Weblogs as a Source of Information about the 2003 Iraq War." *Journal of Global Mass Communication* 3 (2010), 291–301.

Boko Haram

Islamist group predominantly in Nigeria and active since 2002. Formally called *Jamā'at Ahl as-Sunnah lid-Da'wah wa'l-Jihād'*, and also known as Boko Haram ("Western education is forbidden"), the group advocated the imposition of Sharia law and the establishment of an Islamic state in Nigeria. Boko Haram attacks against Nigeria began in 2009 with assaults against military and police garrisons in Borno State in northeast Nigeria. Prior to 2009, Boko Haram limited actions to insurgent attacks in the city of Maiduguri, Borno State, by followers of their leader, Mohammad Yusef.

The extrajudicial execution of Yusef and the associated deaths of hundreds of his followers following a July 2009 raid on Boko Haram's mosque in Maiduguri passed leadership to Abubakar Shekau by 2010, whose leadership grew the group by adding 700 prisoners from a prison break in Bauchi. Following success against the military and police, in 2011 Boko Haram escalated its actions to involve the use of suicide bombers in a series of attacks against military, civilian, and police targets in Borno State, and increased the use of improvised explosive device (IED) attacks including multiple vehicle-borne IED attacks in the Nigerian capital of Abuja against the United Nations headquarters and police headquarters, and the December 25, 2011, bombing of four churches.

In early 2012, the Nigerian government declared a state of emergency in northeast Nigeria and increased military operations in response to rising numbers of Boko Haram attacks. Undeterred by ineffective Nigerian military efforts, Boko Haram fighters spread their movement into northern Cameroon, where by the end of 2012 Boko Haram had captured an estimated seven French nationals in kidnap-for-ransom attacks. Throughout 2012 and 2013 the Nigerian military continued to lose control of large areas, and neighboring states experienced Boko Haram attacks against villages and military forces.

Boko Haram fighters participated in attacks alongside North African Al Qaeda affiliates in Mali in 2012 and 2013. Returning Boko Haram members introduced new tactics to the fight in Nigeria with improvised explosive devices and suicide vests. The Boko Haram splinter group Ansaru attacked a Nigerian column on its way to deploy as members of a United Nations peacekeeping mission. In late 2013, the U.S. Department of State declared Boko Haram and Ansaru terrorist organizations. During this time, Boko Haram developed links with regional terrorist organizations including Al Qaeda in the Maghreb (AQIM). Despite Nigeria, Chad, and Niger forming a Multinational Joint Task Force in 1998, the countries were unable to consolidate funds and military forces to defeat Boko Haram, and throughout 2013 and 2014 each country conducted uncoordinated efforts against Boko Haram within their national borders.

On April 14, 2014, Boko Haram kidnapped more than 276 Christian schoolgirls from the Nigerian village of Chibok, Borno State. Through Boko Haram's use of social media throughout 2014, the Chibok girls attracted worldwide attention to the conflict, generated the #BringBackOurGirls campaign, and defined the group internationally. After the attack at Chibok, Boko Haram increased the number of female suicide bombers, employing them against civilian and military targets. Boko Haram equipped bands of roving, heavily armed young males with captured Nigerian military materiel, including tanks and armored personnel carriers.

The Nigerian military proved unable to stem the Boko Haram attacks through 2014, suffering numerous setbacks, including the loss of the towns of Gwoza, Mubi, and Bama to Boko Haram fighters. Reports of breakdowns of military discipline and desertions in units operating in northeast Nigeria appeared in Nigerian newspapers and social media. Unable to field enough personnel to stem the Boko Haram attacks, the Nigerian military began the ad hoc use of civilian militias, the Civilian Joint Task Force, to protect towns from Boko Haram.

By early 2015 Boko Haram reached its highest level of power and influence in Borno State, claiming to control more than 50,000 square kilometers in northeast Nigeria, and over an estimated two million Nigerians were reported to be displaced internally. The deadliest Boko Haram attack on January 3, 2015, was launched against the town of Baga, Borno State, headquarters of the anti-Boko Haram Multinational Joint Task Force, resulting in more than 2,000 people killed, a report the Nigerian government disputed. The attack attracted attention from Islamic State, depriving the Nigerian government of any claim to have destroyed the group.

By March 2015 the combined forces of the Multinational Joint Task Force had pushed Boko Haram out of Gwoza, Bama, and eleven of fourteen districts previously claimed by Boko Haram. Boko Haram pledged allegiance to the Islamic State of Iraq and the Levant (ISIL) in a Twitter message, rebranding itself as Islamic State's Wilayat Gharb Afriqiyya, or West Africa Province (IS-WAP).

In April 2015 Nigerian forces attacked Boko Haram camps in the Sambisa Forest, giving newly elected Nigerian president Buhari the confidence to declare Boko Haram's defeat in December 2015. Beset by heavy losses in fighters and materiel, internal frictions within Boko Haram's leadership led to a fracture of the terrorist organization. In early August 2016, ISIL announced it had appointed Abu-Musab al-Barnawi as the new leader of the group, a claim that Abubakar Shekau refused to accept. Shekau vowed to fight al-Barnawi while asserting his loyalty to ISIL's leader Abu Bakr al-Baghdadi. Throughout 2016 and into 2018 the two factions continued to fight each other despite both losing control over their respective territories.

John P. Ringquist

See also: #BringBackOurGirls; ISIS Recruitment and War Crimes; Shekau, Abubakar; Vehicle-Borne Improvised Explosive Device (VBIED); War on Terrorism and Social Media

Further Reading

Brigaglia, Andrea. "Ja'far Mahmoud Adam, Mohammad Yusuf, and al-Muntada Islamic Trust: Reflections on the Genesis of the Boko Haram Phenomenon in Nigeria." *The Annual Review of Islam in Africa* 11 (2012), 35–36.

Hentz, J., and H. Solomon, eds. *Understanding Boko Haram*. London: Routledge, 2017.

Ringquist, John P. "War by Tweet, Hashtag, and Media Messaging: Boko Haram's Media Warfare Challenges Nigeria's Information Campaign (2012–2015)." In *Paths of Innovation in Warfare: From the Twelfth Century to the Present*, ed. Nicholas Michael Sambaluk. Lanham, MD: Lexington, 2018.

Smith, Mike. *Boko Haram: Inside Nigeria's Unholy War*. London: I. B. Tauris & Company, 2016.

Zenn, Jacob. "Boko Haram: Recruitment, Financing, and Arms Trafficking in the Lake Chad Region." *Combating Terrorism Center at West Point, Combating Terrorism Center Sentinel* 7, no. 10 (October 2014).

#BringBackOurGirls

A global social media campaign asking for the release of more than 200 Nigerian girls who were abducted by the Islamist terrorist organization Boko Haram in 2014. Considered one of the most important online social mobilization movements, it demonstrated how social media could be used to rally communities, influence public opinion, and foster political action.

On the night of April 14–15, 2014, 276 female students were kidnapped from a government secondary school in the town of Chibok in Borno State, Nigeria, by the extremist group Boko Haram. The girls were sleeping in their dormitories when the armed men attacked the school. Although many girls managed to escape within hours of the kidnapping, mostly by jumping off the vehicles and running off into the surrounding underbrush, 219 girls were taken away and remained missing.

Many of the Chibok girls, aged approximately between twelve and seventeen years old, were Christian. The group Boko Haram, roughly translated as "Western education is forbidden/sinful," was known to have conducted attacks on several schools in the northeast of Nigeria. The abduction of the schoolgirls therefore reiterated Boko Haram's commitment against Western culture, most of all education for women and girls. The captured girls were allegedly forced into sexual slavery, child marriages, and human trafficking or were used in military activities, such as suicide bombings.

In the weeks following the kidnapping, the local population protested against the lack of action of the Nigerian government to save the 219 missing girls. On April 23, 2014, a Nigerian lawyer, Ibrahim Abdullahi, came up with #BringBackOurGirls on Twitter, repeating the same expression he heard from a speech by former World Bank vice president Obiageli Ezekwesili during a UNESCO event in the Nigerian city of Port Harcourt. A few days later, on April 30, Nigerian protesters marched on parliament in the capital city Abuja, calling for action from the government, which was perceived to be slow and inadequate in its response to the kidnapping.

Within weeks, the hashtag #BringBackOurGirls began to trend globally on Twitter and other social media platforms, causing global outrage against Boko Haram and the Nigerian government, becoming for a time Twitter's most tweeted hashtag (the hashtag was mentioned more than four million times in 2014). The organizations Amnesty International and UNICEF backed the online campaign, as well as many world leaders and celebrities such as the then U.S. first lady Michelle Obama, former U.S. secretary of state Hillary Clinton, British prime minister David Cameron, Nobel Prize winner Malala Yousafzai, actress Angelina Jolie, and rappers Wyclef Jean and Chris Brown. #BringBackOurGirls became a global movement for the release of the kidnapped schoolgirls and girls' rights to education. Not only popular online, the story of the abducted girls also spread in international news, and the movement prompted physical protests around the world, with people organizing marches on the Nigerian embassies in cities such as London and Los Angeles.

The campaign #BringBackOurGirls showed how social media can be used to support or encourage real-world change. The publicity created by the campaign put international pressure on the Nigerian government to intervene. In May 2014, the Nigerian president Goodluck Jonathan promised that he would find the girls. Furthermore, the police offered a $300,000 reward for any credible information leading to the rescue of the schoolgirls. The United States also sent surveillance drones and military trainers to Nigeria to help the local government and participated in the rescue mission.

Almost a year later, in March 2015, for the first time in a Nigerian election the sitting president lost his reelection. This defeat was partly attributed to the success of the online campaign to generate public frustration against President Jonathan's inaction. In August 2016, Boko Haram released a video showing some of the schoolgirls. A masked militant said that the girls would be freed in exchange for the release of imprisoned members of the group. In October 2016, the Nigerian government confirmed that twenty-one girls had been freed by Boko Haram following negotiations brokered mainly by the International Committee of the Red Cross. Two other girls were also found in 2016 after escaping the terrorist group, both carrying a baby in their arms. In May 2017, eighty-two of the schoolgirls were released following successful negotiations involving the exchange of five Boko Haram leaders.

However, the campaign also received a certain amount of criticism. Only three months after the popularity peak of the hashtag on social media, it was already losing the attention of the online communities, demonstrating the short temporality of an online movement. In addition, the girls remained missing for at least another two years, and approximately one hundred girls are still believed to be in the hands of Boko Haram. These factors underscore questions about the power of an online movement to truly spur concrete political action.

Overall, the movement demonstrated that using online platforms can help activists drive an issue into the mainstream agenda, not only locally but also internationally. The kidnapping of the 219 girls did not receive much international coverage in the days following the event, but a few weeks after the spreading of the hashtag on digital platforms, mostly because of its popularity among celebrities, the Chibok kidnapping was being discussed in news all around the world. The campaign proved that social media can influence traditional media sources as well as public opinion, and create bridges between people across physical boundaries and cultures.

Karine Pontbriand

See also: Boko Haram; ISIS Recruitment and War Crimes; Shekau, Abubakar

Further Reading

Carter Olson, Candi. "#BringBackOurGirls: Digital Communities Supporting Real-World Change and Influencing Mainstream Media Agendas." *Feminist Media Studies* 16, no. 5 (2016), 772–787.

Chiluwa, Innocent. "'War against Our Children': Stance and Evaluation in #BringBack-OurGirls Campaign Discourse on Twitter and Facebook." *Discourse & Society* 26, no. 3 (2015), 267–296.

Loken, Meredith. "#BringBackOurGirls and the Invisibility of Imperialism." *Feminist Media Studies* 14, no. 6 (2014), 1100–1101.

Olutokunbo, Adekalu Samuel, et al. "Bring Back Our Girls, Social Mobilization: Implications for Cross-Cultural Research." *Journal of Education and Practice* 6, no. 6 (2015), 64–75.

China, Blocking of Social Media by

Ongoing efforts by the government of the People's Republic of China (PRC) to exert control over communication in the digital realm, particularly within the borders of PRC. Three of the key approaches undertaken involve legislation about digital traffic within the country, technical controls on traffic involving users in PRC, and advancing a vision of digital sovereignty on the international stage.

China's participation in the digital realm has grown quickly, even by the meteoric standards of the expansion of computer and mobile technology overall. The first Chinese e-mail was part of an experiment between researchers at the Beijing Institute of Technology and Karlsruhe University in southwestern Germany in September 1987. Blogging first became possible in PRC in 2005; by 2009, the country had more Internet users than the total population of the United States, and by the end of 2011, PRC's digitally active population exceeded 500 million.

Authorities in PRC have proven alert to trends they identify as signifying danger or instability. Its authorities responded with deliberate severity against Tiananmen Square prodemocracy protests in June 1989 in a successful bid to avoid the contemporaneous fate of the Soviet Union, which was collapsing in the face of various challenges including nationalist movements and calls for greater democracy. Authorities in the country have maintained a commitment to controlling information and discourse in order to avoid public conversations that are interpreted as inviting a challenge to the party or the state. In the digital context, this includes infamous restrictions on what information is available online in PRC, including, for example, technical tools put in place so that search engine results for "Tiananmen Square" do not include any hint of the 1989 massacre. Control over information is sufficiently effective that outside estimates of the death toll thirty years after the fact still range wildly between 200 and 10,000.

The social media giants that dominate the rest of the globe, such as Twitter, Facebook, and YouTube, are prohibited in PRC. Instead, domestic social media entities exist and enjoy considerable popularity due to the country's massive population. Douban, a social networking site similar to Facebook launched in 2005 (a few months after Facebook), attracted more than 40 million users in its first six years and has since reportedly reached the 200 million mark. Weibo, a microblogging site launched in 2007, grew to become the hub of a family of social networking sites hosting the digital activity of more than 300 million people in PRC. Weibo's 140-character limit seemingly resembles that of Twitter. However, because of the information capacity of each character, 140 Mandarin Chinese characters can convey a considerably larger message than a comparable number of Latin characters. Linguists suggest that a Weibo message might require 500–600 characters to express in languages based on the Latin alphabet.

Two important events in social media expression and in censorship occurred during 2008. On May 12, a massive 8.0 earthquake hit south-central China near Chengdu in Sichuan province. The disaster killed an estimated 90,000 people and gave way to national mourning. Widespread popular outrage erupted with the discovery that many of the deaths had occurred in the collapses of more than 7,000 underengineered and inadequately constructed schoolrooms. Individuals shared their frustrations via social media with others around the country, and frustration gained traction especially since the country's one-child policy had resulted in many families instantly becoming childless. The protest movement, which was partly facilitated by online discourse, was met by a combination of government actions, including modest monetary compensation for bereaved families, the deployment of riot police to confront demonstrating parents, and legal prosecution of parents and others who persisted in alleging that corruption-induced shoddy building construction had led to thousands of deaths.

The year 2008 also saw the completion of the "Great Firewall," formally known in PRC as the Golden Shield. This system was designed to manipulate online experiences by blocking access to some sites identified as dangerous but to allow other traffic to proceed seemingly as normal. Called an "on-path" system, it reportedly examines both the destinations and the content of Internet communications. Banned sites will frequently be made inaccessible in ways that appear to reflect a problem with the Internet, as for example Google reported in 2011 when users in PRC found that in response to some search terms the computer reported trouble maintaining a connection to the Internet or that a selected link is no longer a functioning webpage. On other occasions, sophisticated filters comb online material and exclude prohibited content from view while allowing the remaining portions of a website to be observed.

Chinese law places the onus of censorship enforcement on service providers, meaning that although the government can get involved in maintaining censorship, entities that choose to do business in PRC are compelled to take part in assistance of censoring policies. This, coupled with the discovery of extensive and prolonged cyberespionage activities targeting Google and its users, resulted in tensions between Google and PRC when the former attempted to expand its business into a lucrative market potentially involving almost 15 percent of the world population.

Domestic web companies are similarly expected to enforce PRC censorship law. Despite its popularity, Douban has removed various types of content. These have included images of Renaissance paintings taken down in 2009 because nude figures in the paintings were asserted to constitute pornography and the elimination of conversations and keywords relating to Hong Kong's iconic Victoria Park the same year in order to counteract memorial demonstrations related to the twentieth anniversary of the Tiananmen Square massacre. Weibo also witnessed extensive self-censorship as the company was prompted to execute PRC policies. Initiatives meant to force users to associate their blogs or other social media accounts with their government-issued ID cards raise the prospect of the government even more easily being able to trace activities it mistrusts with specific individuals for prosecution and punishment.

Users located in PRC, similar to free speech advocates in other countries in which speech is restricted, have repeatedly tried to evade such censorship. Chinese Twitter users, for example, reportedly access the site via proxies, and technologies such as The Onion Router (TOR) were created for the purpose of preserving Internet privacy for users confronting censorship. As of mid-2018, more than 8,000 domain names were banned by PRC for various reasons.

Evidence suggests that in addition to domestic censorship, PRC technologies may include some ability to launch cyberattacks. Websites that provide censored information such as foreign news reports to Internet users in PRC reported suffering a distributed denial-of-service (DDoS) attack in 2015, and digital forensics suggested that these were launched by a system called "the Great Cannon," which essentially uses ordinary Internet traffic in PRC such as commercial activity on the website Baidu and redirects it to contribute to a larger DDoS campaign against a targeted website.

This pattern of technologies conducting censorship and of domestic legal policies compelling servers to support censorship are complemented by voluntary efforts of individuals who monitor information online and report "abuses" by material that meets the government's broad definitions of destabilizing or obscene content. The reported items are then scrutinized and removed. It is reasonable to suspect that social media intelligence (SOCMINT) is a part of censorship efforts.

On the international front, PRC officials have frequently enunciated the concept of Internet sovereignty. This concept, which is met sympathetically in several other countries in which authorities suspect Internet traffic as a source of subversive ideas, aims to establish international norms or even eventually formalized agreements through which PRC's government can claim ownership of Internet traffic in a way evocative of airspace sovereignty. The solidification of such norms would then justify further censorship activities and would reframe the debate about online communication so that those advocating open expression would be significantly curtailed. This agenda may be particularly important to authorities in PRC given recent research suggesting that even temporary access to information may exert lasting impact on a population's interest in regaining greater access.

Nicholas Michael Sambaluk

See also: Operation Aurora; Social Media Intelligence (SOCMINT); The Great Cannon; Twitter, Use of

Further Reading

Chen, Liang. *How Has China's Home-Grown Social Media Wechat Changed the Traditional Media Landscape?* Oxford: Oxford University Press, 2016.

Chen, Yuyu, and David Y. Yang. *The Impact of Media Censorship: Evidence of a Field Experiment in China.* January 2018, https://stanford.edu/~dyang1/pdfs/1984bravenewworld_draft.pdf.

Cheng, Dean. *Cyber Dragon: Inside China's Information Warfare and Cyber Operations.* Santa Barbara, CA: Praeger, 2017.

China's Censorship of the Internet and Social Media: The Human Toll and Trade Impact: Hearing before the Congressional-Executive Commission on China, November 17, 2011. Washington, D.C.: Government Printing Office, 2012.

Cook, Sarah. *Chinese Government Influence on the US Media Landscape.* Washington, D.C.: Freedom House, 2017.

Fischer-Schreiber, Ingrid. *Social Media in China.* Lausanne, Switzerland: UNIL, 2012.

Forbidden Feeds: Government Controls on Social Media in China. New York: PEN America, 2018.

He, Xia, and Rafael Pedraza-Jimenez. "Chinese Social Media Strategies: Communication Key Features from a Business Perspective." *Profession of Information* 24, no. 2 (March–April 2015), 200–209.

Power, Jenny. *Social Media Censorship and the Public Sphere: Testing Habermas' Ideas on the Public Sphere on Social Media in China.* Lund, Sweden: Lund University, 2016.

CompanyCommand.com

An online community of practice forum for U.S. Army captains. Established organically by officers and initially without official approval or support, the forum has been embraced by the Army since 2002. Founded by West Point alumni serving in the Army, CompanyCommand.com began in 2000 as an unofficial virtual space in which company-level officers would converse with each other. The objective was to broaden participation and peer visibility of informal but relevant conversations that regularly occur between peers.

It exists as "opportunistic networks," meaning that interaction between protégés and mentors is informal and shaped by environmental factors. Like its counterpart PlatoonLeader.com, which was similarly established in order to facilitate conversation among lieutenants, the CompanyCommand community grew, and so did its requirements for technical support. Originally a .com project by Army officers but without Army sanction, CompanyCommand was gifted to the Army in 2002 to be supported by West Point infrastructure and exist on Army servers. The Advancement of Leader Development and Organizational Learning (CALDOL) is the Army entity responsible for the effective functioning of the forum. Its informal "charter" in 2007 described it as "voluntary" and "grassroots," and the priority of linking company commanders and fostering conversation and resulting content collectively serve as its guiding principles.

Users have noted that CompanyCommand is a peer-to-peer environment, and this provides an opportunity for open dialog. However, an acknowledged trade-off is that an opportunist network for peer-to-peer communication frequently translates into conversations that end when one or another party gains the information sought or when one or another is drawn to focus attention elsewhere.

Despite these potential shortcomings, the informality of CompanyCommand's structure has encouraged considerable participation among company-grade officers. The forum has supported the sharing of ideas leading to the socialization of military tactics, techniques, and procedures (TTPs) and their application on the battlefield. The success generated by CompanyCommand has also helped prompt greater official acceptance of online communities of practice by the Army.

Nicholas Michael Sambaluk

See also: Army Knowledge Online (AKO); FrontlineSMS; HARMONIEWeb; Platoon-Leader.com; *Small Wars Journal*

Further Reading

Brown, J. S., S. Denning, K. Groh, and L. Prusak. *Storytelling in Organizations: Why Storytelling Is Transforming 21st Century Organizations and Management.* Boston: Elsevier Butterworth-Heinemann, 2005.

Carafano, James Jay. *Wiki at War: Conflict in a Socially Networked World.* College Station: Texas A&M University Press, 2012.

Kimball, Raymond A. "Transformation under Fire: A Historical Case Study with Modern Parallels." *The Letort Papers.* Carlisle, PA: U.S. Army War College, 2007.

Kimball, Raymond A. *The Army Officer's Guide to Mentoring.* West Point, NY: Center for the Advancement of Leader Development and Organizational Learning, 2015.

Palos, Guillermo A. *Communities of Practice: Towards Leveraging Knowledge in the Military.* Thesis. Naval Postgraduate School, September 2007.

Dabiq

A jihadi propaganda journal appearing in several languages, including English. The journal promoted Islamic State of Iraq and Syria (ISIS) viewpoints and provided information to readers about joining and participating in the jihadi organization. Shortly after the death of ISIS proto-founder Abu Musab al-Zarqawi in 2006, the organization established the Al-Furqan Foundation for Media Production, disseminating CDs and DVDs as well as written propaganda materials supporting ISIS. Al-Hayat Media Center, established in 2014, exports ISIS propaganda in English, French, German, and Russian languages. This is part of a larger ISIS initiative to attract attention and recruits who are overseas, particularly among the descendants of immigrants from the Middle East and also among Westerners lacking earlier affiliation with Islam but attracted to the violence and charisma displayed in ISIS propaganda, which consciously imitates the style of professional advertising and entertainment entities.

Al-Hayat quickly formed the journal *Dabiq* as a glossy online publication for export in foreign languages. The journal appeared approximately monthly, with a first issue released in July 2014 and the fifteenth issue appearing in July 2016. The journal proclaimed violent antipathy for monotheist adherents of Christianity and Judaism, polytheist Hindus, and also fellow monotheist Muslims if they were Shia or affiliated with the Muslim Brotherhood. The journal also vocally celebrated the ISIS revival of the practice of slavery.

Analysts suspect that the journal was largely imitative of the Al Qaeda journal *Inspire*, which urged readers to conduct lone wolf attacks against Western countries. In keeping with this conclusion, counterterrorism researcher Haroro Ingram has noted ISIS's distinct predilection toward copying the form and often the substance of other already successful entities. Evidence can be seen in ISIS social media and other forms of propaganda as well as in some of its strategic philosophy.

The journal shares the name of the town of Dabiq in northwestern Syria, which is mentioned in the Islamic hadith and which ISIS asserts will be the site of an apocalyptic battle where jihadi forces will triumph against the armies of "Rome." Although publication of *Dabiq* was discontinued in mid-2016, ISIS continued to use its Al-Bayan radio network to broadcast announcements in English and Russian as well as in Arabic.

Nicholas Michael Sambaluk

See also: Al-Furqan; #AllEyesOnISIS; Dawn of Glad Tidings Campaign; ISIS Recruitment and War Crimes

Further Reading

Atwan, Abdel Bari. *Islamic State: The Digital Caliphate.* Oakland: University of California Press, 2015.

Rogers, Paul. *Irregular War: ISIS and the New Threat from the Margins.* London: I. B. Tauris & Co., 2016.

Salzar, Philippe-Joseph. *Words Are Weapons: Inside ISIS's Rhetoric of Terror.* New Haven, CT: Yale University Press, 2017.

Warrick, Joby. *Black Flags: The Rise of ISIS.* New York: Anchor, 2016.

Dawn of Glad Tidings Campaign

An Arabic language smartphone application established by the jihadi Islamic State (IS) movement, frequently called the Islamic State of Iraq and Syria (ISIS) as a vehicle for projecting propaganda. Reportedly, the application facilitated the retweeting of ISIS messages by the Twitter accounts of those who had also downloaded the Dawn of Glad Tidings app.

ISIS leveraged the app during its campaign in mid-2014 that captured Iraq's second city, Mosul. The city's fall was accompanied by a wave of brutal murders of Iraqi Army stragglers and prisoners and by an orchestrated ISIS propaganda effort to document and disseminate images and messages about the killings. ISIS spread 40,000 tweets the day that it captured Mosul, reinforcing the horror of its regime and seeking to inspire extremism in others beyond the physical grasp of its territories. In addition to retweeting ISIS messages on Twitter, the app also allowed for the sharing among ISIS sympathizers of various types of images, videos, and comments. It appears that the original intent of those designing the app had been to establish a private online forum through which to communicate secretly. The app's use in the context of the capture of Mosul suggests that it was repurposed for use as a propaganda tool.

Although the downloading of most apps involve the sharing or compromising of some pieces of personal information, the Dawn of Glad Tidings app went much farther, including the modification and deletion of flash memory and the app's ability to view WiFi connections. It is unclear exactly how many times the app was downloaded, but estimates range wildly between several hundred and several thousand before it was removed from the Google Play app store.

Nicholas Michael Sambaluk

See also: #AllEyesOnISIS; Facebook, Use of; Hashtag Hijacking; Instagram, Use of; ISIS Recruitment and War Crimes; Twitter, Use of; War on Terrorism and Social Media; YouTube, Use of

Further Reading

Berenger, Ralph D., ed. *Social Media Go to War: Rage, Rebellion and Revolution in the Age of Twitter.* Spokane, WA: Marquette Books, 2013.

Bourret, Andrew, et al. *Assessing Sentiment in Conflict Zones through Social Media.* Monterey, CA: Naval Postgraduate School, 2016.

Irby, Jack B. *The Weaponization of Social Media.* Master's thesis. Ft. Leavenworth, KS: Army Command and General Staff College, 2016.

Schneider, Nathan K. *ISIS and Social Media: The Combatant Commander's Guide to Countering ISIS's Social Media Campaign.* Newport, RI: Naval War College, 2015.

Ullah, Haroon K. *Digital World War: Islamists, Extremists, and the Fight for Cyber Supremacy.* New Haven, CT: Yale University Press, 2017.

Doctrine Man

An online military-themed lampoon of policies and trends in the U.S. military. Reputedly the brainchild of Steve Leonard (U.S. Army Col., retired), "The Further Adventures of Doctrine Man" began as a comic strip in 2008, drawing on personal doodles from 1999 and initially expanding through e-mailed sketches to colleagues and friends. The cartoon's growth in popularity has been cited as exemplifying the U.S. military's relaxing of strictures against military personnel participating in social media. The cartoon significantly expanded its reach in mid-2010, when a Facebook page was established as a platform for sharing Doctrine Man cartoons.

Typical Doctrine Man cartoons involve a four-square sequence of line-and-bubble cartoon figures. Pithy sarcastic comments dealing with Army or sister service issues characterize the vast majority of cartoons. Over a dozen videos titled "Doctrine Man" and done in a complementary style appeared on YouTube in 2011. Although these provided an opportunity for longer, more layered satirical dialogs between simple computer-generated cartoons with computer-generated voices, this series lacked the endurance of the comic strip.

Expansion to a Facebook presence has allowed Doctrine Man not only to reach a wider audience than before, but also to introduce links to current events hyperlinks beside cartoons. Short satirical commentary usually preface the links. Appealing to an audience for the most part personally connected to the military, the

Doctrine Man brand has extended to include paperback book collections of the comics that had first appeared on the Facebook page, first appearing in 2013.

Although the U.S. Army is a central focus of Doctrine Man cartoons, observations are frequently applicable to the sister services, and occasionally cartoons deal specifically with other branches. The niche popularity of Doctrine Man within parts of the military has coincided with changes in how the military has institutionally been brought to view and accommodate the popularity of social media. The earlier existence of cartoons such as the iconic World War II "Willy and Joe" by Bill Mauldin, published in the Army's *Stars and Stripes*, arguably indicates a longer pattern of eventual official acceptance of wry illustrated commentary by and for military personnel. The projection of newer forms of commentary via social media has created the opportunity for such commentaries to more quickly reach a broader audience.

Nicholas Michael Sambaluk

See also: Facebook, Use of; John Q. Public; Web 2.0; YouTube, Use of

Further Reading

Patrikarakos, David. *War in 140 Characters: How Social Media Is Reshaping Conflict in the Twenty-First Century.* New York: Basic Books, 2017.

Qualman, Eric. *Socialnomics: How Social Media Transforms the Way We Live and Do Business.* New York: Wiley, 2009.

Silvestri, Lisa. *Friended at the Front: Social Media in the American War Zone.* Lawrence: University Press of Kansas, 2015.

Doxing

A form of cyber domain vigilantism in which compromising or vulnerable data is derived and disseminated about a victim. Frequently, doxing involves the compilation and release of open source information, although it can also alternatively entail the release of private information.

Doxing specifically aims to publicize information about a targeted individual in order to create vulnerability. As such, it is an activity that means to remove the character of anonymity and obscurity that has traditionally been associated with the digital environment. Research has found that the information-gathering effect of doxing mounts, because "each type of doxing also creates new possibilities to further interfere [and] deanonymization makes it easier to obtain other types of identity knowledge" about a victim (Douglas, 2016, 203). Motives for doxing vary and can include extortion, if the victim is threatened with the divulging of information unless payment is provided, to hacktivism if the objective is primarily to humiliate and publicly damage the victim.

Doxing has been a favorite tool of the hacktivist group Anonymous. As early as 2011, the group began launching doxing actions against diverse entities ranging from purported members of racist hate groups such as the Ku Klux Klan to U.S. federal officials such as the net-neutrality opponent Federal Communications Commission chairman Ajit Pai. The term itself gained notoriety through media accounts of Anonymous tactics.

Some journalists have also reportedly utilized doxing to expose antagonistic figures in their investigative efforts. Recurring cases of mistaken identity have fueled controversy about the ethicality of hacktivist doxing. Similarly, doxing appears to demonstrate an ability to leak information that hurts unintended individuals and groups. Hacktivist doxing that was intended to damage the reputation of Turkish president Tayyip Erdogan in 2016 illustrated this, releasing personally identifiable information about potentially hundreds of thousands of third parties in the process of the attack. Media reports have characterized the use of doxing techniques as "leading to an arms race of financially incentivized, shame-slinging vigilantes" (Ellis, 2017) and have noted that doxing actors have used funding sites including Patreon and WeSearchr to garner resources as digital mercenaries.

North Korea's hack of Sony Pictures Entertainment in 2014 included doxing approximately thirty-eight million of the company's documents, prompting an embarrassing controversy for the company in retaliation for its planned release of a film deemed offensive by North Korean leadership. In the People's Republic of China (PRC), doxing has reportedly been used as a tool by authorities for coercing minor domestic dissidents and shaming other domestic targets. As such, doxing is a coercive weapon in some state digital arsenals as well as being used by nonstate organizations.

Doxing has also been used as a tool for triggering and inciting physical violence. "Swatting," the signaling to law enforcement of a target's accurate personal information and location accompanied by misleading accusations, has been used as a tool to trigger armed action against an unsuspecting target. In 2015, the Islamic State of Iraq and the Levant (ISIL) doxed the personal information of more than one hundred U.S. military personnel for the purpose of inciting sympathizers to murder service personnel in the United States.

Nicholas Michael Sambaluk

See also: Anonymous; Sony Hack, 2014

Further Reading

Douglas, David M. "Doxing: A Conceptual Analysis." *Ethics and Information Technology* 18, no. 3 (September 2016), 199–210.

Ellis, Emma Grey. "Whatever Your Side, Doxing Is a Perilous Form of Justice." *Wired,* August 17, 2017, https://www.wired.com/story/doxing-charlottesville.

Mathews, Roney Simon, Shaun Aghili, and Dale Lindskog. "A Study of Doxing, Its Security Implications and Mitigation Strategies for Organizations." Edmonton, Canada: Concordia University College.

Norris, Ingrid N. *Mitigating the Effects of Doxing.* Thesis. Utica, NY: Utica College, 2012.

Facebook, Use of

A social media giant whose capacity to facilitate networking has attracted interest and use by various entities for strategic purposes including the support of information operations. The social media application allows individuals to create member accounts by providing contact and other information so that they can "friend" other members (often people they know personally) or "like" public figures or organizations for which they already have affinity. Once these linkages are established the app provides an easy and convenient way for friend members to keep in touch through a widening array of communication methods, starting with the posting of publicly readable statements and later through private messages and through the posting of videos.

First founded in 2004 by a circle of undergraduate students from Harvard University, the application's popularity exploded during the mid-2000s, and membership was expanded first to a few other notable universities and then more broadly across institutions of higher learning and subsequently in September 2006 to any interested party thirteen years of age or older. It attracted investment by software giant Microsoft in 2007, became a publicly traded company in 2012 and grew to one billion users, and it acquired the Oculus virtual reality company two years later. Facebook's popularity and its wide accessibility mean that in 2017 the app took in $40 billion in revenue and boasted two billion users to include a virtual "population" larger than any single country at the time.

The app is designed to facilitate communication between users who already share friendships or affinities beyond the digital realm. With respect to conflict, journalists have already commented on how affinity groups have leveraged the app as a method for projecting messages about the need for equipment, as for example by fighters supporting Ukraine in its combat against pro-Russian paramilitary forces in the eastern part of Ukraine. Groups loyal to Kiev have also used the app as a platform on which to display images and messages of fighters who pose for the camera with newly donated equipment and supplies provided by sympathizers. Syrian Kurdish organizations used Facebook to assist in recruitment efforts when mobilizing against the Islamic State of Iraq and Syria (ISIS). Meant to make easy communication between friends, the app is easily adapted to this purpose among populations that share political affinities in wartime.

Facebook's format has always been geared toward a one-to-many mode of communication, and although private messaging has been an option since 2008, the app retains essentially its one-to-many character of communication. The expansion of access to Facebook coincided with the latter phases of the U.S. military occupation of Iraq (2003–2011). Large numbers of military personnel recognized in the app a way to communicate with loved ones at home with an ease and an immediacy unparalleled in the history of warfare. Scholars have suggested that this ease and ubiquity of communication binds deployed soldiers via a virtual tether to their peacetime lives in a way that may prompt new kinds of stresses on combat personnel. Nonetheless, the desire to communicate with friends and family attracted military personnel to adopt and use the app in considerable numbers.

The rise of social media during this period also attracted the attention of authorities in the U.S. military who were concerned about the operational security implications of personnel regularly engaging with social media. The immediacy with which words and images could be posted prompted particular concern. Written posts might include information that could tip off enemies to become aware of impending operations, lulls, conditions on a given base, or other operationally valuable information. Photographic images, such as "selfies" or photos of other personnel, could similarly include information useful to an enemy. Examples could include items in the background, such as vehicles or other equipment whose existence and condition becomes evident to the viewer. The U.S. military briefly attempted to foreclose social media usage by its personnel, and it tried to offer more secure alternatives for communications between military members and their families at home. However, the limitations and ultimate failure of these policies were acknowledged as the drawdown of troops from the Iraq occupation progressed. Carelessness, by terrorists or other antagonists of the United States, in the use of social media such as Facebook has reportedly assisted in the location and targeting of some enemy figures. This underscores the question of operational security when popular social media tools like Facebook are used on the battlefield.

The app's ease of use invites leverage for political activism. Broadly, Facebook has resisted undertaking efforts to do much policing of content beyond removal of reported graphic images and recently examples of evident fake news items intended to propagandize. Colombians in 2008 used the app to organize against the Revolutionary Armed Forces of Colombia (FARC), which had waged revolutionary war and drug-funded terrorism from 1964 until 2017. During the early parts of the Arab Spring in Egypt and Tunisia in 2011, protesters used Facebook to organize demonstrations.

Inversely, states have responded both to and through Facebook in order to retain political power. The Egyptian government banned social media in a vain attempt to stem planning of protests; Bahrain reportedly used data found via Facebook to trace and prosecute dissidents. Iran and the People's Republic of China have banned Facebook and installed their own counterpart social media site that is controlled by the state. North Korea banned the use of Facebook within the country, but in 2010 its government and its state-run media organization set up Facebook pages, presumably to project messages beyond their borders.

Controversy about privacy issues has repeatedly visited the application and its chief architect, Mark Zuckerberg. Its progenitor app used the official ID photos of Harvard students, along reportedly with images "Zuckerberg hacked [from Greek] House websites" in order to construct an array of student photos so that users could rank the attractiveness of their peers. Accused of "breaching security, violating copyrights and violating individual privacy," Zuckerberg was compelled to take down the website, which he called Facemash, in late 2003 (Kaplan, 2003).

The app allows users to select privacy settings and reportedly began exerting greater effort to protect user privacy in 2010. This reportedly coincided with various entities collecting data posted by users. Since the app is free to use, it relies on the sale of ad space and the monetization of data to provide its considerable revenue. It is reported that nation-states have leveraged Facebook's loose and expansive execution of monetization policies to promote information operations.

In March 2018 journalists learned that a political data analysis entity called Cambridge Analytica had garnered, in addition to the data voluntarily contributed by 270,000 Facebook users, information from fifty million additional U.S. Facebook users who had not provided consent and who had not participated in the study. Facebook and Cambridge Analytica then reportedly threatened to sue *The Guardian* newspaper if it broke the story. Facebook's responses following publication of the news first suggested that the data analysis firm no longer possessed the information although subsequent analysis indicates that it may instead have continued to retain access to the data. This pattern indicates a considerable latent vulnerability to organizations that might access and utilize data for information operations in relation to political issues or wars.

Evidence mounted of Russian actions to conduct information operations via Facebook to interfere in the U.S. 2016 presidential election cycle. The apparent objective was to sow dissention in the run-up to the election in order to erode national resolve and potentially U.S. commitment to international alliances and other commitments. As a popular app promoting communication among likeminded members and friends, Facebook proved to be a useful conduit for part of this information operations campaign. Russian sources provided funding and support for diametrically opposed political groups, including some fringe extremist organizations in a "strategy aimed to further incite polarization around hot political issues: race, immigration, religion, and gender" (Boyer and Polyakova, 2018, 10). One anti-immigration group with nearly a quarter of a million members was a creation of this information operations effort, while Russia also provided support to the Black Lives Matter movement and encouraged actions against then-candidate Donald Trump. Coinciding with the Occupy Wall Street anticapitalist movement in the early 2010s, Russia's news agency Russia Today created a Facebook app in order to help protesters network with each other. In the wake of these discoveries, journalists noted that Facebook was unusually forthcoming with federal investigators seeking information about the scope and scale of Russian involvement. Furthermore, analysis suggests that this was not in isolation, since Russian interference is suspected of having also occurred in India's 2014 elections and in various European election seasons including the 2017 presidential election in France.

In a decade and a half of existence, Facebook has demonstrated a tremendous ability to gain and hold popularity among a global user base. These virtues have attracted a continually widening circle of users ever since the initial release of the app, and these have included military personnel. Facebook's one-to-many communication model has proven useful to state and nonstate organizers interested in promoting political causes. The app's legacy of free-to-use connectivity has been intertwined with recurring concerns about the ways it leverages user data into monetizable data that enjoys only questionable protection. This combination of ubiquity and apparent vulnerability has reportedly resulted in nation-states identifying the opportunity to transform a communication network into a landscape for strategically useful information operations.

Nicholas Michael Sambaluk

See also: Operational Security, Impact of Social Media on; Social Media Intelligence (SOCMINT); Twitter, Use of; U.S. Military Use of Social Media; Web 2.0

Further Reading

Bergmann, Max, and Carolyn Kenney. *War by Other Means: Russian Active Measures and the Weaponization of Information.* Washington, D.C.: Center for American Progress, 2017.

Boyer, Spencer P., and Alina Polyakova. *The Future of Political Warfare: Russia, the West, and the Coming Age of Global Digital Competition.* Washington, D.C.: Brookings Institution, 2018.

Bradshaw, Samantha, and Philip N. Howard. *Troops, Trolls and Troublemakers: A Global Inventory of Organized Social Media Manipulation.* Oxford: Oxford University Press, 2017.

Kaplan, Katharine A. "Facemash Creator Survives Ad Board." *The Harvard Crimson,* November 19, 2003, https://www.thecrimson.com/article/2003/11/19/facemash-creator-survives-ad-board-the.

Lin, Herbert, and Jackie Kerr. *On Cyber-Enabled Information/Influence Warfare and Manipulation.* Amsterdam, Netherlands: SSRN, 2017.

Patrikarakos, David. *War in 140 Characters.* New York: Basic Books, 2017.

Paul, Christopher, and Miriam Matthews. *The Russian 'Firehouse of Falsehood' Propaganda: Why It Might Work and Options to Counter It.* Santa Monica, CA: RAND, 2016.

Silvestri, Lisa Ellen. *Friended at the Front: Social Media in the American War Zone.* Lawrence: University Press of Kansas, 2015.

Waldman, Ari Ezra. "Privacy, Sharing, and Trust: The Facebook Study." *Case Western Reserve Law Review* 67, no. 1 (2016), 193–233.

Wardle, Claire, and Hossein Kerakhshan. *Information Disorder: Toward an Interdisciplinary Framework for Research and Policy Making.* Strasbourg, France: Council of Europe, 2017.

FireChat

A software application that allows devices to connect to each other without an ordinary Internet connection. Recent usage has already demonstrated the applicability of the app for nonstate actors such as political dissident groups, and it may prove usable to nonstate combatants as well.

FireChat's purpose is to enable wireless mesh networking. Such networks can incorporate either stationary or mobile devices, with the former being more stable but unmovable and with the latter being portable but liable to technical inefficiencies connected with the additional challenges in maintaining informational routes. Despite the drawbacks, the portability of the latter is distinctly of interest to a number of groups, especially those whose members anticipate relying on frequent mobility. FireChat exists to create this latter type of peer-to-peer connectivity, linking mobile devices to one another.

FireChat first appeared in the spring of 2014, first for Apple iPhones and then quickly also for Android devices. Limitations to FireChat include the fact that it supports text communications but not voice or video; also, its range is believed to be limited to a maximum of something between 70 and 200 feet (approximately 20 to 60 meters). The app's creator does not claim the ability to hop from device to device, so the extent of the mesh may be a single span of about 200 feet rather than a potential chain with devices serving as connecting stations that are spaced closely together. Apple and Android devices may also be incapable of communicating with one another even if both have the app, meaning that an ad hoc mesh network requires a degree of similarity among the member platforms.

Almost immediately, the app's utility for protest movements became apparent. Users in Iraq and in the Hong Kong Special Administrative Region of the People's

Republic of China reportedly began relying on FireChat to continue communications despite actions of both governments to curtail access to the Internet with the intention of limiting demonstrations that the governments did not welcome. The following year, the left-wing government of Rafael Correa appears to have shut down Internet access in order to stifle mounting protest. Officials claimed that Internet outages were the effect of denial-of-service attacks, but analysts believe that such claims were false. Opposition figures encouraged members of their movement to use FireChat, and reported use of the app spiked in Ecuador that year. Catalan nationalist protesters in eastern Spain have also used the app, as have protesters in India when the University of Hyderabad reportedly shut down Wifi access in the face of demonstrations on campus.

It is not yet clear whether FireChat's proven use by demonstrators in political movements might indicate future use as a tactical communications tool by militant nonstate actors. However, improvements to the app may have already included the introduction of end-to-end encryption that prevents third parties and even the app developer itself from accessing messages. FireChat asserted that this capability had been established during mid-2015. If correct, this could constitute yet another relatively secure low-budget communications tool for a range of users, including possibly militant nonstate actors if they possessed enough devices in an area to maintain a mesh network.

Nicholas Michael Sambaluk

See also: FrontlineSMS; Surespot App; Telegram App; WhatsApp

Further Reading

Beachy, Alexander J. *At the Edge: Operating in a Disconnected Low-Bandwidth Environment.* Thesis. Monterey, CA: Naval Postgraduate School, 2015.

Pathak, Parth H., and Rudra Dutta. *Designing for Network and Service Continuity in Wireless Mesh Networks.* New York: Springer, 2013.

Petry, Joshua W. *Let the Revolution Begin, 140 Characters at a Time: Social Media and Unconventional Warfare.* Thesis. Maxwell AFB, AL: School of Advanced Air and Space Studies, 2015.

Santi, Paoli. *Mobility Models for Next Generation Wireless Networks: Ad Hoc, Vehicular and Mesh Networks.* Sussex, UK: Wiley, 2012.

Forum Infiltration

The invasion of online discussion venues, typically for intelligence-gathering purposes or as part of an effort to subvert the actions or attentions of forum members. Online forums are virtual spaces in which like-minded people gather to communicate about a spectrum of interests, from specialty hobbies to political activism, to illicit transactions and planning of criminal acts. Because these forums must be accessible to members and relatively easy to find, forum infiltration is a common form of exploitation by a range of actors. Depending on the target and the purpose, forum infiltration may be employed by law enforcement or by individuals or groups defying legal authorities. Infiltration allows a motivated and skilled actor to use the forum as a platform from which information can be collected, although several other opportunities exist beyond reconnaissance. Groups can be weakened through the seeding of disinformation on a forum to foment discord. Alternatively, forum members can be recruited by an infiltrator for dissimilar causes, or they can be cultivated to participate in various activities in the physical world. Once granted entry to a forum, infiltrators can also move to enable bots to passively collect information. Experts conclude, however, that the more effective forum infiltration cases involve the active use of social engineering to choose targets, gain the trust of participants over an extended period of time, and ultimately even take over leadership of the forum.

Criminal networks, often on the leading edge of technology, use forums as an anonymous ecosystem of contacts and skill acquisition, to exchange illicit materials, gain confederates, and rehearse and plan operations. In 2005, the FBI assigned Agent Keith Mularski to infiltrate the DarkMarket forum, a site where stolen credit cards and PINs were exchanged. By 2006, Mularski had so successfully won the trust of the fractious criminal participants through his mastery of spamming that he was both a site admin, as "Master Splyntr," and had moved the site onto the FBI's server. Law enforcement professionals, as well as trained volunteers, have infiltrated and continuously monitor several kinds of forums. Notable examples include forums used by pedophiles, those used as recruiting sites for violent extremist organizations, and those frequented by criminals either seeking dupes or confederates for scams.

Various bad actors, as individuals or as parts of groups or serving at the behest or acquiescence of nation-states, nonetheless continue to utilize forums and also engage in their own forum infiltration. In 2008, Russian hackers infiltrated pro-Russian forums to recruit allies in attacking Georgian government websites in coordinated acts of sabotage and denial of service. Governmental entities in the People's Republic of China, in Russia, and in Myanmar reportedly operate branches of their intelligence services dedicated to infiltrating forums of political activists. The information that is collected leads to persecution of targeted activists, while also sowing discord and tension. Forum infiltration is difficult to prevent except at the expense of groups walling themselves off and thereby undermining the capacity to attract new members, and few forum administrators have the technological or social expertise to spot infiltrators. In each case, forum infiltration relies on the planned deployment of social engineering to identify an appropriate venue, earn the trust of the participants, and then exploit that virtual social relationship to achieve a planned end, whether the forum was an idealistic "imagined community" or an "offender convergence setting."

Margaret Sankey

See also: Anonymous; Georgia, 2008 Cyber Assault on; Hacktivism; Hashtag Hijacking; Operational Security, Impact of Social Media on; Twitter Account Hunting

Further Reading

Deibert, Ronald, and Rafal Rohozinski. "Liberation vs. Control: The Future of Cyberspace." *Journal of Democracy* 21, no. 4 (October 2010), 43–57.

Goel, Sanjay. "Cyberwarfare: Connecting the Dots in Cyber Intelligence." *Communications of the ACM* 54, no. 8 (August 2011), 132–140.

Leukfeldt, E. R. "Cybercrime and Social Ties." *Trends in Organized Crime* 17, no. 4 (January 2014), 231–249.

Lusthaus, Jonathan. "Trust in the World of Cybercrime." *Global Crime* 13, no. 2 (May 2012), 71–94.

Ottis, Rain. "Proactive Defense Tactics against On-Line Cyber Militias." In *EWIC2010: Proceedings of the 9th European Conference on Information Warfare*, ed. Joseph Demergis. Reading, UK: Academic Publishing, 2010, 233–237.

FrontlineSMS

A stand-alone software developed in 2005 to facilitate information sharing in areas underserved in terms of Internet and cellular phone capabilities. First used by conservationists in South Africa's Kruger National Park, it has been used for monitoring of several elections including in the Philippines, Afghanistan, and Nigeria, and it was also used for emergency management following the 2010 Haiti earthquake.

The U.S. military considered FrontlineSMS as a tactical communications tool because its reliance on a cell phone and a laptop computer opened the opportunity for truly mobile use. Planners anticipated that a laptop might be matched with a group of cell phones to facilitate distribution and use in a combat zone. Unclassified information could in theory be shared either in two-way messaging of 160-character segments or by one machine to a predetermined group of phones receiving a message. Although the system's ease of use and insular character made a built-in resistance to misuse and the potential existed for encryption, the software has experienced more success with emergency response and health workers in remote areas than among military personnel. U.S. military users peaked in the hundreds.

Nicholas Michael Sambaluk

See also: FireChat; HARMONIEWeb; TroopTube; Twitter, Use of; WhatsApp

Further Reading

Backes, Kenneth W. *Leveraging Technology and Social Media for Information Sharing.* Air War College thesis, April 2009.

Elmasry, George. *Tactical Wireless Communications and Networks: Design Concepts and Challenges*. Hoboken, NJ: Wiley, 2012.

Mitchell, Paul. *Network Centric Warfare: Coalition Operations in the Age of US Military Primacy.* London: International Institute for Strategic Studies, 2006.

HARMONIEWeb

An unclassified virtual environment built by the U.S. military expressly for facilitating civil-military information exchange, responding to an identified need of U.S. forces "to partner with non-traditional entities to achieve operational objectives, especially in stability, disaster relief, and humanitarian operations" (Drake, 2008, 4).

U.S. military awareness of the utility in digital collaborative spaces, particularly if they could be made accessible or restricted according to need, propelled interest in the creation of HARMONIEWeb software by 2006. Users would authenticate on a discrete but not secret website, and although the program was known by several nongovernmental organizations, it remained relatively obscure to the wider public.

Among the features of HARMONIEWeb was an automatic translation feature allowing users to read internal webpages in any of thirty-three separate languages and could also provide automatic document translation for documents uploaded by users for viewing by government personnel for nongovernmental organizations or vice versa. A 24–7 support feature was also built in order to address urgent needs encountered by users. Tools within the application included updated imagery from satellite mapping, Microsoft Office Communicator, and Adobe Connect conferencing, which were built into the system. To simultaneously promote sharing and security, the creator of each individual component of HARMONIEWeb was able to control access to that portion of the overall site. Different configurations of pages could be applied to allow users in poorly connected areas to cope with limited bandwidth.

Although HARMONIEWeb offered the opportunity for significantly more secure communication than would be possible via popular public alternative venues for near real-time information-sharing technologies such as Facebook or Twitter, some users argued that this advantage was counterbalanced by challenges to use, including the complexity of the registration process and the system's document tutorial.

One of the high-profile events was the response to the 2011 tsunami that struck Japan, because the response involved humanitarian relief efforts by the U.S. military as well as several nonmilitary entities. In addition to natural disaster scenarios, HARMONIEWeb has also been used in facilitating humanitarian activities in war and postwar zones such as Iraq and Afghanistan, where collaboration and communication between military and nongovernmental organizations are also valuable. Management of HARMONIEWeb was transferred from Joint Forces Command to the Defense Information Systems Agency.

Nicholas Michael Sambaluk

See also: Facebook, Use of; FrontlineSMS; Twitter, Use of; U.S. Military Use of Social Media; WhatsApp

Further Reading

Backes, Kenneth W. *Leveraging Technology and Social Media for Information Sharing.* Air War College thesis, April 2009.

Cryan, William. "HARMONIEWeb." United States Joint Forces Command. No date. https://www.slideserve.com/rod/harmonieweb.

Drake. "Enabling Civil/Government Teaming in HARMONIEWeb." Department of Energy Environment Management Office. October 20, 2008, http://usacac.army.mil/cac2/AOKM/aokm2008/presentations/Wednesday/HARMONIE-BCKS-Brf-20october.pdf.

Elmasry, George. *Tactical Wireless Communications and Networks: Design Concepts and Challenges.* Hoboken, NJ: Wiley, 2012.

Joint Forces Command Briefing by Rear Admiral Wachendorf, JFCOM Chief of Staff, at Industry Symposium 2007 National Defense Industry Association, July 30, 2007. http://www.ndiaghrc.org/Symposium_2007/briefs/02.pdf.

Hashtag Hijacking

The organized commandeering of comment feeds in the social networking and updating service Twitter. Although hashtag hijacking can occur in several contexts and with various motivations, in the context of 21st-century warfare it typically deals with the commandeering of an unrelated comment feed and the imposition of propaganda materials to drown out relevant traffic and replace it with an information operations message.

The Islamic State of Iraq and Syria (ISIS) prominently hijacked a number of Twitter feeds during the apex of its empire and self-proclaimed caliphate, which extended across much of Syria and Iraq between 2014 and 2017. One prominent instance occurred when the Twitter feed #WorldCup2014 about that year's soccer championship games was hijacked by ISIS accounts that posted jihadist messages including images of ISIS fighters kicking a severed human head as if it were a soccer ball. Another example is the hijacking of the hashtag #napaquake in August 2014 that had been established in response to an earthquake in Northern California. ISIS accounts posted text threatening the United States with attacks as well as posting Photoshopped images of the White House on fire. Although many of the examples of ISIS hashtag hijacking coincided with the early establishment and expansion of the organization's territorial holdings in 2014, this did not mark the end of ISIS actions to temporarily commandeer feeds. ISIS sympathizers successfully hijacked the Twitter account of the U.S. military's Central Command in early 2015, creating consternation and embarrassment. In January 2016, ISIS hijacked the hashtag of pop singer Justin Bieber, using the feed as a conduit to reach the entertainer's seventy-four million mostly adolescent and young adult social media followers and present them with a jihadi video titled "Message to the Islamic West."

Although ISIS has proven capable of generating original hashtags of its own, such avenues are primarily of value in disseminating messages to already interested followers. ISIS's aim extends beyond directing messages to its existing sympathizers and additionally includes intimidation of Twitter users who live outside ISIS's physical reach and who would not willingly seek out jihadist propaganda. This may explain the effort that ISIS has placed into hijacking unrelated hashtags.

Business interests, which also face hashtag hijacking threats, have a vested interest in avoiding falling victim to such actions, regardless of the hijackers' motivations. As was the case with the Digorno pizza company in the fall of 2014 when a pizza advertisement #WhyIStayed inadvertently crossed paths with a tweet feed by people sharing their experiences in domestic abuse, hashtag hijacking does occasionally occur by accident. A potentially even more embarrassing case occurred when a mobile wallet app jointly created by three giants in the telecom industry was dubbed ISIS and launched an advertising campaign on Twitter proclaiming the ability of its phones to support of "ISIS." The product's rollout coincided tragically with the most notable territorial and social media offensives of the ISIS jihadist organization in the summer of 2014, and the mobile wallet app was redubbed "Softcard" by September. The question of mathematically detecting a hashtag hijacking has also drawn attention from academia.

As a propaganda tool in 21st century warfare, hashtag hijacking carries formidable potential, allowing a group to temporarily gain access to a large audience and proselytize or intimidate it with tailored propaganda messages, as well as potentially to create embarrassment for a digital victim.

Nicholas Michael Sambaluk

See also: Al-Furqan; Forum infiltration; Twitter, Use of

Further Reading

Maggioni, Monica, and Paolo Magri, eds. *Twitter and Jihad: The Communication Strategy of ISIS*. Milan, Italy: ISPI, 2017.

Poxon, Benjamin W. *Social Media Risk Analysis: How to Use Accepted Risk Assessment Tools to Analyze Social Media Risks in Military Organizations*. Master's thesis. Maxwell AFB, AL: School of Advanced Air and Space Studies, 2017.

Ridland, Andrew. *Seizing the Digital High Ground: Military Operations and Politics in the Social Media Era*. Master's thesis. Norfolk, VA: Joint Forces Staff College, 2015.

Tunnicliffe, Ian. *Social Media: The Vital Ground, Can We Hold It?* Carlisle, PA: Strategic Studies Institute, 2017.

VanDam, Courtland, and Pang-Ning Tan. "Detecting Hashtag Hijacking from Twitter." *WebSci 2016,* Hannover, Germany, 2016.

Instagram, Use of

A leading social media application, popular and in widespread use particularly among military-age populations. Consequently, it has been used not only by personnel to connect with friends but has also been used by people operating in conflict zones and other political violence, such as the 2011 antiauthoritarian movements colloquially known as the Arab Spring. Its use in such scenarios coincides with the prohibition of the app in some countries and fits into a context of concern about operational security (OPSEC) issues occurring with the rise of social media or "Web 2.0." In some cases that usage has been used by digital forensic analysts to pinpoint the presence and positions of forces that have been officially disavowed by nation-states.

Created in 2010 and purchased by the social media giant Facebook in 2012, Instagram allows the online sharing of images among users. Trends are identified through hashtags in a structure reminiscent of the social media app Twitter. By the end of 2013, the app incorporated a private messaging feature. Use of a separate privacy feature enables an account holder to selectively vet potential followers on a case-by-case basis. Although this arguably addresses some privacy concerns held by users, the app's history includes algorithm-driven advertising, illicit activities such as drug dealing by third parties using the app, and recurrent user allegations of privacy violations and censorship. Nonetheless, Instagram is a social media giant and continues to grow, reaching both the 700 million and 800 million threshold for account holders in 2017 alone.

Uploading of new images is possible because of the self-destruction of older data, and algorithms shape user feeds according to popularity. As a result, while some images attain popularity and a few briefly go viral, 70 percent of posts on the app reportedly go unseen. The simultaneous and simulated dual anonymity and community of the app coincides with its popularity among millennial account holders. Nearly 60 percent of Instagram users are eighteen to twenty-nine years old, which itself strongly correlates to the typical age of combatants in wars.

The popularity of social media among soldiers has prompted concern among authorities, and in the years preceding Instagram's creation, the U.S. military attempted to dramatically curtail social media use by service personnel. The largest concern was the social media intelligence (SOCMINT) that might be gathered by alert hostile parties online, and the consequent operational security breaches that an adversary's discoveries would entail. By the time Instagram was unveiled for iOS phones in October 2010 and Androids in April 2012, these restrictions had been walked back.

Events in 2014 unexpectedly demonstrated the validity of some OPSEC concerns. Pro-Russian paramilitary forces have been engaged in fighting since February 2014 in Eastern Ukraine against that country's government and military. Observers have frequently pointed to evidence of covert Russian participation in the conflict, and in one case analysis of online imagery helped lead to proof that a missile from Russia's 53rd Anti-Aircraft Rocket Brigade was used to intercept and destroy Malaysian Airlines Flight 17 in mid-July 2014.

Analysts have found Russian military personnel uploading pictures to Instagram, announcing their firing weapons from the Russian side of the border into Ukraine and other cases in which the geolocation, which is default embedded into most digital photographs, places Russian soldiers' online photos inside Ukraine's borders. Revelations attained through this analysis have resulted in various official denials in Russia, including accusations of account hacking and insistence that any Russian soldiers fighting in Ukraine were doing so of their own accord and were "on vacation" rather than then serving in the Russian army.

Instagram has also proven popular with embedded photojournalists and others who are amid combat but are not themselves combatants. Scholars have observed the controversy that has arisen as a result of the new participants, photo editing tools, and sharing opportunities that Instagram has made possible. The implications of these consequences are likely to become more apparent later in the century.

Nicholas Michael Sambaluk

See also: China, Blocking of Social Media by; Facebook, Use of; Operational Security, Impact of Social Media on; Social Media and the Arab Spring; Social Media Intelligence (SOCMINT); Twitter, Use of; Web 2.0

Further Reading

Alper, Meryl. "War on Instagram: Framing Conflict Photojournalism with Mobile Photography Apps." *New Media and Society* 16, no. 8 (September 2013), 1233–1248.

Rose, Andree E., and Christina M. Hesse. *Current Trends in Social Media and the Department of Defense's Social Media Policy.* Washington, D.C.: Defense Personnel and Security Research Center, 2014.

Santiago, Cesar H. *US Army Public Affairs Officers and Social Media Training Requirements.* Thesis. Fort Leavenworth, KS: Command and General Staff College, 2016.

Truth Is the First Casualty of War: A Brief Examination of Russian Informational Conflict during the 2014 Crisis in Ukraine. Defense Research and Development Canada, 2014.

ISIS Recruitment and War Crimes

Activities which the jihadi Islamic State (IS) movement, frequently called the Islamic State of Iraq and Syria (ISIS), intertwined to make a hallmark of the organization. Although war crimes and brutality feature prominently in one major strand of ISIS propaganda efforts, another strand of the organization's recruitment effort sought to normalize life in ISIS-controlled areas and downplay the culture of brutality that ISIS simultaneously and brazenly advertised.

ISIS's precursor organization was founded in 1999 as a terrorist entity loyal to Al Qaeda. ISIS reportedly began to diverge from other jihadi extremist groups such as Al Qaeda during the fighting in the 2003–2011 U.S. occupation and reconstruction efforts in Iraq. Whereas Al Qaeda figures touted a Salafist strand of extremism while apparently believing that publicizing frequent and overtly grotesque forms of murder was counterproductive, leading figures in the nascent ISIS movement such as Abu Musab al-Zarqawi advocated a different apocalyptic mind-set and insisted on both perpetrating and displaying attacks such as suicide bombings of U.S. forces, nongovernmental organizations, aid workers, and Arab embassies, as well as the beheadings of captured hostages. *Washington Post* reporter Joby Warrick suggested that even Al Qaeda's leaders began to judge that al-Zarqawi was going so far that he would damage the jihadi image globally. Al-Zarqawi was killed in a U.S. air strike in June 2006, but his zeal for conspicuously brutal killings as a form of advertising his movement was carried forward by the group that would come to be known as ISIS. As a result, ISIS propaganda, including much of its recruitment efforts, have gone hand in glove with its war crimes.

ISIS gained momentum in a power vacuum, made possible by a chronic civil war that began in the spring of 2011, concurrent with the prodemocracy Arab Spring movements elsewhere in the region. Syrian rebels initially posed a security problem for the autocratic regime of Bashar al-Assad, but neither they nor al-Assad could quickly consolidate control. As the war continued and the rebels fought on in virtual isolation, al-Assad regained footing and received assistance from Iran, Russia, and Hezbollah. Tension between the anti-al-Assad ISIS and other rebel groups flared in 2013 at a time when the non-ISIS rebels were already losing ground in the war. Hinting at its ambitious intentions, the organization officially changed its name to ISIS in the spring of 2013, but in the context of Al Qaeda, the Taliban, and the larger Syrian civil war, U.S. president Barack Obama openly dismissed ISIS in this phase as a secondary concern.

As a result, ISIS sought to build credibility as a terrorist organization through a combination of military exploits as well as acts of brutality displayed both publicly and projected across social media. It appears that leaders hoped this would provoke fear and rage in enemies and thereby help ISIS military advances gain momentum against panicked enemy forces, while also attracting potential recruits to an openly charismatic terrorist organization.

A high-water mark for the organization in terms of its military prowess and its infamy came in the summer of 2014, with ISIS's swift capture of several areas in Syria from both al-Assad's regime and other rebel groups, and ISIS's abrupt invasion of northern Iraqi and seizure of Mosul. ISIS fighters, numbering no more than

1,500 personnel, swept forward toward the city of Mosul, defended by 25,000 well-equipped Iraqi soldiers. However, ISIS coordinated its advance with a social media campaign that included YouTube videos and Twitter messages advertising atrocities meted out to captured Iraqi troops. The result was an Iraqi Army rout from Mosul, with ISIS fighters murdering stragglers and seizing thousands of military-grade vehicles and extensive equipment. Using an app called the Dawn of Glad Tidings, ISIS parades and murders were featured in the 40,000 Twitter posts that affiliated accounts made the day of the city's capture.

During 2013 and 2014, ISIS primarily propagated its social media message through mainstream platforms, including Twitter, Facebook, and YouTube. These posts were easily found by like-minded social media users. Through actions such as hashtag hijacking, ISIS could commandeer the feeds of completely unrelated topics on Twitter and temporarily inundate conversation with their own message. ISIS activity on Twitter tends to coincide with military victories.

ISIS has also leveraged apps such as Telegraph and the anonymized image and text sharing app JustPaste.it. Whereas mainstream networking platforms like Twitter and Facebook provide opportunity to spread propaganda beyond the terrorist group, anonymized apps lend opportunities to share and plan information between terrorists as they prepare to launch future attacks. Despite difficulty, ISIS has demonstrated an interest in creating similar kinds of anonymized sharing apps on the occasions when the use of legitimate social media apps is foreclosed to them. ISIS tailored its selection of apps to specific purposes, and by using a combination of them it can maximize the physical effect of attacks and then maximize the propaganda impact that these incur.

Images of beheadings and other forms of murder are believed to be posted primarily with Westerners as the target audiences. ISIS has demonstrated a penchant for presenting glossy and high-quality (in a technical sense) social media postings. Counterterrorism researcher Haroro Ingram has noted that "Islamic State media architects are more strategic plagiarists than geniuses" (Ingram, 2015). The images are of theological murder, but "the ISIS playbook [otherwise] looks much like any of the dozens of social-media-marketing 'how-to's' circulated by consultants" (Brooking and Singer, 2016). Although ISIS stylistically copies the trappings of 21st-century professional advertising and entertainment messaging, their communications strategy itself also borrows from 20th-century insurgencies.

Shock, outrage, and fear are concluded to be the reactions ISIS hopes to elicit from its enemies. It also appears to hope to attract two kinds of new recruits in Western countries. This includes second and third generation descendants of immigrants, and also occasionally directionless youth lacking prior affiliation with Islam who might be newly converted to jihadism. New recruits are reportedly urged to act for ISIS in one of two principal ways.

One alternative, particularly when ISIS controlled an empire across parts of Syria and Iraq, had been for recruits to travel from their countries of origin to ISIS-controlled lands and become soldiers. The numbers of ISIS recruits who attempted to travel to Syria, and also the number of people drawn to ISIS ideology, are difficult to judge. The largest number of foreign ISIS fighters are believed to be the 6,000 from Tunisia, with 2,000–2,500 each from Saudi Arabia, Russia, France, and

Jordan. Each of these countries has large Muslim populations that are either indigenous or which have lived in the country for decades, and the proportion of ISIS fighters to overall Muslim population is very small. In terms of ISIS fighters relative to a country's Muslim population, however, the top three countries are reported to be Finland, Belgium, and Ireland. This has spurred conjecture that cultural isolation is an important factor in shaping ISIS recruits. Reports indicate that a number of recruits, including a large proportion of those who had earlier been unaffiliated with Islam and especially female recruits, were surprised to be quickly assigned menial tasks and noncombat roles and that they risked extremely brutal punishment for any infraction of a ponderous list of strictures.

The second alternative ISIS urges is for its new recruits who are outside the Middle East to remain in their home country and plot attacks there. The organization appears to see several advantages in this alternative. First, as ISIS itself loses the ground it had held in Syria and Iraq, transportation to ISIS territory becomes more difficult and dissolution with the movement arguably becomes more likely. Second, ISIS gains a strategic windfall in projecting power through kindling terrorist cells abroad. Although ISIS lacks conventional power projection capabilities such as aircraft carriers or midair-refueled strike aircraft possessed by some of its nation-state opponents, numerous small-scale attacks have occurred in the France, the United States, the United Kingdom, and Canada.

At times, ISIS brutality involves violence that is deeply and almost universally abhorrent to Islamic tenets. A prominent example is the January 2015 immolation murder of Muath Al-Kasasbeh, a Royal Jordanian Air Force pilot whose jet crashed over Syria in December 2014. As a propaganda and marketing tactic, ISIS conducted online competition on Twitter, inviting members to suggest a particularly graphic way to murder the captured pilot. The filmed immolation offended sensitivities across the Middle East as well as Western countries, and in Jordan led to a surge of popular support for increased air strikes against ISIS strongholds. Such episodes suggest that the bloodthirsty character of the organization creates an incontrovertible gap between ISIS and the tenets of mainstream Islam. This separation represents a crucial weakness for the organization, since a hallmark of its combat and recruitment operations have historically been the overt display of extreme violence as a means of attracting attention and recruits as well as of inspiring fear.

Perhaps in an effort to combat this problem, the ISIS message within their fold emphasizes "pragmatic factors such as security, stability, and livelihood" in order to normalize the image of the organization and life within the territories it controls. Concurrent with the seizure of Mosul and the murder of prisoners of war was ISIS's proclamation of a caliphate, constituting an attempt at cementing their appropriation of Islam. ISIS's press releases, civic forum boards, and carefully manipulated documentaries aim to portray life within ISIS territory as normal and wholesome. One otherwise inexplicable example of the normalization strategy is "Cats of Jihad," where ISIS fighters pose their cats with weapons and jihadi paraphernalia in photos uploaded to Instagram. These efforts to make life in ISIS territory seem normal, safe, and almost whimsical seem to be intended specifically to draw and retain recruits. ISIS's monthly online English-language journal *Dabiq* is another example of using technical professionalism to imply normalcy of radical ideas. The journal features Islamist theological statements, narratives about jihad, and information about making explosives.

Western policy makers, who had initially taken little action until late 2014, when North Atlantic Treaty Organization countries began small-scale air strikes targeting ISIS fighters in Syria. The international hacktivist network Anonymous launched efforts to combat ISIS social media propaganda beginning in late 2015. Iraqi forces with significant Iranian support, and Syrian forces with predominately Russian support, recaptured the vast majority of ISIS-controlled territory during bitter fighting in 2017. Although this has dramatically reduced the possibility for ISIS recruits to successfully join the main movement in Syria and has eroded ISIS's ability to mimic a nation-state, ideological recruitment and attacks by lone wolf sympathizers persist.

Nicholas Michael Sambaluk

See also: Al-Furqan; #AllEyesOnISIS; Anonymous; Boko Haram; #BringBackOurGirls; *Dabiq*; Dawn of Glad Tidings Campaign; Hashtag Hijacking; ISIS Trolling Day; Telegram App; Think Again, Turn Away Campaign; Twitter, Use of

Further Reading

Atwan, Abdel Bari. *Islamic State: The Digital Caliphate.* Oakland: University of California Press, 2015.

Brooking, Emerson T., and P. W. Singer. "War Goes Viral." *The Atlantic,* November 2016, https://www.theatlantic.com/magazine/archive/2016/11/war-goes-viral/501125.

Erelle, Anna. *In the Skin of a Jihadist: A Young Journalist Enters the ISIS Recruitment Network.* New York: Harper, 2015.

Ingram, Haroro. "What Analysis of the Islamic State's Messaging Keeps Missing." *The Washington Post,* October 14, 2015, https://www.washingtonpost.com/news/monkey-cage/wp/2015/10/14/what-analysis-of-the-islamic-states-messaging-keeps-missing/?utm_term=.9b2b23f61520.

Irby, Jack B. *The Weaponization of Social Media.* Master's thesis. Ft. Leavenworth, KS: Command and General Staff College, 2016.

Patrikarakos, David. *War in 140 Characters: How Social Media Is Reshaping Conflict in the Twenty-First Century.* New York: Basic Books, 2017.

Rogers, Paul. *Irregular War: ISIS and the New Threat from the Margins.* London: I. B. Tauris & Co., 2016.

Schneider, Nathan K. *ISIS and Social Media: The Combatant Commander's Guide to Countering ISIS's Social Media Campaign.* Newport, RI: Naval War College, 2015.

Ullah, Haroon K. *Digital World War: Islamists, Extremists, and the Fight for Cyber Supremacy.* New Haven, CT: Yale University Press, 2017.

Warrick, Joby. *Black Flags: The Rise of ISIS.* New York: Anchor, 2016.

ISIS Trolling Day

A coordinated social media attack carried out by the hacking group Anonymous against the Islamic State of Iraq and Syria's (ISIS) Twitter, Facebook, and Instagram accounts on December 11, 2015. Founded in 2003, Anonymous, famous for donning Guy Fawkes masks on its various social media outlets, is a decentralized online hacking group that has conducted numerous cyberattacks against government entities, corporations including PayPal and MasterCard, and other social groups such as the Church of Scientology and the Ku Klux Klan (KKK). The group proclaimed ISIS Trolling Day in response to the ISIS-led terrorist attacks in Paris, France, on November 13, 2015, during which ISIS operatives attacked various cafés and restaurants, the Eagles of Death Metal concert at the Bataclan Theatre, and a soccer match between Germany and France at the Stade de France. The attacks, in which the seven assailants died, had resulted in 130 civilian fatalities.

Following the attacks, Anonymous representatives, via YouTube, issued a warning to ISIS, stating the group "from all over the world will hunt you down." The group called for "ISIS Trolling Day" to begin December 11, 2015, in which they asked users to "[show] your support and help against ISIS by joining us and trolling them // do not think you have to be part of Anonymous, anyone can do this and does not require special skills. . . . We ask you to take part of this on Facebook // Twitter // Instagram // Youtube //In the 'Real World.'" Social media users were directed to post hashtags including #Daesh and #Daeshbags with mocking photos and memes meant to troll—to post or write a deliberately provocative message with the intent to provoke an angry and emotional response from the recipient—and to generally degrade ISIS. Daesh is derived from the Arabic acronym for the full name of ISIS (*ad-Dawlah al-Islamiyah fī'l-'Iraq wa-Sham*) and using the word in the Arabic speaking world particularly carries a derogatory connotation as "Daesh" sounds similar to another Arabic verb that roughly translates to "trample down" or "crush something"; ISIS forbids the use of the term by any followers and other Muslims.

ISIS Trolling Day was not the first time Anonymous targeted ISIS. The hacking group also commenced a coordinated social media campaign against ISIS operatives following the attacks on *Charlie Hebdo*, a French satirical magazine, on January 7, 2015, which resulted in the deaths of eight journalists. Although the attack was carried out by militants affiliated with Al Qaeda in Yemen (AQAP), who sought retribution for the magazine's perceived derogatory depictions of The Prophet and Islam, Anonymous issued a statement via its YouTube account that it had "a message for al-Qaeda, the Islamic State, and other terrorists. . . . We, Anonymous around the world, have decided to declare war on you, the terrorists and promises to avenge the killings by 'shut[ting] down your accounts on all social networks.'"

The legacy of the ISIS Trolling Day campaign remains ambiguous, as does many other Anonymous campaigns in general. The hacking group claimed to have disabled thousands of pro-ISIS Twitter accounts while doxing—maliciously publishing private or identifiable information on a certain individual or entity to generate harassment and even direct potential harm from other Internet users—several operatives and recruiters. However, and similar to other Anonymous campaigns, various media outlets, including the British Broadcasting Company (BBC) and *New York Times,* could neither confirm nor verify these claims. Twitter dismissed Anonymous claims outright as well. Several security services and intelligence networks have also criticized actions of Anonymous, claiming the group only hinders investigations and apprehension of terrorists by closing down associated and known social media accounts.

Kate Tietzen

See also: Anonymous; Facebook, Use of; ISIS Recruitment and War Crimes; Twitter, Use of; Web 2.0

Further Reading

McCants, William. *The ISIS Apocalypse: The History, Strategy, and Doomsday Vision of the Islamic State.* New York: St. Martin's Press, 2015.

Olson, Parmy. *We Are Anonymous, Inside the Hacker World of LulzSec, Anonymous, and the Global Cyber Insurgency.* Boston: Little, Brown, 2012.

Rapoport, David C. "Why Has the Islamic State Changed Its Strategy and Mounted the Paris-Brussels Attacks?" *Perspectives on Terrorism* 10, no. 2 (April 2016), 24–32.

Stern, Jessica, and J. M. Berger. *ISIS: The State of Terror.* New York: HarperCollins, 2015.

John Q. Public

An online publication at http://www.jqpublicblog.com, offering commentary about challenges facing Airmen and the U.S. military. The website started as the personal blog of Lt. Col. (Ret.) Tony Carr, which reflects his broad experience during twenty-two years in the Air Force as a C-17 pilot, staff adviser, squadron commander, graduate of the School of Advanced Air and Space Studies, and Distinguish Flying Cross winner with service in Operations Enduring Freedom and Iraqi Freedom. The website commonly features incidents reported to Carr by his extensive network of connections.

The website has gained some influence, as other media have picked up and used stories from the website dealing with emotionally engaging controversies. Prominent topics have included the Air Force efforts to retire the A-10 close air support plane; a general who insulted officers who talked to Congress, accusing them of "treason"; officers accused of sexual assaults; and officers punished for destroying the command climate of their unit by arbitrary, angry outbursts. He has also drawn attention to the difficulties the Air Force faces in retaining pilots, and the hardships of pilots assigned to fly remotely piloted vehicles. With the focus on problems derived from anecdotal reports, critics claim that it conveys a negative tone about life in the U.S. Air Force and lacks substance. However, in addition to providing a venue for whistleblowers, it also provides a forum that advocates humane leadership principles.

On February 17, 2015, Bright Mountain, LLC, a company that manages and monetizes websites through advertising, acquired John Q. Public as part of its focus on selling products for the military and public safety communities. Analysis of the website shows that 30 percent of visitors come from a search engine, with 5,000 people per month searching the term "John Q. Public." It has over 46,000 Facebook fans, who provide 12 percent of its readers. It claims more than 127,000 unique visitors with 200,000 page views. After selling the website, Carr enrolled in Harvard to study international law, where he edits the *National Security Journal.* He also writes for the Center for Defense Information, within the government watchdog group Project on Government Oversight.

Jonathan K. Zartman

See also: Army Knowledge Online (AKO); Doctrine Man; *Small Wars Journal*

Further Reading

Adesnik, David, and Mackenzie Eaglen. "State of the US Military: A Defense Primer." Washington, D.C.: American Enterprise Institute, October 1, 2015.

Fino, Steven. "The Coming Close Air Support Fly-Off." *Air & Space Power Journal* (Summer 2017), 17–38.

Lamothe, Dan. "Retired General Is Subject of Sexual-Assault Investigation." *The Washington Post*, August 28, 2016, A.18.

Losey, Stephen. "Leadership of 'Fear': Fired Wing Commander Berated, Ridiculed Terrified Staff, Investigation Finds." *Air Force Times*, October 4, 2018.

Pawlyk, Oriana "'Good Kill' Targets Air Force Drone Pilots' Disconnect." *Air Force Times*, June 8, 2015, A, 14.

Schogol, Jeff. "Texting Can Kill Your Career." *Air Force Times*, October 26, 2015, A, 14.

Mappr App

Established in 2010, a mobile app allowing the geotagging of images to be shared among a controlled group. The app was launched by a Swedish software company named Cardomapic, in response to a dynamic common among many contemporary and predecessor photo sharing applications and tools within social media.

Digital photography rapidly displaced film photography, first among professionals and then among casual users in the early 21st century, and the trend was facilitated and accelerated by the inclusion of increasingly capable digital cameras onto mobile phones. Digital photography allows the embedding of significant data into the picture file, including information about the time and geographic location in which an image was taken. The inclusion of the latter information is known as the geotag.

On many popular apps and websites, such as Facebook, Twitter, and Foursquare, the geotag information is released with the uploading of the image. As a result, any third party may relatively easily observe the temporal and spatial information associated with an image. In some circumstances, such information can be embarrassing, and in other scenarios the leakage of such information may even entail a breach of operational security. For example, images of soldiers who are ostensibly in one country might be discovered to be deployed to another. Alternatively, the precise locations of a military base might be easily determined by an adversary through the examination of geotags of photographs posted by personnel on social media.

Mappr app's developers, Bart Denny, Jesper Sarnesjo, and Frederik Bromee, designed Mappr to secure geotag information for people who do not want to broadcast their current location to potential stalkers or strangers. "Mappr is useful," it was reported, to users who wish not "to share [their] location with all [their] Facebook friends, either for privacy reasons, or because [they] don't want to clutter the feeds of some friends" (Constine, 2010). Limiting the tagging to a specified list of receivers in each of these larger social media platforms changed the social dynamics and security threat via better privacy tools and made geotagging into a way to promote an activity or place to a select group of interested friends. Additionally, a savvy user can manually tag a photo or activity, allowing a person to "fake" a geotag. Currently, the developers do not collect user geotagging information for sale, and rely on whitelisting with media partners to monetize their program.

Margaret Sankey

See also: Facebook, Use of; Operational Security, Impact of Social Media on; Twitter, Use of

Further Reading

Constine, Josh. "Mappr Lets Users Share Their Location with Facebook Groups." ADWEEK, October 21, 2010, https://www.adweek.com/digital/mappr-share-location-groups.

Marcellino, William, et al. *Monitoring Social Media: Lessons for Future Department of Defense Social Media Analysis in Support of Information Operations*. Santa Monica, CA: RAND, 2017.

Piggott, Jennifer M. *Utilizing Social Media and Protecting Military Members and Their Families*. Maxwell AFB, AL: Air Command and Staff College, 2016.

MySpace, U.S. Military Personnel Usage of

A communication vector via a popular social media platform of the early 2000s, which raised concern within the U.S. military about operational security factors. For military personnel deployed to Afghanistan starting in 2002 and Iraq in 2003, MySpace and other social media constituted a digital counterpart of the letters home from the front lines, records, Vmail, and cassettes used to connect personnel with family members during earlier conflicts. Citing risk to military networks from servicemembers' heavy use of Internet social media, however, the U.S. Department of Defense on May 14, 2007, banned the use of MySpace and a dozen other high-traffic sites, including YouTube, Metacafe, iFilm, StupidVideos, Hi5, Pandora, and Photobucket from Department of Defense computers.

Experts believe the decision to have been part of an effort to maintain operational security, or OPSEC. One of the reasons cited for the ban was the risk to military networks from computer viruses. Although U.S. Army regulations in 2007 restricted soldiers from blogging or sending e-mail with mission-related information and from using social media from official computers, the restriction did not extend to their personal computers. The ban did represent an effort to control the flow of information from the combat zone via the laptops, digital cameras, recorders, and iPods of service personnel. Requirements that soldiers register their blogs and submit content for review also restricted soldiers' ability to transmit unscreened material via the Internet.

John P. Ringquist

See also: Operational Security, Impact of Social Media on; Social Media Intelligence (SOCMINT)

Further Reading

Bittner, Lisa. "How Is Social Media Used by Military Families to Communicate During Deployment?" *Master of Social Work Clinical Research Papers* (May 2014), Paper 239.

Knopf, Christina M., and Eric J. Ziegelmayer. "Fourth Generation Warfare and the US Military's Social Media Strategy." *ASPJ Africa & Francophonie* (October 2012), 3–22.

Operational Security, Impact of Social Media on

The emergence and rise of Web 2.0 is epitomized by the swift prominence enjoyed by social media giants, including perhaps most notably Facebook. The value of social media was readily identified by deployed military personnel, including but by no means exclusively encompassing U.S. personnel in Iraq and Afghanistan combating terrorism. Personnel from several countries conducting missions such as peacekeeping frequently come from societies with the resources, technical familiarity, and traditions of free speech that encourage exploitation of social media tools. Researchers noted the dramatic impact of social media popularity in the mid-2000s and the unprecedented ability for military personnel in a combat zone to communicate with immediacy with friends and family at home.

The arguably revolutionary effect on communications brings accompanying security implications. Of particular concern to the military is the operational security impact that could be exerted, even unintentionally, by postings made by military personnel and observed by adversaries. Glancing references to impending military exercises or other activities could give advanced warning to adversaries or could alert enemies to opportunities for launching attacks on military concentrations or on vulnerable locations. Displaying photos of personnel can divulge clues about who and how many military personnel and what units might be located in an area or conducting certain missions. Seemingly trivial imagery data can provide information about not only the equipment being used but also the types and identities of units and their physical dispositions, thanks in part to the geolocation features that exist as defaults on most digital cameras.

Since apps including giants such as Skype require in user agreements that the app acquire access to text messages and personal information, this information is inherently vulnerable not only to misuse by the app's creators but also exploitation by third parties in the event that this information is either sold to them or is released through leaks or hacking. The exploitation of automatic geolocation data incorporated into various social media apps represents an obvious challenge. The reported average length of a compromise existing without discovery was 229 days in 2014, a marginal improvement over the average of 243 days in 2012.

The harvesting of data available via social media for intelligence purposes is known as social media intelligence (SOCMINT) and contributes a wealth of potentially useful data points. Operational security breaches have already been reported as having impacted aggressive actors. The social media activities of some People's Liberation Army (PLA) members helped forensic analysts trace the headquarters of the PLA units most deeply involved in strategic and industrial espionage and helped lead to indictments in U.S. courts of a small number of these personnel. Bahrain reportedly analyzed citizens' social media activities to identify and track dissidents during that country's Arab Spring political demonstrations in 2011.

Operational security impact exists in cases of kinetic conflict as well. Crowd-sourced analysis of photographs taken of Russian military convoys bringing heavy weapons to pro-Russian paramilitary forces fighting in Eastern Ukraine played an important part in demonstrating that the Malaysian MH17 airliner that was shot down over Eastern Ukraine in July 2014 was struck by a Russian surface-to-air missile and not by Ukrainian forces as was asserted in a Russian disinformation campaign.

The U.S. military's alertness to risks connected to social media are seen in publications that identify it as the "fastest growing vulnerability to the Air Force mission," pointing specifically at personal information that becomes vulnerable, trade-offs enforced with the use of free-to-use apps, and the potential for data mining against military personnel.

Studies surveying small numbers of European military personnel indicated that a slight majority interpreted social media as bringing net opportunity to military establishments, while the remaining respondents answered that social media posed both risks and opportunities and therefore constituted a mixed blessing. The most commonly identified problem was the risk of revealing sensitive information that might endanger the lives of personnel, and the possibility of mission compromise and of distorted messaging was less commonly identified. The principal benefits were assessed to be improvements in the military's ability to message its and its mission to the population in an area of deployment, and almost equally to reflect the mission's value to the civilian population at home. Scholars have observed that militaries lacking a social media presence appear to be more likely to hold optimistic views of the marketing advantages of social media than are held by militaries that have already undertaken a social media presence.

Despite operational security concerns, scholars on the topic have noted messaging opportunities presented by social media. Such applications offer opportunities "for disseminating the national strategic narrative in situations where armed forces have devoted substantial resources to developing their own media outlets" (Hellman et al., 2016, 52). As with the strategic messaging implications, social media impact on operational security depends in large part on the character and flavor of the materials that are shared, and this in turn is related to cultural factors within military organizations and to discipline in the digital realm.

Nicholas Michael Sambaluk

See also: China, Blocking of Social Media by; Facebook, Use of; Instagram, Use of; Social Media and the Arab Spring; Social Media Intelligence (SOCMINT); U.S. Military Use of Social Media; War on Terrorism and Social Media

Further Reading

Garside, Debbie, et al. "Secure Military Social Networking and Rapid Sensemaking in Domain Specific Concept Systems: Research Issues and Future Solutions." *Future Internet* 4 (March 2012), 253–264.

Hellman, Maria, Eva-Karin Olsson, and Charlotte Wagnsson. "EU Armed Forces' Use of Social Media in Areas of Deployment." *Cogitatio* 4, no. 1 (February 2016), 52–61.

Patrikarakos, David. *War in 140 Characters.* New York: Basic Books, 2017.

Silvestri, Lisa Ellen. *Friended at the Front: Social Media in the American War Zone*. Lawrence: University Press of Kansas, 2015.

Solomon, Scott E. *Social Media: The Fastest Growing Vulnerability to the Air Force Mission*. Maxwell AFB, AL: Air University Press, 2017.

PlatoonLeader.com

An online community of practice forum for U.S. Army platoon leaders. First established by individual officers and without official sanction from the Army, PlatoonLeader has been officially supported since 2002. As with its captain-level counterpart forum CompanyCommand.com, PlatoonLeader was designed to utilize digital technology in order to expand on the kinds of peer-to-peer conversations related to the profession of arms and particularly to the experiences, challenges, and solutions involved with being a platoon officer.

Both PlatoonLeader and CompanyCommand are "opportunistic networks," correlating to wide opportunities to connect with peers but intrinsically weak ties between peers. Users have observed that junior lieutenants frequently seek advice and perspective through PlatoonLeader, especially when serving in detachments such as ROTC units that offer only limited interaction with peers. Users can easily navigate to see "topic leaders" and "recent activity" to become familiar with topics of popular interest among their peers. Gifted to the Army in 2002 and subsequently supported by Army servers, the success of PlatoonLeader and its counterpart CompanyCommand have helped foster a deeper official appreciation for the value of communities of practice emerging within the Army.

Nicholas Michael Sambaluk

See also: Army Knowledge Online (AKO); CompanyCommand.com; FrontlineSMS; HARMONIEWeb

Further Reading

Brown, J. S., S. Denning, K. Groh, and L. Prusak. *Storytelling in Organizations: Why Storytelling Is Transforming 21st Century Organizations and Management*. Boston: Elsevier Butterworth-Heinemann, 2005.

Kimball, Raymond A. *The Army Officer's Guide to Mentoring*. West Point, NY: Center for the Advancement of Leader Development and Organizational Learning, 2015.

Kimball, Raymond A. *Walking in the Woods: A Phenomenological Study of Online Communities of Practice and Army Mentoring*. Dissertation, Pepperdine University, April 2015.

Recorded Future

A U.S. and Swedish Internet technology company that specialized in real-time threat intelligence founded in 2009. Threat intelligence refers to an extensive analysis that involves the organization, identification, and collection of information. Threat intelligence traditionally relates recent physical attacks to future threats, whereas cyber threat intelligence (CTI) also relates to how actors recognize and (re)act to threats. As a result, a cyber warfare becomes a conflict between states where threats are directed against the military and other industries for political and economic advantages.

With the recent output of cyber threats, cyberespionages and hacking have become focal concerns for the U.S. military. Outsourcing CTI to entities such as Recorded Future has enabled Internet companies to access and analyze information to ensure cybersecurity in the information age. Recorded Future touts the philosophy that all published information on the web carries a "predictive power." Recorded Future's system uses the term "temporal analytics" to describe how it analyzes open source information in real time. The idea behind real-time analytics is based on the company's aptitude on analyzing the data once it enters the system. However, some experts are skeptical about the effectiveness of the protocols and ideologies of Internet companies, as for example the global nature of the Internet complicates the task of geographically pinpointing a threat.

Companies such as Recorded Future often use blogs as platforms for generating awareness about possible threats of interest to military and law enforcement communities. Such publications simultaneously highlight the analytical prowess of the organization, particularly when subsequent events validate predictions and identified trends. For instance, informational posts on the company's website highlighted activities by hackers in mid-2018 reportedly exfiltrating information about UAV technology to be sold on the dark web. This brief announcement allowed Recorded Future to point to the need for greater cybersecurity precautions for safeguarding sensitive materials, while also potentially drawing attention to the company's own ability to discover and interpret relevant information about security issues.

Information industries like Recorded Future are changing the forms of cybersecurity and cyber laws. But at the same time, more policies and critical discourses are needed to determine whether these industries benefit or detract from the military's cybersecurity. Digital technologies generating a *recorded future* should be professionally and scientifically evaluated further to determine its effectiveness.

Kathy Nguyen

See also: Computer Network Defense (CND); Cyberweapons

Further Reading

Jasper, Scott. E. "U.S. Cyber Threat Intelligence Sharing Frameworks." *International Journal of Intelligence and CounterIntelligence* 30 (2017), 53–65.

Tabansky, Lior. "Basic Concepts in Cyber Warfare." *Military and Strategic Affairs* 3 (2011), 75–92.

Shekau, Abubakar

Leader of the Salafist Islamic terrorist group Boko Haram who is known for leveraging social media to complement the group's operations in northern Nigeria. Born in March 24, 1973, in Tarmuwa Local Government Area of Yobe State in northern Nigeria, Shekau enrolled in a local Islamic school in Borno State, where he studied the Koran and the Hadith (Arabic for "news" or "story" related to The Prophet Muhammad's words, actions, and traditions), forsaking any sciences or math in his education. He is fluent in Arabic, English, Kanuri, Fulani, and Hausa, the dominate language in northern Nigeria. Since Fulani and Hausa are also widely spoken in Nigeria, Cameroon, Chad, Mali, and Niger, this has arguably helped Boko Haram's recruitment and messaging throughout the region.

Shekau eventually studied under Mohammed Yusuf, the puritanical leader and founder of the Yusuffiya sect, also known as the "People Committed to the Propagation of The Prophet's Teachings and Jihad" (*Jama'atu Ahlis Sunna Lidda'awati Wal-Jihad*), or Boko Haram (a mixture of Hausa and Arabic that roughly equates to "Western Education is Forbidden"). Angered over perceived social ills, including police corruption and state brutality, Yusuf blamed Western influences and advocated for a return to Salafist Islam, an ultra-orthodox interpretation of Islamic law that rejects anything deemed incompatible with Koran including Western education, and the creation of an Islamic state in Nigeria. Founded in 2002, Boko Haram gained more members and increasingly radicalized and violent, leading to an important demarcation point for the group in 2009. In late July 2009, Boko Haram supporters and Nigerian security services openly clashed, leaving an estimated 700 dead, including militants and civilians according to various media outlets. Nigerian security services subsequently captured Yusuf and executed him outside police headquarters in Maiduguri in view of the public; the Nigerian government would later claim he was killed in an escape attempt. Following Yusuf's death, his deputy, Shekau, who himself escaped captured, assumed leadership of Boko Haram. Researchers and analysts have suggested this July 2009 attack further radicalized Boko Haram and personally spurred Shekau to seek vengeance for his former mentor.

Shekau continued Boko Haram's advocacy of Salafism and intensified the group's embrace of violence to achieve its means. His disdain for Western education is best characterized by the numerous assaults on schools in northern Nigeria, including attacks on Bayero University in April 2012, Government Secondary School Mamudo in Yobe State in February 2014, and the infamous kidnapping of 276 girls from Government Girls Secondary School of Chibok in Borno State on April 14, 2014, which gained international condemnation and global media attention and prompted the creation of the #BringBackOurGirls awareness countercampaign on Twitter. Shekau has also embraced social media outlets, particularly YouTube, to spread his message against any media, cultural, and religious entities deemed too Western or even too moderate.

In March 2010, Shekau publicly allied Boko Haram with Al Qaeda networks based in Mali, but then also pledged support for leader of the Islamic State of Iraq and Syria (ISIS) leader Abu Bakr Al-Baghdadi in June 2014. By March 2015, Boko Haram's asserted affiliation with both ISIS and Al Qaeda pointed to the ambiguous relationship between the two latter groups; analysts disagree about whether ISIS branched away from Al Qaeda or if the two were never synonymous, despite ongoing conflation in Western media and Al Qaeda's early financial support for the progenitor group that later renamed itself ISIS. In this respect, Shekau's statements suggest that this relationship may be as paradoxical even among jihadist leaders as it is to outsiders. Boko Haram's regional activities, including ransom kidnappings of foreigners, bombings of moderate mosques and Christian churches, and cross border raids and action around Lake Chad triggered the African Union to deploy forces from Cameroon, Central Africa Republic, Chad, Benin, and Niger to join Nigeria in the fight against the group. The Nigerian Department of State Service first offered a bounty of around $156,000 USD for Shekau in September 2011, with the U.S. State Department following suit in June 2012 by adding him to the designated terrorist list; the Americans offered their own $7 million USD bounty for his capture in June 2013.

However, Shekau's whereabouts and even current status remains shrouded in mystery as of May 2018 with various groups, including the Nigerian and Cameroon militaries, alleging his death. He has alleged to have been killed in July 2009, August 2013, November 2014, and even ousted as leader in August 2015. However, after each such instance, Shekau reportedly appears in a YouTube video disputing those claims; his alleged used of a body double complicates confirming either his survival or death.

Kate Tietzen

See also: Boko Haram; #BringBackOurGirls; War on Terrorism and Social Media

Further Reading

Mantzikos, Ioannis. "Boko Haram Attacks in Nigeria and Neighbouring Countries: A Chronology of Attacks." *Perspectives on Terrorism* 8, no. 6 (December 2014), 63–81.

McCants, William. *The ISIS Apocalypse: The History, Strategy, and Doomsday Vision of the Islamic State.* New York: St. Martin's Press, 2015.

Okereke, Emeka. "From Obscurity to Global Visibility: Periscoping Abubakar Shekau." *Counter Terrorist Trends and Analyses* 6, no. 10 (November 2014), 17–22.

Ringquist, John P. "War by Tweet, Hashtag, and Media Messaging," In *Paths of Innovation*, ed. Nicholas Michael Sambaluk. Lanham, MD: Lexington, 2018, 245–255.

Smith, Mike. *Boko Haram: Inside Nigeria's Unholy War.* New York: I. B. Tauris & Co., 2015.

Thurston, Alexander. *Boko Haram: The History of an African Jihadist Movement.* Princeton, NJ: Princeton University Press, 2017.

Small Wars Journal

A free, online publication, founded by Dave Dilegge and Bill Nagle that provides counterinsurgency (COIN)-related, law-enforcement-centered information, published regularly since 2005, financed by the Small Wars Foundation, registered in Bethesda, Maryland. Hochmiller and Muller document the high levels of political influence and connections of contributors to this journal, which facilitates COIN-related knowledge production within an extended, international community of practice (Hochmiller and Muller, 2016). This journal provides ideas that attract a broad variety of professionals: law enforcement officials, academics, politicians, commentators, and policy analysts. David Kilcullen claims that this journal has provided "scholars, warriors and agitators [. . .] a foundation for battlefield success" (Kilcullen, 2010, v). To the extent that *Small Wars Journal* (SWJ) promotes the role of civil society, legality, humanitarian values, and good governance as antidotes to the grievances that motivate some mobilization in an insurgency, it offers a liberal perspective—in the international academic sense of the term.

However, it also promotes a rational-instrumental approach in studying the nature of problems by gathering and assessing data on insurgencies broadly, collating and evaluating the experiences of practitioners, by distinguishing specific forms of collective antistate violence, and ultimately proposing policies in response. This elevates the status and contributions of a transnational community of security experts with an academic approach to violence. This journal also does not neglect the role of ideas and ideology in motivating collective action and violence, as well demonstrated in the book *Jihadi Terrorism, Insurgency and the Islamic State*. The nature of the website prioritizes relevance to current events over well-developed academic theses, yielding a diverse quality of entries that ranges from two-page editorials and book reviews with an occasional superficial note to thorough and substantive articles that integrate the results of protracted research.

Small Wars Journal provides a forum for professionals to debate crucial and difficult issues of strategy at both the operational and grand strategic level. By enabling academics as well as policy makers to propose and critique the contending approaches to complex problems affecting a diverse array of states, it facilitates a shared vocabulary of the problem, and the development of national level military doctrine. Analysts argue that this journal, together with the Warlord Network, has stimulated sufficiently vigorous public debate as to prompt developments in strategic thinking.

The journal developed in the midst of a controversy within the U.S. Department of Defense over the permissibility for members of the military to engage in social media activities. Lawson attributes the growth of this website, in part, to professionals working with General David Petraeus in developing American COIN doctrine, who traveled to Iraq with him to assist in implementing the surge. The people on this team took the position that social media technologies, including blogs, social networks, with photo and video sharing sites, can serve as a valuable weapon in the broader war effort, although other journals such as *Joint Forces Quarterly* and *Military Review* provide similar topic coverage.

In addition to the online journal itself, the editors have published, as of 2018, ten topically organized anthologies of the articles from the journal in books printed by Xlibris in Bloomington, Indiana, most recently *The Hammer of the Caliphate*. These books facilitate more careful analysis of issues by students in a durable, citable form, free from the potential unreliability of Internet access and addresses.

Jonathan K. Zartman

See also: Army Knowledge Online (AKO); John Q. Public

Further Reading

Dilegge, Dave, and Robert J. Bunker, eds. *Jihadi Terrorism, Insurgency and the Islamic State*. Bloomington, IN: Xlibris, 2017.

Dilegge, Dave, and Robert J. Bunker, eds. *Hammer of the Caliphate: The Territorial Demise of the Islamic State A Small Wars Journal Anthology*. Bloomington, IN: Xlibris, 2018.

Hochmuller, Markus, and Markus-Michael Muller. "Locating Guatemala in Global Counter Insurgency." *Globalizations* 13, no. 1 (2016), 94–109.

Kilcullen, David. *Counterinsurgency.* Oxford: Oxford University Press, 2010.

Lawson, Sean. "The US Military's Social Media Civil War: Technology as Antagonism in Discourses of Information-Age Conflict." *Cambridge Review of International Affairs* 27, no. 2 (2014), 226–245.

Newton, Paul, Paul Colley, and Andrew Sharpe. "Reclaiming the Art of British Strategic Thinking." *The RUSI Journal* 155, no. 1 (2010), 44–49.

Wynne, Johanna Thompson. "Wired Fast and Thinking Slow: Cyber Technology and the US Army." Fort Leavenworth, KS: School of Advanced Military Studies, 2016.

Snowden, Edward

A computer expert and hacktivist who stole and publicly released a huge number of top secret documents revealing the vast range of U.S. government surveillance activities, in the most catastrophic security breach to date. He was born June 21, 1983, in Elizabeth City, North Carolina. Snowden dropped out of high school at age fifteen and became a passionate video gamer, working as a webmaster for the website Ryuhana Press that specializes in anime-based games.

In May 2004 he tried to join the U.S. Army Special Forces through a special program for people with incomplete education, but he was given an administrative discharge four months later. He was subsequently hired as a security guard by the University of Maryland's Center for the Advanced Study of Language and moved into information technology, gaining his first security clearance. In 2006, the Central Intelligence Agency (CIA) offered him a job as a communications officer, in spite of lacking any high school or college diploma. After six months of training, he served at the U.S. Embassy in Geneva from March 2007 to February 2009. After conflict with a supervisor over an attempt to break into classified files without authorization, he was allowed to resign and keep his security clearance, with only a derogatory comment on his file.

Retaining his security clearance and with his personnel file protected by privacy laws, Dell hired him and sent him to Japan in June 2009 to teach cybersecurity to Army and Air Force personnel. In October 2009, he began working as a system administrator on a program in which the National Security Agency (NSA) sent classified files to a backup storage facility in Japan. In September 2009, while on the Dell payroll, he made a ten-day trip to India to take an intensive course in sophisticated hacking skills. In February 2011, he applied for a renewal of his security clearance. The government had outsourced the task of conducting background checks to a private company, which did not have access to his CIA file or to his angry Internet postings expressing contempt for government secrecy.

In the summer of 2011, Dell offered him a job in Hawaii, at the NSA regional cryptological center focused on China, where he could work solo as a system administrator, working on the NSA backup system, which did not have any upgraded security protocols or auditing methods. In 2012, he read the NSA inspector general's report on the 2009 surveillance program. Throughout the fall and winter of 2012, he stole tens of thousands of top secret documents, and he would later claim that revelations of illegality from the 2009 report prompted his actions.

On December 1, 2012, he began trying to contact Glenn Greenwald, a high-profile activist journalist who wrote ferocious polemics against U.S. government surveillance. After failing to get Greenwald to set up encryption on his computer and e-mail, he pulled into his scheme Laura Poitras, an activist filmmaker and adversary of the NSA, who was already making a film opposing American government surveillance policies. Snowden used Poitras to recruit Greenwald as a gateway to *The Guardian,* and Barton Gellman who could publicize his claims in the *Washington Post.* So far Snowden had access only to relatively low-level documents, which could serve whistleblowing purposes because they concerned NSA operations in the United States. These did not reveal NSA activities regarding foreign adversaries.

On March 15, 2013, Snowden quit his job at Dell to take another job at Booz Allen Hamilton, which specialized in much more highly classified data, revealing foreign intelligence sources and methods. This job involved a cut in pay and a shift down in status from system administrator to infrastructure analyst, revealing that he targeted the company and the job specifically to steal data that would seriously harm U.S. security and alliance relations.

On April 15, he started on-the-job training in Hawaii and during the next month he stole over a million documents secured in twenty-four different compartments, each protected by different passwords, to which he did not have legitimate access. He deployed a computer program to map the data.

On the evening of May 18 he flew to Tokyo's Narita Airport and on to Hong Kong by the morning of May 20. For more than ten days, he stayed at a residence he had arranged in advance through an unknown party. On June 2, 2013, Greenwald and Poitras met him in Hong Kong, together with another journalist Ewen MacAskill, brought to evaluate Snowden and the documents. On June 6 the first stories by Gellman, Poitras, and Greenwald appeared. On June 9, *The Guardian* newspaper posted a video on its website in which Snowden identified himself and took credit for the revelations. On June 23, with help from Julian Assange, a leading global hacktivist and publisher of WikiLeaks, Snowden flew to Moscow. Russia held Snowden sequestered in the transit zone of the Moscow airport for thirty-seven days, and on August 1, 2013, gave him official sanctuary.

From here, Snowden claimed whistleblower status and received great fame that included nomination for the 2014 Nobel Peace Prize. Although at one point he claimed that he did not bring stolen documents with him to Russia, many newspaper stories later appeared based on documents beyond the 58,000 that he gave to Greenwald and Poitras.

Snowden's actions have provoked a strong debate. His defenders seek to portray him as a whistleblower and person of conscience, who engaged in civil disobedience, sacrificing his reputation, career, and freedom for the collective good of the American people. Snowden's defenders describe his accomplishments in several categories: (1) triggering a debate on the value of privacy in the digital age; (2) raising awareness of the dangers of ubiquitous government surveillance, combined with government secrecy; (3) raising greater distrust of government in general; and (4) motivating the development of an international, ideologically committed coalition, such as the Electronic Frontier Foundation and Freedom of the Press Foundation, pushing for reform of laws governing surveillance.

His opponents argue that the actions he took to claim the whistleblower label serve merely as a disguise for the enormous damage his revelations have inflicted on the security of America and its allies. Epstein argues that terrorists changed their methods of communication after the Snowden revelations, enabling the terrorists who attacked Paris to avoid detection during months of logistical preparations. He also cites the Russian invasion of Eastern Ukraine without Western intelligence picking up any electronic signals of the mobilization.

The Snowden revelations provoked strong international political reactions, represented in the UN General Assembly Resolution of December 2013 entitled "Right to Privacy in the Digital Age." On June 2, 2015, the U.S. Congress passed the USA Freedom Act, which instituted some modest reforms. In other countries, privacy advocates complain states have tended to legalize the existing surveillance practices rather than strictly regulate them.

Jonathan K. Zartman

See also: Assange, Julian; Encryption; Hacktivism; Manning, Chelsea/Manning, Bradley; National Security Agency (NSA); WikiLeaks, Impact of

Further Reading

Epstein, Edward Jay. *How American Lost Its Secrets: Edward Snowden, the Man and the Theft*. New York: Alfred A. Knoph, 2017.

Harding, Luke. *The Snowden Files: The Inside Story of the World's Most Wanted Man*. London: Guardian Books, Faber & Faber, 2016.

Hertsgaard, Mark. *Bravehearts: Whistle-Blowing in the Age of Snowden*. New York: Hot Books, 2016.

Pohle, Julia, and Leo Van Audenhove. "Post-Snowden Internet Policy: Between Public Outrage, Resistance and Policy Change." *Media and Communication* 5, no. 1 (2017), 1–6.

Social Media and the Arab Spring

Networking platforms used during political demonstration and upheaval across the Middle East and North Africa (MENA) marked especially by internal regime changes in Tunisia, Egypt, and Yemen, and the eruption of civil war in Libya and Syria during 2011. These events surprised pundits, analysts, and world leaders alike, who had not anticipated the imminent political and social upheaval. The Arab Spring brought change with such speed and sheer magnitude that it called into question many of the preconceived notions of both the strength of old-guard regimes in the MENA and many contemporary ideas of political scientists. Political concessions were forced in several countries whose regimes successfully withstood protests.

Before the uprisings, many social movement theories neglected the impact of social media technologies on protest movements and commonly relegated them to fashions of the younger generation. Analysts saw protests through a framework of physical collective action to achieve a goal, but they neglected to consider the potential impact of the cyber realm on the process of social mobilization. Early analysis often missed the effect of information spread through the Internet on local antiestablishment movements. Protest organizers used social media to collect and focus grievances against the government, to overcome the inherent hesitancy of individuals to participate in a protest movement. Social media alone did not create the social movements observed in the Arab Spring, but access to interactive media affected the movements' narrative and framing, which dictated success or failure of the various civil efforts to force political concessions.

In today's connected world, Internet access through mobile devices has enabled ordinary citizens far greater access to information than any other period in history. In 2010, as the political unrest in the Arab world started to boil, The Oxford Internet Institute reported that an estimated two billion people enjoyed persistent Internet access across the globe. Specifically, 60 percent of the MENA's population had at least intermittent Internet access during this period. Furthermore, these numbers do not consider the fact that in many Arab countries, multiple people will share Internet devices. The shared nature of these devices exponentially increases the possible number of people with access. Saturation of Internet access does not automatically correlate with the ability to mobilize a successful social movement, however. The United Arab Emirates, Bahrain, Qatar, Lebanon, and Kuwait had some of the highest levels of social media penetration and yet experienced only small levels of unrest during the Arab Spring period.

In the case of the Arab Spring, social media provided a means for assorted civil groups to collaborate. The networks of activists, with a common purpose, survived the government's efforts to disrupt and suppress their activity. By forming a collective protest narrative, different groups shared their experiences and organized joint protests. The formation of a cohesive civil resistance, the "frame" explaining and justifying collective activity, provided a crucial unifying element in holding groups together in a shared commitment. Social media enabled activists to quickly shape the dominant discourse of Arab resistance to the old regimes. This clear, emotionally powerful narrative justifying resistance ensured a wide base of support on the streets.

People in the MENA region, on average, have some of the highest rates of technology adoption and correspondingly have some of the highest rates of government censorship. Government censorship of the Internet, before and during the Arab Spring, followed the historical pattern of repression of civil action groups, where the state adopts strong-arm tactics to limit or disband physical organizations. This type of repression did not work against Internet activists. As information spread and groups formed in the digital realm, group dynamics in the physical realm correspondingly mirrored social media organization patterns. The broader movements in each country lacked a hierarchical structure, and instead a cooperative network developed around anger against the government. Very quickly small local groups not only networked with each other but also connected with seemingly unassociated groups throughout the world, like the Occupy movement in the United States.

From the start of the Arab Spring, governments lost their monopolies on the control of information as Internet-savvy youth promoted their message via social media. This flow of information enabled even groups with long-standing disputes with each other to set aside their differences and coalesce behind a common platform. Aristotle explained this aspect of persuasion as the use of sensus communis, or in today's vernacular, the use of common sense to influence a group through a communal interest. As in the previous explanation of social framing, information released on social media shaped the narrative by using a commonly understood picture that harkened to a shared ethos among the differing organizations.

Although social media helped activists propagate a resistance narrative, this did not automatically create a corporeal means for organization. Already existing civil groups effectively filled this role, with a prime example being Islamic organizations. Social movements need places to meet and connect. In the MENA region, people instinctively go to the closest mosque or teahouse. The mosque and Islamic institutions provided activists the physical and social space to organize resistance against the old regimes. Religion has long been a dominant space for organization and development of social movements. Many of these Islamic groups provided the precursor elements necessary for social mobilization, and the Internet provided the tools for these groups to break the status quo. Social media informed the wider movement, but it also became a space for the revolution in its own right. Social media as a space enhanced the speed of the wider movement, but at times cyber activists saw their objectives overtaken by existing physical movements that did not need them once the flames had been stoked.

Because physical, human relationships provide the essential power of organizations, including states, some analysts dismiss social media's impact on the Arab Spring in the long term. However, formulating effective policies in the current world still requires understanding the nuanced interaction between social media and mobilization. Contemporary social mobilization strategy relies on activism through multiple media formats. The continued spread of devices with Internet access accentuates the importance of understanding the effects of this medium. Correspondingly, in the aftermath of the recent Arab Spring, many governments inside the region and around the world have sought to establish policies to restrict the use of or control social media content. However, the uprisings have shown the error of this strategy.

The Internet has opened a Pandora's box of information, and it cannot be closed. A blocked site only means networked activists will search for ways around those restrictions and will often move their content to servers located in more hospitable countries. In light of the way in which the Internet allows information to metastasize, policy makers should remember that social media is agnostic regarding content. The rapid distribution of information can cause short periods of societal change, producing a temporary surge of stimulus, but established governments have a similar ability to utilize the medium and establish a balance of power.

Sean Nicholas Blas

See also: Facebook, Use of; Twitter, Use of; Web 2.0

Further Reading

Billig, Michael. "Rhetorical Psychology, Ideological Thinking and Imagining Nationhood." In *Social Movements and Culture,* ed. Hank Johnston and Bert Klandermans. Minneapolis: University of Minnesota Press, 1995.

Dutton, W. H., A. Dopatka, M. Hills, G. Law, and V. Nash. *Freedom of Connection, Freedom of Expression: The Changing Legal and Regulatory Ecology Shaping the Internet.* Oxford, UK: Oxford Internet Institute, 2011.

Esposito, John, Tamara Sonn, and John Voll. *Islam and Democracy after the Arab Spring.* Oxford: Oxford University Press, 2016.

Howard, Philip. *The Origins of Dictatorship and Democracy: Information Technology and Political Islam.* New York: Oxford University Press, 2011.

Joffe, George. "The Arab Spring in North Africa: Origins and Prospects." *The Journal of North African Studies* 16 (December 2011).

Noueihed, Lin, and Alex Warren. *The Battle for the Arab Spring: Revolution, Counter Revolution and the Making of a New Era.* New York: Yale University Press, 2012.

Tawil-Souri, Helga. "It's Still about the Power of Place." *Middle East Journal of Culture and Communications* 5, no. 1 (2012), 86–95.

Tsang, Flavia, Ohid Yaqub, Desiree van Welsum, Tony Thompson-Starkey, and Joanna Chataway. "The Impact of Information and Communication Technologies in the Middle East and North Africa." *Rand Technical Report.* Santa Monica, CA: RAND, 2011.

Social Media Intelligence (SOCMINT)

Leverage of the apps used as part of Web 2.0 for intelligence at the strategic, operational, or tactical level. Best understood as a new branch of the long-established art of open-source intelligence, credible successes have occurred in the first decades of the 21st century, although challenges remain.

The history of open-source intelligence (also sometimes called OSINT) dates as far back as the Roman Empire with the existence of media and the sharing of rumors. Although acquisition of classified information can be of profound military value, nation-states and other entities predictably take serious steps to protect such information from divulsion. In many cases, however, unclassified sources can include a great deal of information which, when aggregated with other such sources, can provide intelligence of considerable quality and significance. Prior to the 21st century, popular unclassified sources included newspapers and media broadcasts as well as public records. During the Cold War, the relatively open societies of the noncommunist Western powers and the closed societies of the Communist bloc meant that agents from the latter could conduct part of their surveillance through the monitoring of Western newspapers.

The development of Web 2.0 massively broadens the aperture of open-source intelligence efforts because social media activities are by definition unclassified. Among intelligence professionals, debate has occurred about the relation between SOCMINT and open-source intelligence, and although in general it is felt that SOCMINT is most effectively understood as a subset of open-source material, some assert that SOCMINT also includes information "restricted from public view" and is therefore "more appropriately treated as a mix between open and classified sources" (Antonius and Rich, 2013, 45). In fact, this point underscores the ambiguous relationship between privacy and data disclosed through social media apps. The immense popularity of social media means that massive amounts of activity occur via social media, and among the posts and photos is a wealth of relevant data that can yield significant value when aggregated and analyzed.

In fact, this is central rather than incidental to the creation of many social media apps. The most popular apps tend to be those that are free to use, easy to navigate, and popular among friends or like-minded individuals (and this popularity is frequently influenced in turn by the first two factors). Apps such as Facebook exist in a free-to-use form because the data that users provide, both in their first creating accounts and in their activities through the app, are monetized by the app's owner. Users typically think of themselves sharing information with particular friends or groups, but the app itself is a presence at all these conversations and has an unparalleled ability to aggregate data. Sale of this data and especially of the opportunity to deduce conclusions from it is used by apps and famously by Google, which combines records from its search engine and from its popular e-mail tool and other applications.

SOCMINT is a third-party effort to profit in a strategic sense from the same pattern of shared information. Although SOCMINT actors will not necessarily have access to the panoply of data possessed by the app's owners, the opportunity to aggregate and analyze relevant data has already been recognized as significant.

Deployed military personnel and their families constitute a notable source of open-source intelligence information that is often mostly readily available via social media. Tech-savvy U.S. personnel deployed to Iraq and Afghanistan, communicating with loved ones in the United States, have been a source of potential operational security leaks. Images of personnel on bases or beside equipment could easily if inadvertently divulge information about what units might be present on a base or what activities might be upcoming. Default settings on digital cameras tend to imbed geolocation data into images, meaning that an uploaded photograph provides specific information about an individual's location. Many popular apps are also aimed to incorporate geolocation features that would offer yet more monetizable data for themselves, although the same information could be used by thieves to know when a home is empty or by SOCMINT analysts to know the location of targeted individuals. SOCMINT experts could also interpret from posted words and photos whether a base might be vulnerable to attack or whether their forces might need to withdraw or disperse from an area in order to evade an upcoming battle. The U.S. military reportedly felt enough concern about operational security that it sought at first to prohibit its personnel from engaging with social media, then built more secure alternatives to the dominant social media platforms, and eventually reconciled to the popularity of social media with personnel.

SOCMINT has reportedly been used against U.S. adversaries. Unit 61398 of the People's Liberation Army (PLA) is based in Shanghai and is reported to be a leading element in the cyberespionage campaigns waged by the People's Republic of China (PRC) to acquire economically and strategically significant industrial secrets and other information. SOCMINT is believed to have factored into the analysis conducted by Mandiant, which identified several of the PLA members of Unit 61398 and led to their indictment in a U.S. court. Analysis of social media activity may also have played a part in the tracing, geolocation, and killing of Islamic State of

Iraq and Syria propagandists who had not realized that aspects of their activity provided information about their locations.

SOCMINT has also been used in the conflict in Eastern Ukraine and by nation-states against dissidents and others in the Middle East and PRC. Crowdsourcing of photographic analysis helped confirm that Russia had supplied the antiaircraft missile that was used to shoot down the Malaysian airliner MH17 in July 2014. Some of the vital photos surfaced apparently as a result of loose operational security in a scenario that validates concerns that incautious social media usage risks operational security. In 2011, Bahrain's government used information available via Facebook to identify individuals who participated in political demonstrations, and this information was then used to enable prosecution of dissidents. Analysts note that PRC uses a voluntary force numbering in the tens of thousands to monitor the activities of neighbors online, including on social media. This policing of politically undesirable statements or allusions is another example of SOCMINT.

While apps such as Facebook, which are intended for use among friends, can provide large amounts of SOCMINT valuable to third-party gatherers, analysts have also noted that other apps including Twitter can display useful information about patterns and trends in the attention of various users. Endemic to open-source intelligence gathering is the challenge of finding the useful needle within a haystack in which the useful data might not at first appear very different from the less relevant pieces. One expert described the process as the equivalent to having access to a vast library of information but without an index with which to search in an efficient way. Whether big data technologies will yield a breakthrough to SOCMINT analysts or if the production of chaff data can reliably obscure significant pieces of information in the future remains to be seen.

Nicholas Michael Sambaluk

See also: China, Blocking of Social Media by; Facebook, Use of; Instagram, Use of; Operational Security, Impact of Social Media on; Social Media and the Arab Spring; U.S. Military Use of Social Media; War on Terrorism and Social Media

Further Reading

Antonius, Nicky, and L. Rich. "Discovering Collection and Analysis Techniques for Social Media to Improve Public Safety." *The International Technology Management Review* 3, no. 1 (March 2013), 42–53.

Bejar, Adrian. *Balancing Social Media with Operations Security (OPSEC) in the 21st Century.* Newport, RI: Naval War College, 2010.

Hellman, Maria, Eva-Karin Olsson, and Charlotte Wagnsson. "EU Armed Forces' Use of Social Media in Areas of Deployment." *Cogitatio* 4, no. 1 (February 2016), 52–61.

Mazailli, Alessandro. *OSINT & SOCMINT, Theory and Practice, Intelligence Collection Using Location Tools on Twitter.* Master's thesis. Rome: NITEL, 2016.

Omand, David, Jamie Bartlett, and Carl Miller. "Introducing Social Media Intelligence (SOCMINT)." *Intelligence and National Security* 27, no. 6 (2012), 801–823.

Patrikarakos, David. *War in 140 Characters.* New York: Basic Books, 2017.

Silvestri, Lisa Ellen. *Friended at the Front: Social Media in the American War Zone.* Lawrence: University Press of Kansas, 2015.

Solomon, Scott E. *Social Media: The Fastest Growing Vulnerability to the Air Force Mission.* Maxwell AFB, AL: Air University Press, 2017.

Soft War

A term referring to the leverage of cultural influencers by one nation-state against another in order to force social and ultimately political change against the wishes of the latter state's leadership. Although the exact definition remains contested, the term and the underlying concept borrow from political scientist Joseph Nye's assertions about "soft power," interpreted to constitute intangible factors such as culture, political philosophy, and social or governing institutions.

Soft war essentially denotes a more forceful and belligerent variation of soft power, while still remaining short of physical or official hostility constituting "war." The concept of soft power was developed by neoliberalist scholar Joseph Nye during the 1990s and early 2000s. This coincided with the collapse of the Soviet Union and the establishment of a unipolar world in which the United States stood as an unmatched global superpower, and the latter phases overlapped with the outbreak of the Global War on Terrorism. The persuasive potential of political and cultural values stood at the core of the soft power theory. Such factors, although intangible, could produce considerable favorable shifts in the outlook and policy of antagonistic and self-isolated countries that could be noncoercively co-opted into a community of nations. The idea that cultural export can appreciably shift foreign societies and policies commonly leads to the conclusion that soft power is a tool best suited to use by the United States. Thus, soft war might be thought of as leveraging cultural influences in a more concerted and aggressive fashion, to attain a specific change in another country's society and political posture.

The lack of consensus in defining the term coincides with the fact that various pundits, analysts, and political figures have made accusations in different directions. For example, the Iranian Revolutionary Guard Corps has since 2009 been accusing the United States and its Western allies of waging "soft war" against their regime. Iranian leaders such as Ayatollah Khamenei have specifically accused the United States of engaging in soft war through cultural and propaganda efforts to recontour the Iranian public's commitments to the goals of that country's top leadership. One Iranian official was quoted in by Iranian media during 2014, defining soft war as "America want[ing] to do something to make us stop wanting" (Gholamzadeh, 2015). This may be a reference to the Iranian leadership's interest in becoming a nuclear power, which has been a point of frequent bilateral contention, arguably interrupted by the controversial 2014 Joint Comprehensive Plan of Action.

Ironically, some pundits have also pointed to Iran as prosecuting soft war against the United States. Former U.S. Defense officials in 2016 pointed to the leveraging of "smile diplomacy" by Iran's President Hassan Rouhani and Foreign Minister Mohammad Javad Zarif as an example of Iranian soft war mirroring the soft war efforts that Iranian diplomats have accused the United States of enacting.

U.S.-Iranian relations is perhaps the most popularly identified "soft war," but it is not the only example. Another prominent case is between the United States and North Korea. It has been suggested that the 2014 Sony Pictures movie *The Interview* had the effect of soft war and that it was actively interpreted as a soft war attack by North Korea. The movie depicted a notional plot by which Central Intelligence Agency–recruited journalist-entertainers humiliated and subsequently assassinated North Korean dictator Kim Jong Un. North Korea's ruling Communist Party has invested its six decades in power to portray three successive generations of hereditary leaders as quasi-gods, and the regime's hostile reception of the movie was not a surprise.

The scale and form of North Korean to the film, which it reportedly branded as "terrorism," fits the general conception of soft war. Hackers calling themselves "Guardians for Peace," which have been strongly alleged to have been tied to North Korea, used a modified version of the Shamoon computer virus to exfiltrate and leak confidential electronic records and communications belonging to Sony Pictures in November 2014. Sony reacted by halting the release of the movie, and some in the United States debated whether major hacking actions against large industries constituted attacks on U.S. vital interests that could trigger government response. North Korea officially denied connection to the hacking event, although it also applauded the release of digital materials embarrassing to Sony Pictures. North Korea additionally alluded to the physical violence of the September 11, 2001, terrorist attacks and reportedly menaced moviegoers to avoid the film and residents living near movie houses to evacuate theaters.

Allegations about Russian attempts to intervene in various Western elections, particularly but not exclusively the 2016 U.S. presidential campaign, might also fit into the categorization of "soft war." Forum infiltration and the use of bots and trolls to covertly simulate domestic social and political dialogue are alleged to be frequent Russian actions in efforts to destabilize democratic governance in countries that Russian leaders have identified as hostile to its national ambitions.

Various other forms of aggressive cyberactivity can be associated with the concept of soft war. Many of the most notable examples are connected with political tensions involving Russia. Cyber actions against Estonia during a 2007 diplomatic spat between the small Baltic country and Russia might also plausibly be categorized as soft war, since the hackers working in parallel with the apparent interests of Russia sought to digitally isolate Estonia from the rest of the world and to close down as much as possible the digital aspects of the country's economy. The resulting economic distress was reportedly intended to create leverage that might force the Estonian government to reverse decisions to remove a domestically unpopular statue of a Russian soldier that had been displayed in Estonia early in the Cold War. Although much remains as yet unconfirmed about the Stuxnet virus, it appears to have been meant to stall Iranian nuclear enrichment activities. This may be an example of soft war, although the reluctance of Stuxnet's creator to officially take credit for the action raises the question about the degree to which nonviolent demonstrations of cultural or technical power must be overt in order to be recognized as "soft war." The Shamoon virus, believed to be an Iranian action against computers involved in the Saudi Arabian oil industry in indirect response to the setback initially imposed by Stuxnet, is another action that potentially constitutes soft war.

However, the cyber aspects of the 2008 conflict between Russia and Georgia fit less well with the concept of soft war, because hackers' generally successful efforts to digitally isolate the Caucasus country of Georgia occurred at the same time as physical violence between Russian and Georgian military forces. Therefore the

2008 events in Georgia, including in the cyber domain, arguably resemble "war" more than "soft war." The 2014 cyber actions against Ukraine, called Snake, also resist easy categorization. Ukraine's military was already engaged in battle against paramilitary forces that invaded the eastern regions of the country, and evidence pointed to Russian involvement in support of the paramilitaries. Russia's denial of involvement and the challenges of precise attribution in cyber activities complicates the task of characterizing Snake, but it too may be an example of soft war.

Paradoxically, non-Western countries have been quoted accusing the United States of "using a pretext like elections" to "create doubts and suspicion and line up one group of people against another" (Gholamzadeh, 2015). U.S. sympathy for prodemocracy movements in autocratic systems has often been decried by dictatorial regimes as being efforts to sow dissention and revolution. With the emergence of the term "soft war" in the early 21st century, such accusations adjust in character more than in nature. Skeptics of the utility of soft war have noted that its ability to forcefully persuade can be undercut by the user's national desire to be admired, that countries being acted upon will not necessarily respond passively, and that adversary states may even interpret a larger rival's use of soft power or soft war as a sign of military weakness.

Nicholas Michael Sambaluk

See also: China, Blocking of Social Media by; Estonia, 2007 Cyber Assault on; Forum Infiltration; Georgia, 2008 Cyber Assault on; Shamoon; Snake, Cyber Actions against Ukraine, 2014; Social Media and the Arab Spring; Social Media Intelligence (SOCMINT); Sony Hack, 2014; Stuxnet; U.S. Military Use of Social Media

Further Reading

Blackwill, Robert D., and Jennifer M. Harris. *War by Other Means: Geoeconomics and Statecraft*. Cambridge, MA: Harvard University Press, 2016.

Gholamzadeh, Hamid Reza. "Soft War vs. Soft Power." *Mehr News Agency,* October 14, 2015, https://en.mehrnews.com/news/111041/Soft-War-vs-Soft-Power (accessed April 8, 2018).

Mearsheimer, John J. *The Tragedy of Great Power Politics*. New York: Norton, 2014.

Melissen, Jan, ed. *The New Public Diplomacy: Soft Power in International Relations*. New York: Palgrave Macmillan, 2005.

Nye, Joseph S. *Soft Power: The Means to Success in World Politics*. New York: PublicAffairs, 2005.

Parmer, Inderjeet, and Michael Cox, eds. *Soft Power and US Foreign Policy: Theoretical, Historical, and Contemporary Perspectives*. New York: Routledge, 2010.

Simons, Anna. "Soft War = Smart War? Think Again." *Foreign Policy Research Institute,* April 27, 2012, https://www.fpri.org/article/2012/04/soft-war-smart-war-think-again.

Soft War: A New Episode in the Old Conflict between Iran and the United States. Philadelphia: University of Pennsylvania, 2013.

Surespot App

An open source instant messaging tool first released in 2014, whose designed emphasis on security reportedly attracted its use by personnel affiliated with the Islamic State of Iraq and Syria (ISIS). The app is a notable example of how the secure communications technologies emerging in the early 21st century offer not only faster and more convenient ways to communicate, but also constitute unprecedented opportunities as commercial-off-the-shelf (COTS) tools for militants. Applications such as Surespot provide a degree of add-on communications security without entailing the investment in a specialized crypto phone.

Surespot uses end-to-end encryption, whereby the users can communicate secure even from eavesdropping or surveillance by the app provider. This design, coupled with the app itself being open source, has won Surespot praise from the Electronic Frontier Foundation as a relatively secure method by which users can exchange messages. Although a free-to-install software, the app is sustained through a combination of donations and the sale of added functionality such as secure voice messaging.

These secure features imply strong possibilities as a cheaply used commercial means of maintaining operational security while communicating military information. Such opportunities are of particular interest to nonstate organizations such as terrorist groups, because they have a special interest in swift and secure communication and because they may lack the resources with which to develop tailored technologies that are possible for the military and intelligence entities of nation-states. Media reports in the United Kingdom in mid-2015 indicated that more than one hundred personnel connected to the terrorist group ISIS had used the Surespot app in the previous seven months. Evidence has reportedly also suggested the use of the app by at least one ISIS sympathizer in the United States, and analysts believe that the terrorist group sees such apps as an avenue through which to gain exposure to wider audiences.

ISIS fighters in Syria have used the app as a means of communicating directly with potential sympathizers overseas. Blog posts by ISIS recruiters have also been found that specifically directed potential recruits to four electronic methods by which to reach the terrorist group and receive further guidance on how to join and fight. In addition to Twitter, Kik, and Tumblr, the blog directed recruits to use Surespot. Other ISIS members are known to have used Surespot to have communicated to recruits advice about how to make explosive devices.

Nicholas Michael Sambaluk

See also: Crypto Phone; ISIS Recruitment and War Crimes; Operational Security, Impact of Social Media on; Telegram App; War on Terrorism and Social Media; WhatsApp

Further Reading

Graham, Robert. "How Terrorists Use Encryption." *CTC Sentinel* 9, no. 6 (June 2016), 20–26.

Nguyen, Hoaithi Y. T. *Lawful Hacking: Toward a Middle-Ground Solution to the Going Dark Problem.* Thesis. Monterey, CA: Naval Postgraduate School, 2017.

Scaperotto, Albert. *The Foreign Fighter Problem: Analyzing the Impact of Social Media and the Internet.* Thesis. Maxwell AFB, AL: School of Advanced Air and Space Studies, 2015.

Voelz, Glenn J. *The Rise of IWar: Identity, Information, and the Individualization of Modern Warfare.* Carlisle, PA: Strategic Studies Institute, 2015.

Winter, Charlie. *Documenting the Virtual "Caliphate."* London: Quilliam, 2015.

Telegram App

A cloud-based instant messaging smartphone application. Telegram and other applications seek as an add-on software to provide a degree of security in communication without involving the investment in a specialized crypto phone. Although technologists debate how much anonymity it might actually provide, its qualities as a private means of communication have led to its use by terrorist organizations such as the Islamic State of Iraq and Syria (ISIS) and have also prompted some dissent-wary governments to block access to it.

Telegram was originally release in the spring of 2013, and within its first eighteen months, it grew from a user population of 100,000 to one of fifty million. The user population nearly quadrupled again between the end of 2014 and 2017. Conversations between individuals or among groups can be conducted via messaging that is cloud stored. Accounts are connected to telephone numbers, although the numbers can be rapidly swapped so that the same account can be connected to an individual who rapidly exchanges phone numbers. Accounts can also be deleted on command or after six months of inactivity. Intuitively, these characteristics are of interest to covert planners and activists, as well as to terrorist groups. "Online encryption messaging platforms such as Telegram have allowed Islamic State operatives in caliphate territory to keep in touch with recruiters and guide plotters in Europe" (Heil, 2017, 9).

As a result, several governments have sought either to access messages or to block the app. Russia, Iran, and Germany are reported to have intercepted messages sent via the app, and Iran, Pakistan, and the People's Republic of China have blocked access to the service. The app's Russian-based developers relocated to Germany and later Dubai, possibly in connection with adverse attention from the Russian government.

Although the technology offers considerable privacy and has reportedly attracted large numbers of users in South Korea, where telecommunications surveillance was dramatically upgraded in 2014, many cryptography experts argue that Telegram in its standard format does not provide true communications security. The app's creators have asserted that Telegram exceeds rivals such as WhatsApp in offering security to users. Although a two-factor authentication variant for secret chats came into being at the end of 2014, Telegram ordinarily operates on a much less secure single-factor authentication basis. Without providing an endorsement, the Electronic Frontier Foundation reportedly concluded that the secret chat feature was secure and that the ordinary app provided considerably more modest privacy. Furthermore, although group conversations can rapidly be edited or retracted by any user, the conversations are also held in cloud storage, and this detracts from the app's security.

ISIS is known to have recommended, in announcements to its sympathizers via Twitter, the use of Telegram as a tool by which its members and sympathizers could communicate and plan attacks. This included planning for the coordinated attacks in Paris in November 2015, which killed 130 people, and later the videoed murder of a priest in the Normandy region of France in 2016. The planning for a 2016 truck attack against a Berlin Christmas market was also found to have been conducted using the app. Telegram has closed down more than seventy channels known to be used by ISIS, but use by the jihadist group had nonetheless continued, and the app's creator has refused requests to close down the app entirely. The year 2017 saw Indonesia lift its ban on the app and Russia openly contemplate imposing a ban on the app within that country, pointing to the ongoing use of Telegram by terrorists.

Nicholas Michael Sambaluk

See also: Crypto Phone; Encryption; ISIS Recruitment and War Crimes; Surespot App; War on Terrorism and Social Media; WhatsApp

Further Reading

Burke, Jason. "The Age of Selfie Jihad: How Emerging Media Technology Is Changing Terrorism." *CTC Sentinel* 9, no. 11 (November/December 2016), 16–22.

Doherty, Jim. *Wireless and Mobile Device Security.* Boston: Jones & Bartlett Learning, 2016.

Heil, Georg. "The Berlin Attack and the 'Abu Walaa' Islamic State Recruitment Network." *CTC Sentinel* 10, no. 2 (February 2017), 1–11.

Rhee, Kyung-Hyune, and Jeong Hyun Yi, eds. *Information Security Applications: 15th International Workshop, WISA 2014, Jeju Island, Korea, August 25–27, 2014.* New York: Springer, 2014.

Thuriasingham, Bhavani, et al., eds. *Security and Privacy in Communication Networks: 11th International Conference SecureComm 2015, Dallas, TX, USA, October 26–29, 2015.* New York: Springer, 2015.

Think Again, Turn Away Campaign

A widely criticized U.S. Department of State social media campaign hinging on a Twitter account @ThinkAgain_DOS. Frequently referred to as "Think Again," it was intended to counteract the Islamic State of Iraq and Syria (ISIS) propaganda on social media and dissuade people who were considering allegiance to the jihadi terrorist organization.

@ThinkAgain_DOS was established in December 2013 as an English-language Twitter account. Although overtly a U.S. government enterprise and therefore an example of white operations rather than of covert propaganda, the engagement was apparently meant to be less rigid and more responsive and more widely read by potential ISIS recruits than might be the case with other U.S. government information outlets. The project had precedent in earlier U.S. government efforts at dissuading terrorist recruits online; by 2011, parallel efforts had been undertaken in Arabic and Urdu. The @ThinkAgain_DOS initiative came in response to signs of a ramp-up of ISIS organizing its social media and propaganda entities. As such, the State Department enterprise began operation some months before ISIS formally unveiled its al-Hayat Media Center or the glossy English-language online jihadist journal *Dabiq* in mid-2014, concurrent with ISIS offensives across much of Syria and sweeping deep into northwestern Iraq.

Despite its comparatively early start and enthusiastic presidential support for the State Department project, the "Think Again" campaign steadily became bogged down in criticism from U.S. journalists, who pointed to lack of evidence of the project's success in persuading viewers from reconsidering their interest in ISIS. Journalists also particularly criticized the unsophisticated character of the U.S. messaging, which contrasted unfavorably with the technical competence of ISIS messaging in its own propaganda and recruitment projects. "They're trying to reach these kids, but it's backfiring," because "it's like the grandparents yelling to the children, 'Get off my lawn'" (Miller and Higham, 2015), according to one counterterrorism consultant.

Despite problems, "Think Again" arguably managed to land a viral propaganda punch with its *Welcome to ISIS Land* video, which was viewed more than 800,000 times on YouTube. Critics conceded the popularity of the video but contended that going viral did not necessarily equate to dissuading potential terrorists. Whether the video's views, which dramatically dwarfed any other video produced by "Think Again," even came from people considering terrorism is unclear. The @Think-Again_DOS account currently carries a message dated March 23, 2016, directing viewers to the Global Engagement Center at @TheGEC, but the latter account has been suspended.

Nicholas Michael Sambaluk

See also: Al-Furqan; #AllEyesOnISIS; *Dabiq*; Dawn of Glad Tidings Campaign; ISIS Recruitment and War Crimes

Further Reading

Miller, Greg, and Scott Higham. "In a Propaganda War against ISIS, the U.S. Tried to Play by the Enemy's Rules." *The Washington Post,* May 8, 2015, https://www.washingtonpost.com/world/national-security/in-a-propaganda-war-us-tried-to-play-by-the-enemys-rules/2015/05/08/6eb6b732-e52f-11e4-81ea-0649268f729e_story.html?utm_term=.3e02923586ad.

Patrikarakos, David. *War in 140 Characters: How Social Media Is Reshaping Conflict in the Twenty-First Century.* New York: Basic Books, 2017.

Siame, Lyson, et al. *Defeating ISIS by Winning the War of Ideas.* Research report. Maxwell AFB, AL: Air Command and Staff College, 2017.

Ullah, Haroon K. *Digital World War: Islamists, Extremists, and the Fight for Cyber Supremacy.* New Haven, CT: Yale University Press, 2017.

TroopTube

A social media video sharing platform built for the U.S. Army as an alternative to YouTube for video uploading. However, its emphasis on security severely curtailed the interest of personnel in using it, and the site was shut down three years later. It was first launched in late 2008. This coincided with the latter phases of the military occupation of Iraq (2003–2011), in which considerable numbers of personnel were deployed and wanted to communicate with family by using the increasingly available social media platforms, which gained popularity as the occupation continued. Prominent among these was YouTube, a video posting website.

Although social media platforms allowed unprecedented amounts of communication between deployed personnel and their friends and families, operational security concerns mounted. The U.S. military struggled with numerous and potentially significant operational security breaches, and in 2008 it decided to prohibit military members from uploading videos onto the highly popular YouTube. In order to help make the ban enforceable without provoking a drop in moral, the military aimed to provide workable alternatives that would allow personnel to take advantage of advances in communication technology that characterized the early 21st century, but to do so in controlled ways that would not compromise secure information and thereby put the lives of deployed personnel at risk.

Analysis has concluded that TroopTube was "not a bad idea" (Bruhl, 2009, 22), but major factors were evident from the outset that ultimately doomed TroopTube as a viable alternative to commercial counterparts. One was the uphill challenge of standing up a new application in hopes that it can compete, even within the finite circle of potential users such as the U.S. military, with an alternative that had already gained widespread acceptance, familiarity, and use by a massive population that appreciated its ease of use. A second, related reason that TroopTube could not compete was that it required use of a Controlled Access Card (CAC) and registration of an account to use it at all, rather than needing registration only for the uploading of videos, as is the case with YouTube. Third, videos are censored in order to ensure that classified material is not intentionally or unintentionally divulged.

Although this sets a prima face context of operational security, it also complicates use in ways that tended to drive potential users away from the application and to YouTube, which stood as a readily available and more user-friendly alternative. Yet another shortcoming involved the limited size of videos, since TroopTube uploads were limited to a maximum of 20 MB or 5 minutes of video. Contemporary YouTube, even during the early phases of rise in its popularity, allowed videos of up to twice this length. The YouTube prohibition ended in early 2011, and the military shut down the application at the end of July 2011.

Nicholas Michael Sambaluk

See also: Army Knowledge Online (AKO); FrontlineSMS; HARMONIEWeb; Operational Security, Impact of Social Media on; YouTube, Use of

Further Reading

Bruhl, Jakob C. *Soldiers in the Blogosphere: Using New Media to Help Win the War for Public Opinion.* Research paper. Maxwell AFB, AL: Air Command and Staff College, 2009.

Nunn, Stephen W., and Leah Y. Wong. *Knowledge Management for Shared Awareness.* San Diego: Space and Naval Warfare Systems Center Pacific, 2013.

Silvestri, Lisa Ellen. *Friended at the Front: Social Media in the American War Zone.* Lawrence: University Press of Kansas, 2015.

Trusted Flagger (YouTube)

A crowdsourced organization of volunteers that marks content violating YouTube's Community Guidelines and assists in the policing of the site. Although any user can flag objectionable videos for review, disturbing videos, including those recruiting for violent extremist groups, featuring child abuse, and spotlighting acts of performative violence like beheadings, continued to appear on the site. In 2012, content managers noticed that some users flagging objectionable content performed far better than others—that 90 percent of the things they reported as violating standards were validated by YouTube's Trust and Safety staff and removed, while most flaggers were validated 30 percent of the time.

According to YouTube, the program provides four premium tools for trusted flaggers. First, it offers a "bulk-flagging" function that allows a flagger to report batches of videos simultaneously. Second, these flaggers may take part in private forums that discuss YouTube's policies and enforcement procedures. Third, these flaggers have exclusive access to decisions from YouTube about flagged content. And finally, the program prioritizes input from these flaggers for review by content moderators. Trusted flaggers do not have the authority to remove content.

To be eligible for this program, a flagger must have a valid YouTube account, contribute frequently and accurately, and abide by the company's Community Guidelines. Individuals, government agencies, and nongovernmental organizations (NGOs) that meet these criteria may take part in the program, and volunteers are subject matter experts from groups like the Anti-Defamation League, Institute for Strategic Dialogue and Rape, and Rape, Abuse & Incest National Network, as well as conscientious individuals. A secondary program, YouTube Heroes, was created to reward and incentivize users who accurately add subtitles to videos, flag objectionable comments, and run help forums. Businesses and for-profit organizations are not eligible to participate. And trusted flaggers are not allowed to use their position to advertise products or services. If trusted flaggers abide by YouTube policy, they can not only use the enhanced tools but also receive special benefits called "perks." These bonuses include testing new products, attending exclusive events, and connecting directly to YouTube personnel. Most trusted flaggers do not receive a salary from YouTube.

Trusted flaggers' primary responsibility is to identify and highlight content that violates YouTube policies. Some flaggers also use their tools and direct connection to YouTube employees to identify unjust penalties, improper account terminations, and other problems. Trusted flaggers must have in-depth knowledge of policies, features, and restrictions, making their expertise valuable to users who are unable to reach YouTube directly. Some trusted flaggers offer pro bono support and advice on platforms such as Reddit and Twitter.

Although some users appreciate the services trusted flaggers provide, others have expressed suspicion about these unpaid "super users." Since government agencies can become trusted flaggers, provided they follow YouTube's guidelines, some users worry about monitoring and censorship. Others are concerned that these mostly anonymous flaggers may wrongfully target rule-abiding users or accounts. The most common criticism is that YouTube does not release much information about the program, and trusted flaggers are not supposed to talk about their responsibilities. This lack of transparency raises concerns about issues like data privacy, freedom of expression, and corporate accountability.

James Tindle and Margaret Sankey

See also: Forum Infiltration; Web 2.0; YouTube, Use of

Further Reading

Conway, Maura. *Violent Extremism and Terrorism Online in 2016: The Year in Review.* Budapest, Hungary: Network of Excellence for Research in Violent Online Political Extremism, 2016.

House of Commons Public Affairs Committee. *Radicalisation: The Counter-Narrative and Identifying the Tipping Point*. London: House of Commons, 2016.

Parker, Brandon. "Virtually Free Speech: The Problem of Unbridled Debates on Social Media." *Intuition: The BYU Journal of Psychology* 12, no. 2 (2017), 107–118.

Twitter, Use of

An online social networking and updating service whose intense traffic and participation has prompted it to become a contested digital landscape, as a recruiting avenue for extremist groups, an organizing network for political movements, and an online environment for proxy rivalries parallel to physical conflict.

Twitter was launched in the summer of 2006 to serve as a platform for, in the words of cofounder Jack Dorsey, "'a short burst of inconsequential information,' . . . that's exactly what the product was'" ("Twitter Creator Jack Dorsey," 2009). Popularity at an entertainment and interactive media conference in 2007 helped boost a precipitous rise in popularity and usage, and by the time Twitter was a decade old, it counted 319 million active users (a global participation comparable in number to the total number of people in the United States). Brief announcements (called "tweets"), initially 140 characters that were expanded in 2017 to a 280-character limit, can be posted, and users can opt to follow particular accounts by name (called "handles" and signified with "@") or can read and participate in discussion of particular topics according to topic name (called "hashtags" and signified with "#").

A market research firm's study found that 40 percent of Twitter activity was meaningless chatter, a nearly comparable amount was conversational content, that forwarding of information accounted for nearly 10 percent of traffic, that self-promotion constituted 6 percent, and that news and spam contributed the remaining 3 percent each. The meteoric popularity of the social networking service inexorably introduced several kinds of controversial traffic. These in turn have prompted Twitter itself to begin censoring trending hashtags considered to be offensive or abusive. Twitter has also instituted and suspended a verification program that had been aimed to forestall some users from creating accounts in order to simulate and defame other, usually public, personalities.

Conflict in the physical domains increasingly bled into the digital domain in the early 21st century, including into Twitter. As early as August 2009, both Twitter and the social networking site Facebook suffered a denial-of-service attack, taking Twitter offline for several hours. It was subsequently found that the target had been a pro-Georgian user and that the timing had been in reference to the first anniversary of the 2008 South Ossetia War. Twitter as a whole had been a collateral but temporary casualty of this attack. Although Twitter as a whole has generally been operational (up 98 percent of the time in 2007), events such as the hacking of some official media accounts in 2013 do occur.

Twitterbots, computer programs that automatically generate or retweet posts, have been estimated to account for as many as twenty million account profiles. Many of these false accounts are used by advertisers in order to artificially build a following for a product or service in hopes of persuading real consumers of a trend. However, political organizations and militant groups have adopted similar tools as part of their efforts to spread propaganda.

Twitter has announced efforts to suspend the accounts of users who promote extremism, having closed down hundreds of thousands of such accounts. Skeptics have observed that the ease with which extremists can create new accounts and can establish new Twitterbots means that this is at best a short-term approach rather than a solution to the challenge of propaganda usage by extremist entities. Similarly, actions such as blocking anti-Semitic tweets from being accessed in Israel do nothing to end the hatred that creates the posts.

By 2011, signals appeared about the utility of Twitter for political conflict, and since war is an extension of policy conflict this hinted at the potential weaponized use of Twitter. Twitter was one of the ways in which prodemocracy protesters across several Middle East nations shared information and planned demonstrations during the "Arab Spring" events of December 2010–December 2012, although some contemporary studies suggested that only a tiny fraction of the populations of the involved countries were active on Twitter. Meanwhile, in Afghanistan, the U.S.-Afghan-NATO coalition's @isafmedia was arguing in cyberspace against two Taliban accounts about the veracity of reported violence and about the impact on the Afghan population. Contemporary U.S. accounts seemed smug that @isafmedia possessed twice as many followers as both Taliban accounts combined. A spate of violence in Gaza during 2014 led antagonists to highlight their own plight with competing hashtags #GazaUnderAttack and #IsraelUnderFire.

The use of Twitter as a tool for warfare became vividly obvious due to other events in 2014. The jihadi Islamic State (IS) movement, frequently called the Islamic State of Iraq and Syria (ISIS), launched an invasion of northern Iraq and quickly captured the city of Mosul. On the road to Mosul, ISIS forces worked to coordinate their physical invasion with a digital counterpart. As ISIS fighters captured territory, they quickly massacred captured Iraqi soldiers and civilians they suspected of being unsympathetic to the ISIS jihad. These murders were filmed and rapidly posted along with other propaganda messages. Inundated with frightening signals in social media of a vaunted ISIS force ruthlessly murdering those who stayed to fight or failed to flee, the Iraqi Army garrison of Mosul, 25,000 strong, disintegrated in the face of an invading force estimated to be no stronger than 1,500. When ISIS forces seized the city, they murdered Iraqi Army stragglers, commandeered more than 2,000 former Army vehicles, and paraded these newest feats on social media. The propaganda campaign during the Mosul invasion was spearheaded by the Twitter hashtag #AllEyesOnISIS.

As ISIS fought to consolidate its hold on newly captured territories, it also continued to push its message digitally to the world beyond its physical empire. Part of these efforts included hashtag hijacking, wherein account holders and Twitterbots posted a surge of messages spuriously associated with already popular topics. One particular example involved the ISIS commandeering of hashtags related to the World Cup during the global soccer tournaments. Taking advantage of the opportunity to foist their message on unsuspecting Twitter users who expected to see posts commenting on soccer, readers were instead confronted with messages about ISIS and images of ISIS fighters kicking a severed head, which an ISIS Twitter account identified as the kind of soccer ball used in their territories.

Shortly thereafter, in March 2015, a jihadi movement in Nigeria called Boko Haram changed its name to the Islamic State in West Africa Province (ISWAP). The change appears to have been made both to gain credibility through association with the then-victorious ISIS movement and also to express solidarity with their brand of jihad. ISWAP leader Abubaker Shekau proved a frustratingly resilient figure, adept at leveraging a combination of physical violence and social media to destabilize Nigeria and expand jihadi holdings there.

The most globally infamous single action by ISWAP was the kidnapping of 276 female students from a high school at Chibok in northeastern Nigeria. ISWAP ostentatiously boasted about its ability to raid the town and kidnap the girls, and it mocked the government for its inability to prevent or counter these actions. An international social media effort called #BringBackOurGirls successfully raised awareness of the girls' plight, but repatriation efforts brought little result. Between mid-2016 and the start of 2018, barely one hundred of the girls had been rescued, released, or had escaped.

The #BringBackOurGirls campaign points to the limitations of Twitter's use in war. Like all other tools, it is not a panacea. Twitter could socialize news and it could elicit compassion, but this alone was not sufficient to secure the return of the kidnapped girls. However, this campaign, like the 2011 debate between @isafmedia and the Taliban Twitter accounts, shows that Twitter is not necessarily an instrument useful only to jihadi forces specifically or to aggressors generally. Other campaigns further demonstrate efforts to use Twitter as part of a social media toolbox for countering extremist messages, for targeting ISIS figures, and also its potential for promoting positive purposes within the United States.

Among these efforts was a U.S. State Department Twitter hashtag called #ThinkAgainTurnAway that aimed to dissuade those Twitter users who were considering fighting on behalf of ISIS, although the apparently limited effectiveness of this effort prompted significant criticism. Another occasional but spectacular application has been for U.S. forces to capitalize on particularly compromising posts made by unwary ISIS figures. Even savvy users have slipped, such as by posting grizzly images that nonetheless retain geolocation data that allows U.S. experts to pinpoint a jihadi figure's location and strike it. Furthermore, some in the U.S. military have suggested that social media could be used more effectively for visibility and recruiting, and Twitter could be an important avenue for such an initiative in the future.

The years 2016 and 2017 saw heightened concerns and accusations about the leveraging of Twitter for geopolitical or violent purposes. This included widespread contention during much of the 2016 U.S. presidential race and afterward, featuring evidence pointing to Russian activities meant to foment and nurture discord and unrest within the United States. Mid-2017 witnessed a spike in Twitter hostility between Ukraine and Russia, three years after paramilitaries with demonstrable affinity to Russia seized the eastern part of Ukraine. In November 2017, news came to light that numerous fake and inactive Russian and Ukrainian Twitter accounts had been identified two years earlier by an accounts manager at Twitter and flagged as a possible problem. Indeed, fake and inactive accounts are latent and mobilizable "populations" in a Twitter environment. Neglecting to close the accounts, Twitter soon thereafter fired the employee who had brought the problem to light. Reportedly, Twitter was not considering the possibility that such accounts might be used for propagating messages in information warfare, but Twitter was focused on avoiding actions that might reduce the number of users (and thus the implicit relevance) of the service.

The first decade of Twitter has demonstrated that the social networking and updating field is a prime landscape for propaganda and messaging operations, particularly when these can showcase military exploits or can magnify the significance of smaller military actions. These years have also indicated that the social media landscape, prominently including Twitter, remains in significant aspects ungoverned and open to the domination of the more digitally savvy adversary with the most applicable objectives.

Nicholas Michael Sambaluk

See also: #AllEyesOnISIS; #BringBackOurGirls; Facebook, Use of; Hashtag Hijacking; Instagram, Use of; ISIS Recruitment and War Crimes; ISIS Trolling Day; Shekau, Abubakar; Twitter Account Hunting; War on Terrorism and Social Media; YouTube, Use of

Further Reading

Berenger, Ralph D., ed. *Social Media Go to War: Rage, Rebellion and Revolution in the Age of Twitter.* Spokane, WA: Marquette Books, 2013.

Bodine-Baron, Elizabeth, et al. *Examining ISIS Support and Opposition Networks on Twitter.* Santa Monica, CA: RAND, 2016.

Bourret, Andrew, et al. *Assessing Sentiment in Conflict Zones through Social Media.* Monterey, CA: Naval Postgraduate School, 2016.

Gregory, Robert H. *Turning Point: Operation Allied Force and the Allure of Air Power.* Master's thesis. Ft. Leavenworth, KS: Army Command and General Staff College, 2014.

Irby, Jack B. *The Weaponization of Social Media.* Master's thesis. Ft. Leavenworth, KS: Army Command and General Staff College, 2016.

Patrikarakos, David. *War in 140 Characters: How Social Media Is Reshaping Conflict in the Twenty-first Century.* New York: Basic Books, 2017.

Schneider, Nathan K. *ISIS and Social Media: The Combatant Commander's Guide to Countering ISIS's Social Media Campaign.* Newport, RI: Naval War College, 2015.

"Twitter Creator Jack Dorsey Illuminates the Site's Founding Document. Part I." *Los Angeles Times,* February 18, 2009, http://latimesblogs.latimes.com/technology/2009/02/twitter-creator.html.

Ullah, Haroon K. *Digital World War: Islamists, Extremists, and the Fight for Cyber Supremacy.* New Haven, CT: Yale University Press, 2017.

Twitter Account Hunting

A term connected with efforts to identify and eradicate "troll" accounts on social media platforms, notably Twitter. Automated Twitter "bots" have commercial uses in marketing as well as in more militant contexts, where trolls sow discord and are associated with forum infiltration, hashtag hijacking, and the spread of destructive misinformation, often to covertly promote a political agenda.

Although account hunting has been reported in countries with autocratic systems in which dissent is ipso facto interpreted as latent rebelliousness, the most publicly known work toward effective account hunting has been directed at sympathizers of terrorist groups such as the Islamic State of Iraq and Syria (ISIS) and other accounts suspected of fomenting discord by disseminating misinformation or racist sentiment.

Frequently, such accounts are bots, set to automatically post statements without requiring actual involvement from a human user. The threat of bot-conducted misinformation campaigns and forum infiltration has prompted creation of a bot-hunting prize competition by the Defense Advanced Research Projects Agency (DARPA) and a number of related research efforts within academia. DARPA has shown interest in automated systems that can use algorithms to dramatically accelerate the search for and identification of bots.

A winning project developed by Sentimetrix, an analytics company, developed an algorithm capable of identifying all thirty-nine bots within a 2014 forum on vaccinations in a space of barely two weeks, although the algorithm also flagged a false positive. A team from the University of Southern California correctly identified all the bots without any false positives, although its process required an additional week.

As with other forms of competition, improvements in hunting bots is expected to prompt increased efforts by the creators and users of such bots to hide and work more insidiously. The algorithms tested in DARPA's competition worked in different but disclosed ways. Sentimetrix's system was trained to watch for behavior noticed in bots that had been active during India's 2014 elections, including grammatical peculiarities, verbiage resembling natural language computer programs, and extended periods of posting activity that would be difficult for a human being to duplicate.

Reports have noted that, although bot-hunting systems are improving, research into best practices is "revealing their hand in a way that allows bot-makers to design strategies to specifically defeat these algorithms" ("How DARPA Took On the Twitter Bot Menace," 2016). If bot-hunting programs continue to attain success using these presumptions, it seems likely that bot designers will respond by building into their future bots characteristics that obscure these telltale clues. Designers of bot-hunting algorithms have also presumed that bots tend to inflate their own apparent relevance by linking them to one another, but analysts have pointed out that decoupling bots and preventing them from following each other as readily could be a quickly accomplished countermeasure against bot hunters relying on that assumption.

Although account hunting efforts largely focus on research into ways to facilitate the accurate and swift detection of bots, not all of the traffic that hunters seek out is automated. Research has also proceeded into understanding the ways the patterns of human-used account trolls, including by ISIS sympathizers, various nationalist revanchists, racists, and other extremists. The research group Recorded Future has noted that ISIS supporters posting online have often responded to the closure of their social media account by setting up a new account and posting further messages that are similar in content and sometimes even in script. Additionally, the new accounts sometimes even resemble their predecessors' Twitter handles. In the short and even medium term, this may be a useful tool for the makers of hunting algorithms to identify such accounts.

Another approach tailored as a counter to human-used accounts is to build hunter bots that will identify and engage the human troll and absorb the troll's attention and energy. One project, called @Assbot, finds human-used accounts and targets them by splicing and regurgitating randomized pieces of other tweets. Trolls confronted by @Assbot have sometimes engaged in online arguments "with" the bot, unaware that it is not another human user. Obviously, similar bots could easily be directed by extremists almost as easily as they have already been tested against unwary extremists. Skeptics have also noted that increased use of bots to combat hostile accounts could, like the winner of the DARPA competition, inadvertently flag and target false positives. The proliferation of bots on both sides could pose new complications in the collection of usable social media intelligence by militaries and intelligence agencies.

Nicholas Michael Sambaluk

See also: Forum Infiltration; Hashtag Hijacking; Social Media Intelligence (SOCMINT); Twitter, Use of

Further Reading

"How DARPA Took On the Twitter Bot Menace with One Hand Behind Its Back." *Technology Review,* January 28, 2016, https://www.technologyreview.com/s/546256/how-darpa-took-on-the-twitter-bot-menace-with-one-hand-behind-its-back.

Prier, Jarred. *The Command of the Trend: Social Media as a Weapon in the Information Age.* Maxwell AFB, AL: School of Advanced Air and Space Studies, 2017.

Sanovich, Sergey. *Computational Propaganda in Russia: The Origins of Digital Misinformation, Working Paper No. 2017.3.* Oxford: Oxford University Press, 2017.

Subrahmanian, V. S. "The DARPA Twitter Bot Challenge." *Computer* 49, no. 6 (June 2016), 38–46.

Wei, Wei, et al. "The Fragility of Twitter Social Networks against Suspended Users." https://pdfs.semanticscholar.org/4793/c9a440d677aa568adc41d2b7c532f60caff6.pdf.

U.S. Military Use of Social Media

Since 2007, the U.S. military has drastically changed its approach regarding the use of social media, from an official ban of social networking sites to a complete opening toward the new media. Social media can be defined as new means of communicating and sharing information online either between two or more individuals or organizations. They are now being used by the military personnel for individual and private purposes, but they are also part of the broad communication strategy and of the Information Operations (IO) of the U.S. military.

In May 2007, the U.S. Department of Defense (DoD) blocked recreational social networking sites such as YouTube and MySpace from its computer networks, in connection with concerns about the social media intelligence (SOCMINT) value that might be derived by adversaries reading posted materials. The filtering action was directed by the Joint Task Force-Global Network Operations (JTF-GNO),

DoD's operational arm tasked with implementing Internet modifications and security enhancements. In its Operational Directive Message (ODM) 059-07 called "IAP Access Control List (ACL) Security Filter Update," the JTF-GNO directed DoD service providers to control the access of specific websites. Although thirteen sites used for networking and music or photo sharing were banned by the directive, soldiers and the other employees were still allowed to access the sites through personal computers or at private Internet cafés on military bases.

The rationale behind the web-filtering policy was the necessity to limit both excessive bandwidth consumption and the potential risk of social networking sites serving as a channel for malicious code. The ban received criticism as it was perceived as a way to control the information sent by the soldiers on the battlefield, and also because of its potential negative effects on the morale of those deployed around the world.

With the exception of this directive and its different iterations, there was no DoD-wide formal policy addressing social media and Internet access prior to 2010. Therefore, between 2007 and 2010, the different branches of the U.S. military initiated their own in-house policies on social networking. For instance, in January 2009, the Army created a new Online and Social Media Division at the Office of the Chief of Public Affairs in order to open up the lines of communication with the public. According to Lieutenant Colonel Kevin Arata, then director of the new division, it was important for the Army to reach across all generations and demographics so that everyone could obtain information about the Army, and social media were considered the best tool to further the conversation with the public. In April 2009, the Air Force Public Affairs Agency published a pamphlet entitled *New Media and the Air Force*, which encouraged the members of the Air Force to use social media to communicate their interests and their areas of expertise. The document clearly stated that all Airmen were communicators and that using social media would be a good way to ensure the truthful representation of the Air Force online.

A few months later, in August 2009, in the hope of establishing an official social media policy for DoD, Deputy Defense Secretary William J. Lynn III directed a comprehensive study of social media sites. The DoD review's objective was to weigh the benefits of social networking and other new media platforms against potential security vulnerabilities. It was finally in February 2010 that the department issued a memorandum—DTM 09-026 on the "Responsible and Effective Use of Internet Capabilities"—that created a policy to allow access to social networking services (SNS) such as Facebook, Twitter, Flickr, and YouTube across all DoD components. The memorandum also provided guidelines for military use of social media sites and recognized that Internet-based capabilities were integral tools for operating and collaborating within the Department of Defense and with the general public. It was then stated by DoD officials that while information sharing may seem like a lack of security, the department could no longer afford to do just one or the other. Noteworthy, the Principal Deputy Assistant Secretary of Defense, Price Floyd, announced the new policy via his Twitter feed.

After the release of the official DoD policy, the different branches of the military issued their own guidelines to explain how social media should be used and what should and should not be posted online. For example, the Office of the Chief of Public Affairs of the Army released in 2011 the *US Army Social Media Handbook*, an extended iteration of a previous version issued in 2010. The guide explicitly described the kind of sensitive information that soldiers should refrain from mentioning, such as unit locations, deployment dates or equipment specifications and capabilities, but also encouraged the organization to establish and maintain a presence online.

Currently, social media are being recognized by the U.S. military as an invaluable connecting tool, with each branch having its own official presence on social networking platforms such as Facebook, Twitter, and YouTube (the Army's Facebook page has more than 4.5 million followers). It has been admitted that social networking platforms are a useful way of improving the morale of the deployed members as it enables them to stay in touch with their family. One of the most cited examples is the ability for the children to do homework with their parent who is at the other end of the world via Skype. A few research studies have indeed been conducted on the impact of social media on couples and families during deployment, and they have mostly demonstrated that these communication tools are considered helpful toward maintaining healthy relationships.

In addition, networking platforms are a low-cost advertising tool, especially for recruitment. With millions of people following the different pages, the U.S. military has access to a much broader audience and can connect easily with potential recruits. The military can leverage social media to promote its activities and operations not only through the official platforms but also directly through the soldiers who share their personal stories. Overall, social media are considered useful for the military for recruitment, retention, and support of personnel.

Moreover, social media are now an important part of the U.S. military communication and media strategy toward both the American and the international publics. Having a strong online presence can help shape public discourse according to the military's strategic objectives. For instance, social media are a powerful tool to spread the military's key themes and messages as well as to address negative news stories and redress the military's image if needed. As part of the military's Information Operations (IO), social media can also be used as a tactical tool, to increase situation awareness through greater information sharing such as during a crisis, but also to influence a foreign population's perception. This type of activities refers to psychological operations (PSYOP), a subcategory of IO, which are defined by the military as planned operations to convey selected information and indicators to foreign audiences to influence their emotions, motives, objective reasoning, and ultimately, the behavior of their governments, organizations, groups, and individuals. PSYOP are considered an important aspect of diplomatic, military, and economic activities, and social media are a useful tool as they can easily reach large masses. Research aims to determine how actions of popular platforms influences behaviors, and relevant findings would be relevant both to the detection of an adversary's operations via social media or to the conduct of one's own. Finally, social media are also being used by the military as part of their intelligence-gathering operations. Conducting open-source intelligence operations on social media enables the military to easily collect information on different targets and improve their analysis of adversaries.

In summary, the U.S. military's use of social media has progressively evolved since its filtering policy of 2007. If information has always been a central aspect of the conduct of war, it is even more so in the information age. Conflicts in the 21st century are being orchestrated around influence campaigns and information operations, and social media, as an important element of the media environment, are being strategically used by the U.S. military across a range of operations for the successful achievement of its objectives.

Karine Pontbriand

See also: Blogs; MySpace, U.S. Military Personnel Usage of; Social Media Intelligence (SOCMINT); War on Terrorism and Social Media; Web 2.0

Further Reading

Department of Defense Personnel Access to the Internet: Report to Congress in Response to Request on Page 323 of Senate Armed Services Committee Report Number 110–77. Washington, D.C.: Department of Defense, 2007.

Directive Type Memorandum 09–026—Responsible and Effective Use of Internet-Based Capabilities. Washington, D.C.: Office of the Deputy Secretary of Defense, 2010.

Knopf, Christina M., and Eric J. Ziegelmayer. "Fourth Generation Warfare and the US Military's Social Media Strategy: Promoting the Academic Conversation." *Air & Space Power Journal–Africa and Francophonie* 3, no. 4 (2012), 3–23.

Perry, Chondra. "Social Media and the Army." *Military Review* (March–April 2010), 63–67.

The Use of Web 2.0 in the Department of Defense. Washington, D.C.: Department of Defense, 2009.

US Army Social Media Handbook. Washington, D.C.: Office of the Chief of Public Affairs Online and Social Media Division, 2011.

War on Terrorism and Social Media

An interconnection between a long-term 21st-century conflict pitting the United States and various governments around the world against an array of salafa-jihadist organizations on the one hand and a wave of communication tools optimized for the digital realm on the other. The conflict and the communication trend achieved prominence early in the 21st century and developed simultaneously, and predictably each exerted impact on the trajectory of the other.

Although Al Qaeda and like-minded jihadist organizations launched lethal but sporadic terrorist attacks prior to the 21st century, their activities drew pronounced attention following the September 11, 2001, coordinated hijackings of four U.S. airliners and their use to crash into iconic American buildings such as the World Trade Center in New York City and the Pentagon in Washington, D.C. The U.S. response involved the waging of a global hunt for the Al Qaeda organization of Osama bin Laden as well as affiliated groups expressing solidarity with his terrorist franchise. This response initially focused on rooting out Al Qaeda elements and their state sponsors in the Taliban regime that controlled most of Afghanistan between 1994 and 2002.

In what arguably was an effort to maintain the initiative, forestall the development and deployment of suspected weapons of mass destruction, and simultaneously displace a dictatorial regime with a robust and democratic successor state, the United States followed its early successes in Afghanistan with an invasion of Iraq that toppled Saddam Hussein's regime in three weeks. Dismissal of Iraqi military personnel and the dispersal of armed and largely antagonistic former soldiers proved to be a serious and early misstep in occupation policy. Another was an underlying U.S. lack of appreciation for the degree to which successive Iraqi wars—and especially the Saddam Hussein regime itself—had left the country in shambles, from its physical infrastructure to its civil society. The brief campaign of March–April 2003 was subsequently followed by a tumultuous military occupation lasting until the lapse of a status of forces agreement in December 2011.

Social media came into its own especially during the period of the Iraq occupation. Although the concept of "Web 2.0" dates to 1999, many of the icons of connectivity were launched a few years afterward. In social media, this included Facebook in 2004 and Twitter in 2006, while Skype (an Internet tool for voice and video chat) was created in 2003.

The Iraq occupation fueled political controversy in U.S. domestic politics and the virtual presence, via social media, of deployed personnel created a dynamic largely unparalleled in modern military history. Although the physical burden of the occupation was borne by military personnel constituting a scant fraction of 1 percent of the U.S. population, social media helped bring imagery of those personnel home with unprecedented immediacy. U.S. military bureaucracy responded with reluctance to the impact that social media usage seemed likely to incur on operational security, both in terms of endangering personnel and in presenting opportunities for enemies to predict and hamper U.S. military activities. It may also have impacted the character of the emerging antiwar movement in the United States, which while vitriolic toward national policy makers demonstrated a palpably conscious effort to avoid harassment of military personnel, in stark contrast to the approach that some antiwar protesters took toward service personnel during the U.S. conflict in Southeast Asia, which the United States waged particularly in Vietnam from 1965 to 1973.

Armed conflict erupted during the occupation which in fact resembled a civil war within Iraq. The small coalition that the United States organized in 2003 witnessed an exodus the following year following upticks in violence within Iraq and spectacular terrorist attacks against mass transit in Madrid, Spain. The years 2005 and 2006 saw further violence, including especially graphic forms of murder conducted by an organization known as Al Qaeda in Iraq. Its leader, a Jordanian national using the nom de guerre Abu Musab al-Zarqawi, sought to incorporate social media and traditional media into a larger propaganda effort supporting his jihadist agenda. Interestingly, this created a rift between Zarqawi's forces and those of mainstream Al Qaeda, whose leaders objected less to murderous activities than to the degree to which Zarqawi emphasized the open advertisement of violence and directed a large proportion of his efforts at Muslims he considered apostate rather than focusing his violence against the Americans.

Scholars disagree about whether Zarqawi's group splintered from Al Qaeda through this disagreement in 2006 or if the two movements were temporarily kindred but never synonymous. It is believed that Zarqawi received very early financial support from bin Laden, and Zarqawi's association with Al Qaeda and his prewar activities in Iraq formed one part of the U.S. government argument favoring invasion of Iraq in the first place. Zarqawi intended to project violence via social media in order to intimidate enemies and inspire sympathizers to become recruits, and his movement, which would proclaim itself the Islamic State in 2010 and again in 2013, marked a new trend in terrorism and its use of emergent media.

Zarqawi was killed in a U.S. air strike in June 2006, but his organization survived his death and the death of his cosuccessors in a subsequent air strike in April 2010. The organization survived in portions of Iraq and leveraged a civil war in Syria to gain further strongholds, allowing the group to claim to be the Islamic State of Iraq and Syria (ISIS). It prepared a bold social media campaign to complement its swift seizure of lands across northern Iraq in the first half of 2014. The media campaign, called the Dawn of Glad Tidings campaign, included creation of a mobile phone app that transformed a sympathizer's device into a bot that retweeted messages on behalf of ISIS. These messages included #AllEyesOnISIS. Such Twitter handles would constitute useful ways to attract attention from users who were already interested or sympathetic with ISIS. To reach and intimidate a larger global audience, ISIS accounts engaged in activities such as hashtag hijacking so that totally unrelated topics, such as #WorldCup2014 used by soccer fans to follow the global matches occurring at the same time as the dramatic ISIS conquests, were subjected to a barrage of jihadist social media activity.

Analysts believe that this propaganda campaign was meant to frighten the larger world with an inflated specter of the omnipresence of ISIS. The campaign was remarkably successful at controlling global attention in 2014. Countries did belatedly respond to the territorial conquests of ISIS, to the danger that the movement posed as it drew sympathizers to travel to Syria and northern Iraq to become fighters, and to the military capabilities of a movement that had abruptly captured extensive amounts of military hardware from fleeing Iraqi national army units and oil production facilities across portions of both Iraq and Syria.

An international coalition of more than sixty countries formed from 2015 onward to confront ISIS. It was an ironic assortment with some strange bedfellows, including not only Iraqi and Kurdish forces but also U.S. and Russian airpower and ground units closely aligned with Iran. Experts note that the battlefield reverses suffered by ISIS, which by 2016 lost almost all of the territory it had held at its peak, appear to have undercut the effectiveness of ISIS propaganda efforts. The amorphous hacktivist entity Anonymous waged a campaign of ridicule that it called "ISIS Trolling Day," taking place on December 11, 2015, as ISIS forces were being bludgeoned and the movement was showing signs of faltering as a state. Analysts are divided on the net effectiveness of a U.S. information campaign called "Think Again, Turn Away," which aimed to dissuade would-be ISIS recruits from becoming jihadists. Social media companies such as Twitter also worked to remove content and accounts that expressed support for ISIS, and by 2016 this effort was also making an effective contribution.

Although ISIS's fortunes were still strong, some groups outside the Middle East vocalized support and affinity for the movement. ISIS boasted of these pledges of support in the eighth issue of its digital propaganda magazine *Dabiq* in March 2015. The journal was one part of a media effort coordinated by an ISIS branch called al-Furqan. Sympathizers in Libya and Tunisia were matched by the allegiance pledged by Boko Haram, a militant group mostly located in northern Nigeria. Abubakar Shekau, leader of Boko Haram, proved to be a wily adversary for Nigerian military and government officials since he demonstrated an ability to use social media as a propaganda tool to project messages that were illustrated and punctuated by kinetic attacks, such as the seizure of nearly 300 female Christian hostages in April 2014.

Although Facebook and Twitter remain the giants in the social media realm, other examples of Web 2.0 technologies appeared in subsequent years, some of which offer capabilities of use to nonstate combatants. Among these are Telegram and Surespot, which use encryption to provide greater privacy to the user than other communication alternatives. ISIS identified Telegram by name in its tweets to sympathizers, recommending the app as a secure means of communication for planning attacks. Planning of the ISIS attacks in Paris in November 2015 as well as attacks in France and Germany in early 2016 was conducted via Telegram. British analysts reported that Surespot was being used by more than one hundred ISIS fighters during a similar time frame. Still other apps, such as FireChat that provides a limited mesh network communication capability, is known to have been used by political demonstrators in the Middle East and Asia but has not yet been utilized by terrorists for communicating in the austere environment of a battlefield.

As of mid-2018, the most territorially and kinetically powerful jihadist movements seemed to have been quashed or to have gone to ground. However, the readiness with which ISIS and Boko Haram incorporated social media into the context of their violence and the short-term benefits they enjoyed as a result suggest that leaders of future extremist movements likely took note.

Nicholas Michael Sambaluk

See also: Al-Furqan; #AllEyesOnISIS; Boko Haram; #BringBackOurGirls; *Dabiq*; Dawn of Glad Tidings Campaign; Hashtag Hijacking; ISIS Recruitment and War Crimes; ISIS Trolling Day; Operational Security, Impact of Social Media on; Shekau, Abubakar; Surespot App; Telegram App; Think Again, Turn Away Campaign; Twitter, Use of

Further Reading

Bejar, Adrian. *Balancing Social Media with Operations Security (OPSEC) in the 21st Century*. Newport, RI: Naval War College, 2010.

Close, Joshua. *#Terror: Social Media and Extremism*. Maxwell AFB, AL: Air Command and Staff College, 2014.

Graham, Robert. "How Terrorists Use Encryption." *CTC Sentinel* 9, no. 6 (June 2016), 20–26.

Patrikarakos, David. *War in 140 Characters*. New York: Basic Books, 2017.

Prier, Jarred. "Commanding the Trend: Social Media as Information Warfare." *Strategic Studies Quarterly* 12, no. 4 (Winter 2017), 50–85.

Silvestri, Lisa Ellen. *Friended at the Front: Social Media in the American War Zone.* Lawrence: University Press of Kansas, 2015.

Watts, Clint. *Messing with the Enemy: Surviving in a Social Media World of Hackers, Terrorists, Russians, and Fake News.* New York: HarperCollins, 2018.

Weimann, Gabriel. *New Terrorism and New Media.* Washington, D.C.: Wilson Center, 2014.

Winter, Charlie. *Documenting the Virtual "Caliphate."* London: Quilliam, 2015.

Web 2.0

The term referring to the Internet landscape of the early 21st century, highlighting the rise of user-generated content. Information architecture consultant Darcy DiNucci coined the term "Web 2.0" in January 1999, pointing specifically to an impending transition in the character of the web. The dominant format of 1990s web content consisted of posted information on various independent websites, which had been growing rapidly in number since the experimental ARPANet gave way to the World Wide Web in 1989. DiNucci pointed to a shift in which web users would increasingly come to create content as well as access content provided to them via webpages. The term gained little traction until 2004, and two years later *Time* magazine identified its Person of the Year based on the impact of web users' online presence due to the shift toward a Web 2.0 virtual landscape.

Web 2.0 is characterized by the ubiquity of social media giants, particularly Facebook and to an extent YouTube, but other important examples include sites such as Twitter, Instagram, and MySpace. These sites are built around facilitating users' ability to create accounts and post material that is then visited by peers; businesses and other organizations, including government agencies, have since also created accounts and a presence on these sites, and the monetization of Web 2.0 platforms relies on the traffic and on the sale and exploitation of personal information garnered through the platforms' ability to track user traffic as well as other data.

The rise of Web 2.0 creates both a new landscape in which conflict extends in the 21st century, and it also establishes new repositories of vast amounts of information that is of strategic as well as economic interest to different entities worldwide. Social media has proven to be a valuable and frequently utilized avenue for the spreading of political messages as well as of other speech. Web 2.0 technology influenced the course of "Arab Spring" antiauthoritarian movements in 2011. Countries already notorious for their censorship have responded to the rise of the Web 2.0 with concern that has sometimes translated into the establishment of fenced-in web landscapes in which a population's access to outside information can be controlled, monitored, and curtailed.

Propaganda campaigns and other political messaging actions have made deliberately high-profile efforts to garner attention. Some examples include the Boko Haram propaganda campaigns mocking the Nigerian military for its inability to destroy it as a jihadi rebel group. Much of the rest of the world responded with horror when Boko Haram kidnapped nearly 300 female Christian students from the town of Chibok and led them into various forms of slavery; an anti–Boko Haram messaging campaign #BringBackOurGirls emerged in the wake of the April 2014 kidnappings. #Kony, a social media awareness movement begun in 2012 with the aim of facilitating the arrest of the Ugandan rebel cult leader and convicted war criminal Joseph Kony, achieved more than 120 million views on social media platforms, most notably on YouTube and Vimeo. As of the spring of 2018, however, neither of these movements had achieved total success.

The Islamic State of Iraq and Syria (ISIS) enthusiastically leveraged social media during its infamous invasion of Mosul in northern Iraq in 2014. Its Dawn of Glad Tidings propaganda campaign and use of the Twitter handle #AllEyesOnISIS, as well as subsequent use of hashtag hijacking to draw Twitter users unsuspectingly to ISIS pages, were part of a propaganda campaign that sought to intimidate adversary populations while attracting and recruiting potentially sympathetic web users outside the Middle East via a complex misinformation campaign alternatively to either travel to the Syrian and Iraqi lands then controlled by the jihadi quasi-state or instead to launch terrorist attacks as homegrown terrorists in their countries of origin. Opponents of ISIS have included the hacktivist movement Anonymous, which used social media platforms to ridicule ISIS in December 2015.

Controversy, speculation, and investigation into the leveraging of Web 2.0 in the 2016 U.S. presidential election point to other forms of strategic opportunities being harnessed for potential political objectives. Trolling and the deliberate kindling of online and in-person confrontations by artificially seeding Web 2.0 platforms with hate speech are believed to be a method used by nation-states whose leaders are interested in fomenting dissention overseas. Public attention to this issue has dramatically increased following the 2016 elections in the United States, but evidence exists of earlier tampering as well, such as in Indian elections during 2014. Many experts and others attribute these misinformation efforts to the Russian government. Other analysis of the events from 2016 has led to discovery that the leading Web 2.0 platform, Facebook, had for years been aware of politically motivated exploitation of data that had ostensibly been eligible only for scholarly analytics.

Given the extensive amount of information that users willingly if uncomprehendingly provide to Web 2.0 platforms, the opportunity for this information to be bought or stolen by state and nonstate entities for cyberwar purposes suggests the potential impact of the Web 2.0 landscape being part of soft war and of other forms of warfare in the 21st century. Carelessness with postings and information willingly provided to the world via Web 2.0 platforms reputedly contributed to the identification of hackers serving in the People's Liberation Army Unit 61398. The potential risk to operational security dovetails with operational security concerns by many of the world's militaries.

Nicholas Michael Sambaluk

See also: #AllEyesOnISIS; #BringBackOurGirls; Blogs; Boko Haram; China, Blocking of Social Media by; Dawn of Glad Tidings Campaign; Facebook, Use of; ISIS Recruitment and War Crimes; ISIS Trolling Day; MySpace, U.S. Military Personnel Usage of; Operational Security, Impact of Social Media on; People's Liberation Army Unit 61398; Social Media and the Arab Spring; Soft War; The Great Cannon; TroopTube; Twitter, Use of; U.S. Military Use of Social Media; War on Terrorism and Social Media; YouTube, Use of

Further Reading

Ayers, Cynthia E. *Rethinking Sovereignty in the Context of Cyberspace*. Carlisle, PA: Center for Strategic Leadership, 2016.

Herrera, Linda. *Revolution in the Age of Social Media: The Egyptian Popular Insurrection and the Internet*. London: Verso, 2014.

Niekerk, Brett van, and Manoj Maharaj. "Social Media and Information Conflict." *International Journal of Communication* 7 (2013), 1162–1184.

Patrikarakos, David. *War in 140 Characters: How Social Media Is Reshaping Conflict in the Twenty-First Century*. New York: Hachette, 2017.

Prier, Jarred. "Commanding the Trend: Social Media as Information War." *Strategic Studies Quarterly* 11, no. 4 (Winter 2017), 50–85.

Rid, Thomas. *Cyber War Will Not Take Place*. Oxford: Oxford University Press, 2013.

Zeitzoff, Thomas. "How Social Media Is Changing Conflict." *Journal of Conflict Resolution*. https://www.zeitzoff.com/uploads/2/2/4/1/22413724/zeitzoff_how_social_media_jcr.pdf.

WhatsApp

Designed by two former Yahoo employees, Jan Koum and Brian Acton, a software application dating from 2009 to take advantage of the increasing ubiquity of cell phones. Originally, the appeal of the app was to make long-distance phone calls easily, but end-to-end encryption allows users to send messages, photos, videos, and conduct free phone calls privately, a feature that has increasingly become a point of contention between the company and national governments. In 2014, Facebook acquired WhatsApp for $19 billion, and the company has been working to monetize the app through the addition of an enterprise customer service app for which it will charge businesses to interact on the platform with customers. In the United States and Western Europe, WhatsApp usership skews to young adults, who access it more often than Facebook, and frequent use correlates significantly with extroversion.

Globally, however, WhatsApp has more than a billion and a half users, many in developing countries where people can access very inexpensive data plans from the hilltops, even in remote rural areas. There, the appeal is to an older, sometimes illiterate population, who leapfrogged from no digital access to smartphones, rather than to a computer in between. Their unfamiliarity with "fake news" and propaganda makes them vulnerable to viral videos with catastrophic consequences, such as antivaxx messages in Brazil, which contributed to a yellow fever epidemic, Myanmar condoning the use of the app for spreading hate messages against the Rohingya minority, and Filipino president Rodrigo Duterte's videos of encouragement for vigilante actions against drug dealers. In India, viral videos warning against child snatchers led to the vigilante mob murders of twenty-nine people.

Because of its privacy features, WhatsApp is a favorite tool of people conducting illicit transactions. The Islamic State of Iraq and Syria (ISIS) bragged of using the app to arrange movement of their assets, including the sale of sex slaves. The November 2015 Paris and April 2017 Stockholm terrorist attacks were planned and conducted using the app's messaging features. The company resolutely refuses to allow governments access to conversations, resulting in it being blocked in the

People's Republic of China. It was temporarily blocked in Turkey and in Sri Lanka, and in 2016 Brazilian authorities arrested Facebook's vice president for Latin America for not complying with directives to turn over information. In India, WhatsApp has agreed to limit the ability to forward message to more than five people at a time, run newspaper ads and send out street theater troupes to educate people about misinformation, but not to intervene in the content sent. This policy guarantees the app will remain in favor for those desiring privacy, or secrecy.

Margaret Sankey

See also: Facebook, Use of; FrontlineSMS; Surespot App; Telegram App; Zello app

Further Reading

Colraine, James. *Encrypted Messaging Apps in the Age of Terrorism and Snowden: Savior or Safe Haven?* Master's thesis. Washington, D.C.: Georgetown University, 2016.

Rastogi, Nidhi, and James Hendler. *WhatsApp Security and Role of Metadata in Preserving Privacy.* Troy, NY: Rensselaer Polytechnic Institute, 2016.

Shehabat, Amad, et al. "Encrypted Jihad: Investigating the Role of Telegram App in Lone Wolf Attacks in the West." *Journal of Strategic Security* 10, no. 3 (2017), 27–53.

WikiLeaks, Impact of

Complex and controversial results emanating from the release of more than one million items alleged to be classified U.S. government materials onto a website. WikiLeaks was created in 2006 by Australian computer programmer Julian Assange, and while its first years were comparatively uneventful, it attained global fame starting in 2010 with the release of documents and video files leaked by personnel whose work in the U.S. military and as U.S. intelligence agency contractors provided access to vulnerable data.

Following the first wave of releases traced to a then U.S. Army enlisted soldier Bradley Manning (since becoming Chelsea Manning), the U.S. government responded by laying out a strategy to attempt to maintain more reliable control over information possessed by the intelligence community. Predictably, the three pillars of the new strategy involved ensuring that access to classified information was restricted to those with both the authorization and the need to know, that technical aspects of protection from misuse were practiced, and ongoing monitoring of use to detect signals of misuse. Defense Department employee suggestions included expanding the role of voice over Internet protocol in order to reduce the number of people who might have direct access to data. This example points to the interrelation between the three pillars and how policy or technology changes in one could lend assistance in another.

When this plan was presented to Congress in March 2011, the report noted that "we have had 'bad apples' who have misused . . . information before and, unfortunately, we will see them again" (Stone, 2011, 6) but that a balance between protection of information and sharing within appropriate circles remained necessary. Indeed, subsequent events did follow, including an insider threat by Edward Snowden. His role as an intelligence contractor with Booze Allen Hamilton had given access to troves of further data. Later still, a sophisticated combination involving the hacking of e-mail information, its submission to WikiLeaks for distribution by a third party, and the leveraging of compromising data into political messaging via Internet trolling emerged to raise political and security controversy in the United States.

The U.S. government struck back at WikiLeaks after the 2010 releases by constraining the organization's funding. This approach ran parallel to one part of the strategy against terrorist groups, since the United States after the September 11, 2001, terrorist attacks launched by Al Qaeda recognized that preventing monetary donations to terrorist groups might choke off the resources needed for future attacks and moved to hinder the transfer of money to groups associated with terrorism. In the case of WikiLeaks, financial entities such as MasterCard, Visa, and PayPal were persuaded to eliminate tools that had earlier enabled account holders to contribute money to WikiLeaks. Anonymous, an amorphous entity that targets a range of state and nonstate groups that it considers oppressive or engaged in censorship, retaliated against these companies with distributed denial of service (DDoS) as a retribution for their cooperation with the U.S. government.

Sympathizers of the WikiLeaks organization argue that the site is dedicated to public transparency, while U.S. government officials including Central Intelligence Agency (CIA) head Mike Pompeo has suggested that WikiLeaks' exfiltration and publicizing of purportedly classified records makes the site resemble a spy agency. "WikiLeaks walks like a hostile intelligence service and talks like a hostile intelligence service" (Pompeo, 2017). In addition to welcoming leaks from individuals such as Chelsea Manning/Bradley Manning and Edward Snowden, CIA Director Pompeo asserted in 2017 that WikiLeaks "encouraged its followers to find jobs at CIA in order to obtain [and exfiltrate] intelligence" (Pompeo, 2017) and that the organization has simultaneously courted support from autocratic countries that share an interest in eroding U.S. intelligence capabilities.

Analysts have outlined the array of mutually reinforcing tools that Russia employs in order to advance its information agenda internationally. This includes the broadcasting of tailored news pieces via regime-controlled media outlets Russia Today (called RT) and Sputnik, which deploy strategically tailored items alongside ordinary reporting in order to lend credibility to the former and to attract viewership whose political outlooks can be finessed to suit the regime. More covertly, troll accounts in social media propel discussion and tension in ways that erode public trust and cohesion in a target country. WikiLeaks materials provide the fodder for much of this effort, because "before the trolls begin their activities on social media, the cyber warrior hackers first provide hacked information to Wikileaks," which constitutes "a place to spread intelligence information through an outside organization" (Prier, 2017, 68). Troll accounts could subsequently point to documents on WikiLeaks and provide links to documents to encourage traffic to the site. Experts have observed that this process resembles the ways in which Soviet espionage officers would leverage third parties such as universities to inadvertently serve as corroborating third parties and thereby reinforce the apparent credibility of a piece of disinformation.

WikiLeaks' most well-known informants have enjoyed varying fortunes. Manning served seven years of a thirty-five-year sentence for espionage and computer fraud in connection with the information exfiltrated to WikiLeaks, and the remainder of the sentence was commuted by President Barack Obama three days before he left office in January 2017. Snowden found ongoing temporary sanctuary in Russia starting in 2013, although Russian officials declared that cessation of his activities were a precondition to a permanent asylum status.

Different sources provide conflicting visions even regarding the impact that the site's many purported leaked documents have exerted. Since the released documents reportedly constitute items that are averred to be classified documents, examination and distribution of the documents would involve unapproved access to classified material. In the midst of WikiLeaks and other disclosures, and controversies about negligence by senior officials such as former Secretary of State Hillary Clinton and former Army General and Central Intelligence Agency chief David Petraeus, even achieving consensus about the impact of unauthorized use and release of documents has proven a polarizing and politicized issue.

Hacks of the Democratic National Committee coincided with the 2016 election campaign season and are strongly believed to have been the work of Russian cyberespionage specialists. The materials were released to the global public via WikiLeaks, lending credibility to the theory that Russia's intelligence apparatus utilizes WikiLeaks in order to publicize documents that it intends to leverage for strategic purposes. Supporters of then-candidate Hillary Clinton have insisted after her election defeat in 2016 was primarily the result of these actions. Forensics of Russian hacking and trolling activities that reportedly leveraged Russian-provided WikiLeaks materials indicates that the overarching Russian objective was to sow discord and dissention so as to detract from national unity. The rationale for promoting discord might have been that distracting the United States from international commitments, such as its leading role in the North Atlantic Treaty Organization, would facilitate other moves to intimidate other Alliance members and destabilize the organization as a whole. However, the relationship between hacking efforts and an already controversial 2016 election have contributed to a politically charged and polarized dynamic in the United States.

Although WikiLeaks has frequently and most (in)famously targeted U.S. government documents, other countries have also been found in the crosshairs. French socialist leader (and later president) Emmanuel Macron's campaign suffered the exfiltration and release via WikiLeaks of 9 gigabytes of campaign-related e-mails in May 2017.

Parties interested in the release and use of documents have dismissed governmental assertions that such use is illicit, and they have sometimes further rejected suggestions that improperly disclosed material may be useful to various state or nonstate adversaries of the United States. Often, counterarguments rest on the counterallegation of rampant overclassification by some agencies in an effort to emphasize the relevance of otherwise banal documents. Public interrogation of the topic involves a counterproductive degree of tautology: overclassification cannot be confirmed by outside parties who are ignorant of security concerns that may be discussed in classified materials, but gaining the information necessary for a realistic examination of documents would logically presuppose access to them.

Some have suggested that governmental policy is failing to keep pace with technological developments and that inadequate options and protections exist for individuals who seek to work "within channels" to propose authorization for information disclosures. According to this argument, faulty procedures and protections stymie efforts to promote authorized disclosure and thereby tempt would-be whistleblowers to redefine their remaining alternatives as being either to maintain secrecy or to disclose illicitly. To the extent that it points to flaws in the existing structure as the main cause of illicit disclosure, this outlook is similar to arguments citing overclassification in government agencies.

Others in government have written that one of the longer-term results of WikiLeaks-style disclosures is likely to entail a reluctance on the part of officials, particularly in diplomatic positions, to share information with policy makers in a free and complete way, aware that their reports may be compromised and their candor used against their country and against themselves personally.

For the past decade, WikiLeaks has intermittently dominated international headlines, and the information that it has released has given rise to extensive debate about how information breaches impact national security and about the nature of classification and information sharing. The most famous inflection points for its breaches followed the Manning exfiltrations in 2010, the Snowden exfiltrations in 2013, and the hacking activities linked to Russian intelligence in 2016. As an organization, WikiLeaks claimed to act in an effort to create governmental transparency. However, the effects of its actions are the subject of considerable debate, in part because it has appeared especially eager to publicize the secrets of the world's more open and democratic nations and to do so in ways that may assist autocratic nations as well as undemocratically inclined nonstate actors. Analysts have noted also that WikiLeaks has potentially served as a proof of concept for other brands of illicit information dissemination in the coming years of the 21st century.

Nicholas Michael Sambaluk

See also: Anonymous; Assange, Julian; Manning, Chelsea/Manning, Bradley; Snowden, Edward

Further Reading

Bernard, Gregory M. *Whistleblowing in a WikiLeaks World: A Model for Responsible Disclosure in Homeland Security.* Thesis. Monterey, CA: Naval Postgraduate School, 2012.

Elsea, Jennifer K. *Criminal Prohibitions on the Publication of Classified Defense Information.* Washington, D.C.: Congressional Research Service, 2013.

Miller, Bowman H. "The Death of Secrecy: Need to Know . . . with Whom to Share." *Studies in Intelligence: Journal of the American Intelligence Professional* 55, no. 3 (September 2011), 13–18.

O'Loughlin, John. "The Perils of Self-Censorship in Academic Research in a WikiLeaks World." *Journal of Global Security Studies* 1, no. 4 (November 2016), 337–345.

Pompeo, Mike. "Director Pompeo Delivers Remarks at CSIS." *CIA News & Information,* April 13, 2017, https://www.cia.gov/news-information/speeches-testimony/2017-speeches-testimony/pompeo-delivers-remarks-at-csis.html.

Prier, Jarred. "Commanding the Trend: Social Media as Information Warfare." *Strategic Studies Quarterly* 11, no. 4 (Winter 2017), 50–85.

Starcovic, Philip J. *Using Voice Over Internet Protocol to Create True End-to-End Security.* Thesis. Monterey, CA: Naval Postgraduate School, 2011.

Stone, Corin R. *Statement of Record before the Senate Homeland Security and Governmental Affairs Committee: "Information Sharing in the Era of WikiLeaks: Balancing Security and Collaboration."* Washington, D.C.: Office of the Director of National Intelligence, 2011.

YouTube, Use of

Application of the predominant Internet video website for purposes related to military conflict. Created in February 2005 and acquired by the search engine company Google a year and a half later, YouTube is a California-based video sharing website that by 2018 stands as the second-most popular website on the Internet. A diverse array of material is added, from vlogs to videos meant for entertainment to informational pieces and beyond. By 2017 uploads amounted to 400 new hours of content per minute. The site collects lucrative advertising revenue, a portion of which it may share with the hosts of particularly often-trafficked pages. YouTube Red is a subscription service unveiled in late 2015 as an alternative for users to pay YouTube in order to avoid advertisements, and separately some pages require subscription fees. Viewership stands at the core of YouTube's business model, and viewership is in large part sustained by the continual influx of new content. Unauthorized postings of copyrighted materials and identified examples of harassment or extremism (as defined by YouTube) are among the videos liable for removal.

A panoply of videos relate in some way to military affairs, from videos of TED Talks discussing military robots or the significance of the World War II Norden bombsight, to uploads of military-themed documentary television programs, to videos of historic or modern battle footage, to vlogs by veterans, to channels discussing different kinds of weapons. Several think tanks as well as professional military education institutions such as the Army War College and Naval War College possess YouTube channels hosting video recordings of talks about historic and current events at an unclassified level.

Video content commands a particular visceral power and therefore is a potentially valuable avenue for messaging that is meant to promote political views, including in the context of wars. YouTube has largely sought to avoid provoking tensions that might adversely affect viewership and has taken steps to encourage viewership among the widest possible swath of the global population.

As a result, YouTube has designed its interface so that it is available in seventy-six language versions. YouTube has also repeatedly acted to ameliorate the concerns of various political and state entities, most often by locally censoring particular content. For example, a national version of YouTube for Turkey is subject to Turkish law regulating content and has curtailed local availability of videos that Turkish authorities deem insulting to that country's founder, Mustafa Kemal Ataturk. On other occasions, countries such as Libya, Morocco, and Turkmenistan have blocked access to YouTube within their countries. The People's Republic of China (PRC) blocked the site for five months starting in late 2007 and then began an ongoing block of YouTube starting in March 2009. PRC's actions may fit within a larger context of that country asserting Internet sovereignty by blocking sources of outside influences that national authorities mistrust and instead replacing those sites with Chinese alternatives.

In February 2008, Pakistan's government unilaterally moved to block local access to YouTube in response to the existence of videos displaying an inflammatory Danish cartoon depiction of the Islamic Prophet Muhammad. Pakistan's effort to block YouTube access spilled beyond the country's own networks and resulted in a global collapse of YouTube for about 2 hours. The potential for such images to trigger violent reaction is unclear, although in 2010 an event ridiculing Muhammad in Garland, Texas, was met by a failed jihadist attack conducted by an ISIS sympathizer.

The purported role of YouTube videos in inciting violence is debated and a source of controversy. An inflammatory low-budget video called *Innocence of Muslims* was met by YouTube blockages in Afghanistan, Bangladesh, Russia, and Sudan in September 2012; this was the video that Barack Obama administration officials in the United States blamed for sparking violence in Egypt and Libya. In the latter country, an organized assault on the U.S. consulate in Benghazi culminated in the first fatality of a U.S. ambassador in four decades. Political controversy ensued within the United States as subsequent evidence indicated that the assault on the consular compound had been planned in advance of the video's posting. It was then suggested that administration officials had disingenuously associated the attack with the YouTube video in order to draw attention away from the inadequate security posture in Benghazi and to politicize speech that the administration saw as dangerous anti-Islamic radicalism.

Militants have recognized how videos can be used advantageously. One way militants have shown interest in using videos is as how-to instructional pieces on topics such as fabricating bombs and other explosive devices; intended for the eyes of dispersed sympathizers and recruits, these videos are meant to remotely facilitate attacks. Another use for videos is to display the results of ambushes and bombings, frequently accompanied by religious songs with jihadist political themes, called nasheen, to advertise the movement's military exploits.

Videos of both types are actively removed from mainstream social media websites, and YouTube acted in late 2010 to curtail such videos from its website. In November 2010, it removed some of the videos of Anwar al-Awlaki, a U.S. national known for his aggressive recruitment and motivational addresses on behalf of jihadist causes until he was killed by munitions deployed from a U.S. remotely piloted aircraft the following September. In December 2010 YouTube added a feature by which users could flag videos as carrying a terrorist message, and flagged videos could then be examined and potentially removed from the website. Reports assert that YouTube may have begun cooperating with governmental surveillance programs around this time.

School shootings in 2018, occurring as part of an evident trend dating to the Columbine High School massacre of April 1999, prompted several societal shifts in the tenor of the gun debate in the United States, and YouTube announced upcoming changes in policy regarding videos involving the sale of "certain firearms accessories" or "instructions" related to the modification of firearms to permit the simulation of automatic fire (YouTube), since automatic weapons have been generally prohibited in the United States since 1968. Concern among gun rights groups for the fate of historically oriented channels dealing with military history seemed to ease during the late spring of 2018 when these policies did not immediately result in wholesale elimination of military history channels from YouTube, but the long-term impact of this policy, including any additional impact on videos on modern arms that might be of interest to nonstate actors, remains to be seen.

Nicholas Michael Sambaluk

See also: China, Blocking of Social Media by; TroopTube; Trusted Flagger (YouTube); U.S. Military Use of Social Media; War on Terrorism and Social Media; Web 2.0

Further Reading

Anden-Papadopoulos, Kari. "US Soldiers Imaging the Iraq War on YouTube." *Popular Communication* 7, no. 1 (January 2009), 17–27.

Andre, Virginie. "The Janus Face of New Media Propaganda: The Case of Patani Neojihadist YouTube Warfare and Its Islamophobic Effect on Cyber-Actors." *Islam and Christian-Muslim Relations* 25, no. 3 (January 2014), 333–356.

Mirrlees, Tanner. "The Canadian Armed Forces 'YouTube War': A Cross-Border Military-Social Media Complex." *Global Media Journal—Canadian Edition* 8, no. 1 (2015), 71–93.

Siegel, Alexandra, and Joshua A. Tucker. "The Islamic State's Information Warfare: Measuring the Success of ISIS's Online Strategy." *Journal of Language and Politics* 17, no. 2 (2018), 258–280.

YouTube. "Policies on Content Featuring Firearms." *YouTube Help,* https://support.google.com/youtube/answer/7667605?hl=en.

Zello App

Developed as "Loudtalks" by Alexey Gavrilov, and purchased by Bill Moore of Zello in 2012, an app that allows users of any smartphone to use their device as a walkie-talkie on 2G, 3G, 4G, and GPRS/EDGE networks. This facility to connect users one-on-one or in groups of common interest in the style of a party line or police scanner to share voice, text, and photos, came to prominence in the United States in Hurricanes Harvey and Irma as a way to coordinate disaster response. The volunteer Cajun Navy and other rescue groups used it to monitor, locate, and respond to trapped people, while a group of midwives used it to offer medical advice to those trapped far away from professional help. Official response channels like 911 and FEMA were quickly overwhelmed, prompting people to crowdsource information, creating maps of inaccessible areas and even organizing social pressure on pastor Joel Osteen to open his Houston megachurch to people fleeing the flooding. Disturbingly, emergency management studies found that people were more likely to believe and respond more readily to people on a civilian network than official warnings. In 2017, Latvians used Zello to coordinate the nationwide search for Ivan Berladins, a missing five-year-old.

Protesters worldwide have also been using Zello since 2012. In the Ukraine, Venezuela, Hong Kong, and Turkey, activists with the app had similarly crowdsourced maps that allowed them to set up barricades, evade the police, and communicate their political message to one another and the general public. In Venezuela, the Zello group reached 450,000 people, the largest utilization the company knows about to date. Russia responded to Zello and other social media apps by passing a 2014 law requiring "information distribution brokers" to keep records of messages and share them with law enforcement, which Zello refused to do, pointing out that only the most recent message is retained, which is erased as soon as a new one comes in. When governments have attempted to block the app, the company itself assists in recoding the app to continue functioning, as they did during Venezuelan protests in 2014. Unfortunately, these features that make Zello useful for activists also work in the hands of terrorists, including the Islamic State of Iraq and Syria (ISIS) and the planners of the April 2017 Stockholm attacks and the May 2017 bombing of Ariana Grande's concert in Manchester, United Kingdom.

Margaret Sankey

See also: FrontlineSMS; Surespot App; Telegram App; War on Terrorism and Social Media; Web 2.0; WhatsApp

Further Reading

Al-Akkad, Amro, and Christian Raffelsberger. *How Do I Get This App? A Discourse on Distributing Mobile Applications Despite Disrupted Infrastructure.* State College, PA: International ISCRAM Conference, University Park, 2014.

King, Larry J. *Social Media Use during Natural Disasters: An Analysis of Social Media Usage during Hurricane Harvey and Irma.* Orlando: International Crisis and Risk Communications Conference, 2018.

Social Media Bibliography

Atwan, Abdel Bari. *Islamic State: The Digital Caliphate*. Oakland: University of California Press, 2015.

Bergmann, Max, and Carolyn Kenney. *War by Other Means: Russian Active Measures and the Weaponization of Information*. Washington, D.C.: Center for American Progress, 2017.

Bradshaw, Samantha, and Philip N. Howard. *Troops, Trolls and Troublemakers: A Global Inventory of Organized Social Media Manipulation*. Oxford, UK: Computational Propaganda Research Project, 2017.

Brunty, Joshua, and Katherine Helenek. *Social Media Investigation for Law Enforcement*. New York: Routledge, 2013.

Campen, Alan D., and Douglas H. Dearth, eds. *Cyberwar 3.0: Human Factors in Information Operations and Future Conflict*. Fairfax, VA: AFCEA, 2000.

Carafano, James J. *Wiki at War: Conflict in a Socially Networked World.* College Station: Texas A&M University Press, 2012.

Carter, Ashton. "News Transcript: Stennis Troop Talk." April 15, 2016, https://www.defense.gov/News/Transcripts/Transcript-View/Article/722859/stennis-troop-talk.

Cross, Michael. *Social Media Security: Leveraging Social Networking While Mitigating Risk*. Waltham, MA: Syngress, 2014.

DiStefano, Adam T. *The Weaponization of Social Media: A Guide to Protecting Your Brand from Cyber Criminals Using Social Media*. Thesis. Utica, NY: Utica College, 2017.

Erbschloe, Michael. *Social Media Warfare: Equal Weapons for All*. Boca Raton, FL: CRC, 2017.

Erbschloe, Michael. *Extremist Propaganda in Social Media: A Threat to Homeland Security*. Boca Raton, FL: CRC, 2019.

Forbidden Feeds: Government Controls on Social Media in China. New York: PEN America, 2018.

Gertz, Bill. *iWar: War and Peace in the Information Age*. New York: Threshold, 2017.

Gupta, Ravi, and Hugh Brooks. *Using Social Media for Global Security*. New York: Wiley, 2013.

Hellman, Maria, Eva-Karin Olsson, and Charlotte Wagnsson. "EU Armed Forces' Use of Social Media in Areas of Deployment." *Media and Communication* 4, no. 1 (2016), 51–62.

Klimburg, Alexander. *The Darkening Web: The War for Cyberspace*. New York: Penguin, 2017.

Knopf, Christina M., and Eric J. Ziegelmayer. "Fourth Generation Warfare and the US Military's Social Media Strategy." *ASPJ Africa & Francophonie* 3, no. 4 (2012), 3–22.

Lehrman, Yosef. "The Weakest Link: The Risks Associated with Social Networking Websites." *Journal of Strategic Security* 3, no. 2 (Summer 2010), 63–72.

Marcellino, William, et al. *Monitoring Social Media: Lessons for Future Department of Defense Social Media Analysis in Support of Information Operations*. Santa Monica, CA: RAND, 2017.

Mitnick, Kevin. *The Art of Invisibility: The World's Most Famous Hacker Teaches You How to Be Safe in the Age of Big Brother and Big Data*. New York: Little, Brown, 2017.

Montagnese, Alfonso. *Impact of Social Media on National Security*. Rome: Center for Military Strategic Study, 2012.

Olson, Parmy. *We Are Anonymous: Inside the Hacker World of LulzSec, Anonymous, and the Global Cyber Insurgency*. New York: Little, Brown, 2012.

Patrikarakos, David. *War in 140 Characters*. New York: Basic Books, 2017.

Patterson, George. *Review of Social Media and Defence*. Canberra, Australia: Australian Defense Forces, 2011.

Paul, Christopher, and Miriam Matthews. *The Russian "Firehose of Falsehood" Propaganda Model: Why It Might Work and Options to Counter It*. Santa Monica, CA: RAND, 2016.

Petry, Joshua W. *Let the Revolution Begin, 140 Characters at a Time: Social Media and Unconventional Warfare*. Maxwell AFB, AL: School of Advanced Air and Space Studies, 2015.

Prier, Jarred. "Commanding the Trend: Social Media as Information Warfare." *Strategic Studies Quarterly* 12, no. 4 (Winter 2017), 50–85.

Rid, Thomas. *Cyber War Will Not Take Place*. Oxford: Oxford University Press, 2013.

Sambaluk, Nicholas Michael, ed. *Paths of Innovation in Warfare: From the Twelfth Century to the Present*. Lanham, MD: Lexington, 2018.

Sanger, David E. *The Perfect Weapon: War, Sabotage, and Fear in the Cyber Age*. New York: Crown, 2017.

Schneider, Bruce. *Data and Goliath: The Hidden Battles to Collect Your Data and Control Your World*. New York: Norton, 2015.

Shehabat, Ahmad, Teodor Mitew, and Yahia Alzoubi. "Encrypted Jihad: Investigating the Role of Telegram App in Lone Wolf Attacks in the West." *Journal of Strategic Security* 10, no. 3 (2017), 27–53.

Silvestri, Lisa Ellen. *Friended at the Front: Social Media in the American War Zone*. Lawrence: University Press of Kansas, 2015.

Singer, P. W., and Emerson T. Brooking. *LikeWar: The Weaponization of Social Media*. Boston: Houghton Mifflin Harcourt, 2018.

Soldatov, Andrei, and Irina Borogan. *The Red Web: The Kremlin's War on the Internet*. New York: Public Affairs, 2015.

Solomon, Scott E. *Social Media: The Fastest Growing Vulnerability to the Air Force Mission*. Maxwell AFB, AL: Air University Press, 2017.

Tsang, Flavia, et al. *The Impact of Information and Communication Technologies in the Middle East and North Africa*. Santa Monica, CA: RAND, 2011.

Ullah, Haroon K. *Digital World War: Islamists, Extremists, and the Fight for Cyber Supremacy*. New Haven, CT: Yale, 2017.

United States Congress. *China's Censorship of the Internet and Social Media: The Human Toll and Trade Impact*. Washington, D.C.: Government Printing Office, 2011.

Van Niekerk, Brett, and Manoj Maharaj. "Social Media and Information Conflict." *International Journal of Communications* 7 (2013), 1162–1184.

Van Till, Steve. *The Five Technological Forces Disrupting Security: How Cloud, Social, Mobile, Big Data and IoT Are Transforming Physical Security in the Digital Age*. Cambridge, MA: Butterworth-Heinemann, 2018.

Waldman, Ari Ezra. "Privacy, Sharing, and Trust: The Facebook Study." *Case Western Reserve Law Review* 67, no. 1 (2016), 193–233.

Wardle, Claire, and Hossein Derakhshan. *Information Disorder: Toward an Interdisciplinary Framework for Research and Policy Making*. Strasbourg, France: Council of Europe, 2017.

Weimann, Gabriel, and Bruce Hoffman. *Terrorism in Cyberspace: The Next Generation*. New York: Columbia University Press, 2015.

Section 3

Technology

INTRODUCTION

Although digital technologies and methods have garnered frequent and legitimate attention for the impact that cyberwar and the leveraging of social media can exert, a vast array of physical technologies continue to play extremely important roles in warfare into the 21st century. War continues to be inherently about pursuing policy goals and compelling an enemy's acceptance via force. Physical technologies remain entirely relevant.

Prominent scholars of military affairs have observed that although an "important factor in affecting the results of conflict" (Black, 2013, 264), technology's effectiveness depends also on factors such as tactics and strategy and tactics, which themselves adjust because of efforts "to exploit the new weapons capabilities" (Dupuy, 1984, 287) or alternatively to evade the impact of an adversary's weapon. MacGregor Knox and Williamson Murray add that "revolutions in military affairs *always* [italics added] occur within the context of politics and strategy," while technology itself typically "has functioned . . . as a catalyst" (Knox and Murray, 2001, 180, 192).

It needs to be remembered that military hardware can be quite rugged and also often expensive, and this means that the introduction of new weapons does not automatically cause the disappearance of examples of older systems if they retain some degree of effectiveness. Two decades into the 21st century, many of the weapons systems that were fielded in the 1991 Gulf War continue to appear in combat, sometimes in upgraded forms, and much of this hardware remains potent on the battlefield. Anecdotes from Afghanistan even report the use of World War II–era Soviet-made submachine guns and British rifles, and forces fighting in the Syrian civil war are occasionally seen carrying first-generation German assault rifles alongside infrequently appearing bolt action rifles. Not all 21st-century weapons were produced after the year 2000. In fact, history shows a pattern in which leading military powers discard or export surplus materials, which continue to find use for decades. This pattern endures in the 21st century. Therefore, "the technologies of the past are not simply behind us. They are instead a part of the world that we inhabit" (Sambaluk, 2017, "Offsets").

Consequently, two notable trends deserve attention, because understanding them explains the selection of many technology topics examined in this volume. The first trend is the dispersion of once-emergent technologies, particularly of technologies that had originally been the preserve of superpowers during the Cold War. The second is a strategic decision in the United States responding to the impact of this diffusion.

Diffusion of arms has been the norm rather than the exception, but until the 20th century the United States was rarely a first-rate power with first-rate weapons. However, the two world wars catalyzed geopolitical transitions with important consequences for military technology and for the U.S. relationship to arms development. Britain, which began the century as the global empire par excellence, was thrown into fiscal decline and strategic retraction, and this trend overlapped with the emergence of the dictatorial regimes that triggered the Cold War. With its industrial strength actually increased by World War II, the United States in the late 1940s found itself in an unparalleled superpower status, in a unique military position as the world's only nuclear power, and simultaneously invested in increasing global strategic commitments.

Throughout the second half of the 20th century, the United States maintained a military preeminence that it had first ever entertained only in 1945. Crucial strike technologies, first an array of nuclear weapons and then increasingly an assortment of increasingly precise weapon delivery technologies, exemplified this high-tech superiority. Fundamentally, nuclear weapons made possible unprecedented destruction of a target, and precision guidance systems if directed against known targets facilitated unparalleled accuracy in weapons delivery. In both cases, the United States leveraged these technological facilitators to project military power as a means of achieving strategic deterrence against its adversaries (especially the Soviet Union), taking aggressive steps unacceptable to democratic powers during the Cold War. Investment progressed in several related technology areas, such as aviation, submarine technology, and computerization, since capabilities in these areas enabled the striking power of nuclear weapons and precision technology.

In several other areas, U.S. weapon technology was equaled or surpassed by rival systems during much of the Cold War era. Until the development of the Abrams tank in 1980, U.S. tanks left much to be desired and did not exceed the value of rival systems from the Federal Republic of Germany, Israel, and the Soviet Union. The semiautomatic Garand rifle, superlative in World War II, marked the only time that U.S. rifle technology was not either inferior to rival types or emulative of foreign designs until the reliable refinement of the M-16 family starting in the mid-1960s. A wave of big-ticket weapon technologies, such as the Patriot missile-defense system, Bradley Fighting Vehicle, Abrams tank, Apache attack helicopter, and Blackhawk utility helicopter, first appeared in the late 1970s and early 1980s. These helped redress longstanding qualitative deficits in conventional combat capabilities. Deployed just after the Cold War with incredible success against Iraqi forces occupying Kuwait in 1991, this generation of weapons systems vividly outperformed and destroyed Soviet-made counterparts. The millennium ended with a global sense of the dominant position of U.S. military technology in all aspects of warfare, from ground vehicles like the Abrams to aircraft like the F-117 Nighthawk stealth attack plane to the global positioning system (GPS) and other satellites that connected users to information.

Throughout the Cold War, other nations had intermittently demonstrated or implied technologies that rivaled or seemingly surpassed contemporary U.S. technology. News of the first Soviet nuclear test, the Soviets' achievement of the first space satellite, the suspected Soviet development of nuclear-powered aircraft, and hints of Soviet research into potentially militarized activities in space caused political pressure within the United States to redouble its research and development into new technology, including military hardware.

During the Cold War, U.S. planners had constantly eyed Soviet developments and urged development of ever-more-capable technologies with the hope that these could be fielded in time to maintain a qualitative margin over the expected leaps in Soviet technical abilities. Advocates of U.S. military technological development in the 21st century frequently assert that a different dynamic has emerged, since now many countries deploy some weapons only a few years behind U.S. models. The dynamic appeared also in fighting between Israeli ground forces and Hamas fighters in 2006, when the latter group was reputedly able to order up-to-date weapons and effective body armor online, erasing much of the technological edge Israeli troops had earlier been presumed to have. This dynamic is less new than it is unfamiliar to planners unversed in history. European technology races included the multinational rifle revolutions in the 1880s and 1890s and the dreadnaught construction race of the early 1900s, though U.S. involvement was minimal and peripheral.

Cold War rivalry between the United States and the Soviet Union also precipitated a flood of arms, ranging from semiobsolete surplus to nearly modern hardware, from the two superpowers to their presumptive allies both in western and eastern Europe but also across what was then known as the Third World. The collapse of the Soviet system by 1991 brought a diffusion of Soviet hardware across former Soviet republics and meanwhile called into question the maintenance of a large U.S. military. Diffusion of the underlying scientific and technological principals behind Cold War weapons means that "legacy" style systems continue in use by soldiers of an ever-widening diversity of nations.

For example, North Korea's early missile development in the 1980s was derivative of Soviet types, which were themselves impacted at first by German vengeance weapons (Chertok, 2005, 237). Analysis in the early 2000s noted that the construction of sheltered and hidden military facilities caused estimates of its capabilities "ambiguous, incomplete, or erroneous" (Pinkston, 2008, 44), although "North Korea ha[d] already shown that the United States missile defense system is not a deterrent to [its] development" of increasingly long-range rockets (Gipson, 2007, 3). The following decade witnessed maturation of North Korean technology, including missile tests suspected to have intermediate range capability sufficient to target U.S. bases, and also a reported expansion of North Korea's nuclear stockpile.

Concern about North Korean nuclear and missile capabilities can be seen in a larger context in which nuclear weapons command enduring relevance despite having first been pioneered during the Cold War. Despite first appearing on battlefields a century ago, chemical weapons sporadically appear in use and chemical weapons stockpiles remain salient issues in destabilized conflict zones such as Syria. Controversy surrounding the 2003 Iraq War and the Joint Comprehensive Plan of Action's arrangement in 2015 and cancellation in 2018 have arguably distracted attention away from an apparent theme of concern about the proliferation of nuclear weapons in the early 21st century.

Diffusion and technology transfer play complex roles in weapons development beyond the nuclear context. Many of the iconic U.S. military systems of the 21st century stand on the shoulders of precursor technologies. U.S. unmanned aerial vehicles (UAVs), for example, range from systems like the Hellfire missile-armed MQ-9 Reaper that is able to study enemy movements "for as long as 10 hours before engaging" (Blom, 2010, 118) to the hand-launched RQ-11 Raven used for tactical video surveillance. However, the panoply of U.S. UAVs draws on the Vietnam-era Ryan 147 Lightning Bug, which "flew over 34,000 surveillance missions across Southeast Asia" (Rowley, 2017, 6). Nor is technology transfer solely one-directional entropy from the U.S. perspective: sale of Lightning Bugs to Israel has been associated with Israeli interest in UAVs, and their development and sale of the Pioneer to the U.S. Navy led to the Pioneer's use for correcting naval gunfire during the 1991 Gulf War (Blom, 2010, 71–72, 88–89).

Research avenues that have been explored and dismissed in the past can sometimes provide technical foundations for later research and other times provide a political shield for subsequent examination by other nations. Cold War–era U.S. programs that "are decades out of date . . . or have been cancelled" (Saalman, 2014, 7) have nonetheless been cited by spokespeople of the People's Republic of China as a justification for their own technological studies in the 21st century. In some cases, stymied research can be later brought to fruition because of other seemingly unrelated breakthroughs in materials or techniques, but elsewhere technological hurdles can remain thorny challenges for decades. The outcome of such reinvigorated research efforts may be mixed and is difficult to predict. "In the absence of realistic expectations, innovation efforts can be dealt destructive . . . blows," either by excessive expectations that lead to disappointment or to underestimations that "stifle[s] interest and investment" (Sambaluk, 2018, 287–288).

Since U.S. strategy throughout the Cold War was founded on the ability to project power and to compensate for the larger numbers of a potential Soviet enemy, the prospect of a weapons technology breakthrough by any adversarial power triggers alarm for many analysts and planners. Reportedly of commensurate concern to U.S. planners is the rise of adversaries with technologies that nearly match the capabilities of U.S. counterparts. Analysts suggest that sufficiently modern systems, combined with either superior numbers or more regional sets of commitments and priorities, could thwart U.S. power projection and deterrence strength in a given area. Experts have pointed to several tech categories, such as "stalker satellites" (Chow, 2017) as well as antiaccess/area denial weapons, which could be used essentially to interrupt or invalidate the technological prowess of a superpower. Experts have observed that rising regional powers can pose credible challenges to the United States even with technologies that fail to match its power projection capabilities, and a similar point that the People's Republic of China could "pos[e] problems without catching up" was made at the millennium's outset (Christensen, 2001).

Jeremy Black's observation that "technological change is designed in response to perceived needs" (Black, 2013, 246) remains true, and the second trend is itself prompted in response to wariness of the strategic implications of the trend of diffusion of Cold War technology. Selecting iconic technologies and categorizing nuclear predominance as a "First Offset" and the combination of latter Cold War technologies epitomized by stealth and precision guidance as a "Second Offset," planners have urgently declared interest in establishing a "Third Offset."

Research into next-generation projectiles encompasses a broad assortment of approaches and technologies. Much of this concerns improved performance for bombardment and air defense. Despite proving an elusive goal for three decades, electromagnetic railguns continue to inspire interest. In this system, electromagnetism propels a warhead at approximately Mach 6–7; moving three times the speed of a conventional shell, a comparable railgun projectile carries far greater kinetic force. On warships, railguns would draw on the ship's existing power systems and would eliminate the space and risk associated with propellant explosives being aboard ship. The need to rearm warships with complex new weapons systems is a short-term disadvantage of railguns. This prompted interest in hypervelocity projectiles that could be fired from guns currently employed on Navy cruisers and destroyers and whose performance is inferior to railguns but exceeds conventional powder-propelled shells (O'Rourke, 2016, 15).

Several nations are believed to be working on hypersonic flight research to pioneer new missiles. While hypersonic glide vehicles glide on the atmosphere at extreme speeds, hypersonic cruise vehicles travel with the internal propulsion of a high-speed rocket motor or jet engine. Analysts suggest that hypersonic missile technology could be designed for bombardment or for air defense purposes, and that in the former role their extreme speed and unpredictable flight paths will "substantially increase the threat for nations with otherwise effective missile defense systems" (Speier et al., 2017, 10).

Contemporary research into lasers notably addresses anti-ship missile threats, and these reportedly may one day include hypersonic missiles. Although laser power, "atmospheric disturbances" (Speier et al., 2017, 14), and line-of-sight factors mitigate the potential performance of lasers and form "an upper limit on the ability of an individual laser to deal with saturation attacks" (O'Rourke, 2016, 41), analysts note the potentially fast engagement times and ability to engage maneuvering targets as advantages for future laser interception systems. Experimental U.S. deployment of the Laser Weapon System (LaWS) starting in 2014 met with success.

Other research concerns the health impact and the combat effectiveness of smaller caliber ammunition. Efforts to reduce the environmental and health effects of corrosive and other harmful material components at storage facilities and at training sites are directed toward creation of "green" ammunition, while research into guided small arms aims to improve the probability of hitting targets and of avoiding collateral damage.

Analysts anticipate dramatic developments in many technological fields during the present century, and alongside advocates of "offset" technologies are specialists calling for U.S. research in order to establish precedent in emerging high-tech landscapes. Autonomous technologies, potentially deployed in swarms, hint at the interconnections between fields like computerization and the digital realm with various physical technologies. Advocates of research argue that "the US has an opportunity to help influence the forms autonomy will take and the purposes it will serve," allowing work toward the Third Offset to "also promote US national interests consistent with its principles and values" (Lewis, 2017, 42). This perspective of computer research into autonomy fits interestingly with assertions that U.S. decisions about deployment of cyberweapons potentially sets precedent for later adopters whose technology runs slightly behind. It conflicts with the advice of arms control advocates who, writing about hypersonic missiles, argue that "there is probably less than a decade available to substantially hinder the potential proliferation of hypersonic missiles and associated technologies" (Speier et al., 47) and that while "unilateral actions against missile proliferation will have limited effectiveness" (Speier et al., 2017, 36), U.S. leadership in multinational efforts could yield greater effect.

Advocates of a Third Offset thus agree on the need for innovation and for U.S. leadership in determining paths of future technology development, but different analyses prioritize different technology areas and sometimes even disagree about whether U.S. efforts should focus on development or abstention from certain technology categories.

Technology crossover between military and civilian applications also helps shape the technological landscape in the early 21st century. Many Third Offset advocates assert the urgent need to adopt products from the more dynamic civilian realm, whose research and development in some key areas outpaces that of government institutions. This point is sometimes associated with additive manufacturing or "3D printing." Analysts have suggested that additive manufacturing already exhibits a pattern of increasing flexibility and accessibility and that although "3D printing has been compressed into a much shorter time period, it has followed practically identical phases" (Veronneau et al., 2017, 10). Additive manufacturing landmarks already include fabrication of houses, food, and mechanism components for a range of items including for firearms components; potential additive manufacturing avenues with military applications also include sophisticated body armor.

One recurring point involves the frustration attendant to current acquisition structures and cycles. A system designed "to ensure high quality" so that systems can "last many decades" and remain effective, while also providing for accountability about contracting procedures, has been described as "increasingly incompatible with the current environment" (Lewis, 2017, 10). Rapid prototyping and experimentation may well require diverging from "standard requirements processes" (Van Atta et al., 2016, 65), but studies indicate that governmental models are not uniquely inefficient with resources (MacMillan). Scholars have also noted that "the military have been much more important in the development of technology than the civilian innovation-centric picture has allowed" (Edgerton, 2007, 158).

Slippage and deferment in scientific and technological research funding are blamed for a recurring pattern of delays in futuristic aircraft concepts, until they are ultimately overtaken by current needs. Jeremy Black points to "the customary danger of present military and political needs crowding out future options" (Black,

2013, 229). The cycle renews with the emergence of another new generation of futuristic concepts to be doomed in their turn, until "today's Air Force is flying and applying band-aids to airplanes with 1970s-era aero, structural and propulsion technology" (Ernhard, 2009, 50).

Although historical study shows that the Cold War was not as simple or straightforward as is sometimes suggested, the diversity of actuated and potential threats in the 21st century does force planners and technologists to prepare to meet a range of challenges, either through an array of weapons systems or through a smaller set of more versatile platforms. When adversaries pose underanticipated challenges, a military is often left with a hammer and searching for a nail. Analysts have noted that overwatch missions in Iraq were flown by available warplanes. Cold War–era preparation against an adversary with fourth-generation fighters meant that those were F-16s and F-15s, flying missions that a specialized armed reconnaissance aircraft "able to fly from more airfields and provide more coverage in benign environments" (Ernhard, 2009, 82) could accomplish the same task for a tenth the cost.

Planners have to remember, however, that an adversary with fourth-generation fighters could quickly destroy planes designed to be efficient in highly permissive environments where the enemy lacks air defenses. Furthermore, potential adversaries will study a major power's acquisition decisions and adjust in order to pose a threat that is effective against the resulting platforms and doctrines. This is precisely what Chinese strategists called for in the mid-1990s after observing U.S. success in the 1991 Gulf War. These considerations may have contributed to the development of the F-35 Joint Strike Fighter and help account for some of the advocacy outside the Air Force for multirole aircraft historically. Cost is another factor that can drive voices within the Air Force to add multirole traits to a high-priority system; this can be seen historically, and the impetus can be seen in the $89.8 million unit cost of the F-35 (Gertler, 2018, 21).

It has been suggested that the "'revolution in military affairs' is over" because "present and future opponents and allies of the United States know what US forces can do" (Knox and Murray, 2001, 190). Since both modernity and obsolescence are relative, 21st-century conflicts are fought with an array of emerging technologies alongside once-vaunted weapons that are sometimes called legacy systems. Their impact in warfare is in large part a function of how they are used.

FURTHER READING

Black, Jeremy. *War and Technology*. Bloomington, IN: University of Indiana, 2013.

Blom, John David. *Unmanned Aerial Systems: A Historical Perspective*. Ft. Leavenworth, KS: Combat Studies Institute, 2010.

Chertok, Boris. *Rockets and People, Volume 1*. Washington, D.C.: National Aeronautics and Space Administration, 2005.

Chow, Brian G. "Stalkers in Space: Defeating the Threat." *Strategic Studies Quarterly* 11, no. 2 (Summer 2017), 82–116.

Christensen, Thomas J. "Posing Problems without Catching Up: China's Rise and Challenges for US Security Policy." *International Security* 25, no. 4 (Spring 2001), 5–40.

Dupuy, Trevor N. *The Evolution of Weapons and Warfare.* Fairfax, VA: Da Capo, 1984.

Edgerton, David. *The Shock of the Old: Technology and Global History since 1900.* Oxford: Oxford University Press, 2007.

Ernhard, Thomas P. *Strategy for the Long Haul: An Air Force Strategy for the Long Haul.* Washington, D.C.: Center for Strategic and Budgetary Assessments, 2009.

Gertler, Jeremiah. *F-35 Joint Strike Fighter (JSF) Program.* Washington, D.C.: Congressional Research Service, 2018.

Gipson, Issac G. *The Effectiveness of the US Missile Defense Capabilities as a Deterrent to the North Korean Missile Threat.* Thesis. Monterey, CA: Naval Postgraduate School, 2007.

Knox, MacGregor, and Williamson Murray. *The Dynamics of Military Revolution, 1300–2050.* Cambridge: Cambridge University Press, 2001.

Lewis, Larry. *Insights for the Third Offset: Addressing Challenges of Autonomy and Artificial Intelligence in Military Operations.* Arlington, VA: Center for Naval Analysis, 2017.

MacMillan, Ian. "Fighter Jets, Supercars, and Complex Technology." *Strategic Studies Quarterly* 11, no. 4 (Winter 2017), 112–133.

O'Rourke, Ronald. *Navy Lasers, Railgun, and Hypervelocity Projectile: Background and Issues for Congress.* Washington, D.C.: Congressional Research Service, 2016.

Pinkston, Daniel A. *The North Korean Ballistic Missile Program.* Carlisle, PA: Strategic Studies Institute, 2008.

Rowley, Gary D. *Armed Drones and Targeted Killing: Policy Implications for Their Use in Deterring Violent Extremism.* Washington, D.C.: National Defense University Press, 2017.

Saalman, Lora. *Prompt Global Strike: China and the Spear.* Independent Faculty Research, 2014.

Sambaluk, Nicholas Michael. "Chariots and Fire: Part 1, Offsets, Revolutions and History." Air University Television, 2017. https://www.youtube.com/embed/8kBqUw2bURU?rel=0&modestbranding=1.

Sambaluk, Nicholas Michael, ed. *Paths of Innovation: From the Twelfth Century to the Present.* Lanham, MD: Lexington, 2018.

Speier, Richard H., et al. *Hypersonic Missile Nonproliferation: Hindering the Spread of a New Class of Weapons.* Santa Monica, CA: RAND, 2017.

Van Atta, Richard H., et al. *Assessment of Accelerated Acquisition of Defense Programs.* Alexandria, VA: Institute for Defense Analysis, 2016.

Veronneau, Simon, et al. *3D Printing: Downstream Productive Transforming the Supply Chain.* Santa Monica, CA: RAND, 2017.

Additive Manufacturing (3D Printing)

A method of manufacturing where raw material is melted and built up in layers to produce a desired object. Traditional manufacturing methods involve sculpting a block of material until the desired final product is achieved. This process can be slow and wastes a large amount of material. Although research into additive manufacturing dates back to the 1980s, the 3D printing applications grew tremendously from in the early 2010s forward. Unlike previous developments, the target audience for this new wave of technology was hobbyists and "at home" inventors. Any person with the competence to construct a 3D model on a computer could "print" that model from the comfort of their own desktop. Engineers and students could engage in "rapid prototyping," being able to quickly and affordably experiment with physical versions of their projects.

Two main types of 3D printers were developed for the rapid prototyping and hobbyist market. The first involves a mostly open box—the back is closed, and the bottom is the location of the build plate—with an extruder head mounted within it on a three-axis rail system. One of two types of plastic, most often either acrylonitrile butadiene styrene (ABS) or polylactic acid (PLA), are supplied in a spooled form to the extruder head. Each 3D printer comes with its own printing software, which breaks down a 3D model into layers and sets a work path for the extruder head. The second type of 3D printer is based on stereolithography. This method makes use of a laser to harden selected areas of a liquid plastic resin, which is located in a pool at the bottom of the printer. The build plate is located upside down and progressively rises out of the pool as each layer of the model is completed. The benefits of stereolithography are that the printer has fewer moving parts, and the amount of support material needed for overhangs on the part being printed is minimized.

The widespread success of the desktop 3D printer reinvigorated the idea of using additive manufacturing in industry. The result was larger machines that could handle stronger/heavier materials than plastic, such as metal. In 2014, the National Aeronautics and Space Administration (NASA) announced that it was beginning work on a 3D printer that could function in zero-gravity environments. The destination set for this device was the International Space Station (ISS). When launching a vehicle into space, every ounce of weight matters. To complicate the issue further, it is hard to choose what spare parts get sent to the ISS, especially when the decision could mean the difference between life and death. For the same amount of weight, however, the spare parts can be replaced by raw material for a 3D printer. This solution guarantees that the parts astronauts have are the ones they need. In July 2018, Lockheed Martin presented the 3D printed fuel tank caps that it created for its LM 2100 series of satellite systems. The achievement demonstrated both the company's faith in additive manufacturing and the possibilities for the overall future of manufacturing. The fuel tank caps were constructed from titanium, which was cheaper and easier for Lockheed Martin to purchase in the form required for 3D printing. In addition, the dimensions of each cap were huge—46 inches in diameter. Additive manufacturing produced a large-scale, vital component from an industrial-grade material.

Myriad applications in plastics have been complemented by strides in additive manufacturing of organic products. As early as 2015, doctors were using 3D printers to create replica organs of specific patients on which to practice complex surgeries. Only one year later 3D printed body parts began to emerge. Tissue structures—such as noses and ears—have been successfully accepted by a living body. However, the filament used was still plastic-like. By 2017, actual biological filament was starting to be utilized, making the production of full organs possible. Sample cells from a patient are cultured in order to grow more cells, then combined with hydrogels—Jell-O-like cellulose material—which serve as support material. Scans of a patient's organs are taken and used as the design instructions for the printer. This same technique has already been used to produce working ovaries for mice, which have resulted in healthy offspring.

The benefits of additive manufacturing for the future of the military lie in medical care, logistics, and tactical innovation. Following NASA, armed forces could replace their supply chains for spare parts with metal-based 3D printers. If a part is needed for a weapon, piece of equipment, or vehicle, the maintenance department need only print the model for that part and install it.

In the field, forces may experience situations for which they have not been prepared. It may be determined that a special piece of equipment is required, and there now exists the ability to develop and create this solution at the base closest to the operation. What could take years going through the procurement process—in which government makes formal request of a solution, contractors attempt to coordinate with the military customer on exact specifications and incorporate the final product into existing production facilities, yielding delivery of a product—can now happen in the amount of time needed to model and print the tool. The total cost would only be the salary of the military engineers and the materials needed for production. The new tool can be quickly shared with other units by sending the digital part file electronically.

Biological additive manufacturing can benefit injured personnel. Medical facilities at bases of deployment can provide specialized medical treatment using 3D printers by producing patches for organ and skin injuries. The low cost and wide availability of 3D printers also poses a new security challenge, as citizens have begun using them to produce untraceable firearms. In August 2018, eight states sued the federal government to stop the spread of blueprints for these guns.

Elia G. Lichtenstein

See also: Defense Innovation Unit Experimental (DIUx); Defense Technical Information Center (DTIC)

Further Reading

Barnatt, Christopher. *3D Printing: The Next Industrial Revolution.* Self-published, CreateSpace, 2013.

Begley, Leslie D. *Increasing Capabilities and Improving Army Readiness through Additive Manufacturing Technologies.* Carlisle, PA: Army War College, 2017.

Lipson, Hod, and Melba Kurman. *Fabricated: The New World of 3D Printing.* Hoboken, NJ: Wiley, 2013.

Zimmerman, Brock A., and Ellis E. Allen III. *Analysis of the Potential Impact of Additive Manufacturing on Army Logistics.* Monterey, CA: Naval Postgraduate School, 2013.

Anti-Satellite Weapon (ASAT)

Weapons designed to eliminate the functionality of satellites. There are three types of conventional ASATs. The first is a co-orbital ASAT, which is carried into orbit by a space-launched vehicle. The second is a direct ascent ASAT, which is a missile launched from the ground, air, or sea. These missiles destroy the satellite without entering orbit. The third is a directed energy ASAT. These can be alternatively deployed on the ground or in space.

The United States was the first to explore ASAT technology, during the 1950s and 1960s in the context of the Cold War. Program 505 modified Nike-Zeus missiles for use as ASATs. The program was successfully tested in 1963 and became operational shortly afterward. A second ASAT program named Program 437 used a modified Thor IRBM and became operational in 1964. By 1972 both programs had been canceled due to budget cuts. Between 1977 and 1987 the United States developed the ASM-135 missile launched by an F-15 aircraft. In 1997 the United States successfully tested the first directed energy ASAT. The United States used the Mid-Infrared Advanced Chemical Laser (MIRACL) at an aging satellite, MST 13. During the test, the United States discovered that a lower powered laser used to align the MIRACL actually had already blinded the satellite. Although the test showed MIRACL was not needed to disable a satellite, it proved the effectiveness of directed energy ASATs.

During the Cold War the Soviets developed their own ASAT program known as Istrebitel Sputnikov (IS). IS became operational in 1973 and was upgraded to the IS-M system by 1976, to be replaced by the IS-MU in 1991. The systems developed by the Soviets relied on conventional explosives for destroying orbiting targets. In the 1980s the Soviets developed two ASAT programs, the Kontakt air-launched ASAT carried by MiG-31s, and an airborne laser known as the A-60. In the 1990s the Russian Federation started development of an ASAT weapon capable of reaching geostationary orbit (GEO).

The People's Republic of China (PRC) has been at the forefront of research and development of ASATs throughout this period. Starting in 1991 the Chinese began developing ASATs as a response to the successful use of satellites by the United States during the 1991 Gulf War and apparent U.S. reliance on satellite technology. On January 11, 2007, China successfully shot down the Yun 1C weather satellite with a DF-31 ballistic missile from its base in Xichang. The missile traveled 715 miles from the launch site to successfully strike the satellite. The successful test led other countries, such as the Russian Federation, to restart their ASAT programs. In 2008 the United States successfully launched a modified SM-3 missile to shoot down a malfunctioning satellite. Additionally, the PRC started to develop lasers with the capability of damaging or disabling U.S. satellites. In September 2006 the PRC reportedly fired lasers at U.S. spy satellites flying over their country with the purpose of damaging the optical features of the satellite. The PRC is believed to have made regular use of lasers as a method of countering foreign intelligence-gathering satellites. Much of China's space program since the 2000s has focused on the development of ASATs and their employment against U.S. satellites. It is believed that the PRC now has enough ASATs to disable all American satellites in low earth orbit (LEO).

A number of nations are starting to view ASAT weapons as a secondary option to offset space power. First, only a few nations have the technology to develop ASATs. Additionally, ASAT tests have proven to be problematic for nations attempting to maintain control of space. U.S. and Soviet ASAT tests created a significant increase in space debris. Estimates of the increase in space debris as a result of the test during this period is at least over 10 percent. The National Aeronautics and Space Administration (NASA) estimated that debris from the 2007 test by the PRC accounted for 25 percent of all space debris in existence. The explosion from the satellite ejected debris at speeds of 700 to 1,400 miles per hour. The debris field stretched from 3,800 kilometers in orbit to 200 kilometers in altitude. Space debris are a significant threat to satellites. Debris can move at speeds as fast as 30,000 kilometers per hour. At those speeds very small debris can do significant damage to any spacecraft. In 2008 two U.S. satellites had to evade space debris caused by the 2007 test. In 2009 two more satellites also had to change their course to avoid the debris. Projections are that much of the 35,000 debris objects created by the test will remain in orbit for approximately one hundred years or more.

A nontraditional method for taking down satellites is the employment of cyber warfare. Cyberattacks against satellites offer a number of benefits. First, cyberattacks can disable a satellite instantaneously. Direct ascent ASATs have the disadvantage of having a time delay from launch to the moment of impact. Second, cyberattacks do not create the debris field caused by conventional ASAT attacks—thus preventing the debris field from disabling an actor's own satellites. Additionally, cyber warfare is much harder to track. Cyberattacks have the capability of destroying satellites without revealing the attacker.

Currently, the only ASATs that can be used without causing a significant space debris field are directed energy ASATs. This makes them preferable to nations intending to use ASAT technology. Another option going forward is the use of cyberattacks to disable satellites. These two options remain the most optimal since they are the least likely to cause collateral damage to other satellites orbiting.

Luke Wayne Truxal

See also: Directed-Energy Weapon; Electromagnetic Pulse (EMP); Global Positioning System (GPS); Solid State Laser (SSL)

Further Reading

Chow, Brian G. "Stalkers in Space: Defeating the Threat." *Strategic Studies Quarterly* 11, no. 2 (Summer 2017), 82–116.

Hughes, James. "Chinese Ballistic Missile Developments." *The Journal of Social, Political, and Economic Studies* 32, no. 2 (2007), 153–161.

Koplow, David A. "ASAT-isfaction: Customary International Law and the Regulation of Anti-Satellite Weapons." *Michigan Journal of International Law* 30, no. 4 (Summer 2009), 1187–1272.

Strauch, Adam. "Still All Quiet on the Orbital Front? The Slow Proliferation of Anti-Satellite Weapons." *Univerzita Obrany. Ustav Strategickych Studii* 14, no. 2 (2014), 61–72.

Anti-Ship Cruise Missile (ACSM)

Low-flying cruise missiles designed to destroy surface vessels at sea. They are different from Anti-Ship Ballistic Missiles (ASBMs) in that they are smaller, do not have as large jet plume, fly at lower altitudes, and do not reach supersonic speeds. This makes them difficult to detect and shoot down. ACSMs can be launched using land, air, and sea platforms.

The most notable use of ACSMs came during the 1982 Falklands War. The Argentine Air Force used the Exocet missile to great effect against the United Kingdom's naval forces. On May 4, 1982, Argentinian Super Etendard aircraft attacked the HMS Sheffield using Exocet missiles successfully striking the ship. The HMS Sheffield eventually sank on May 10, 1982. Once again on May 25, 1982, Argentinian pilots using ACSMs were able to sink the SS Atlantic Conveyor.

After the Falklands War, the People's Republic of China (PRC) realized the importance of ACSMs going forward. The PRC began to equip their aircraft with more ACSMs to counter American naval power. Part of that country's Long Wall doctrine is to use ACSMs and ASBMs as a part of an integrated missile strategy to wage a sea denial campaign against the U.S. Navy. Currently the main PRC ACSM is the HY-2 "Silkworm." This missile has a range of 300 miles and can be launched by either land or air.

As the PRC's ACSM technology has improved, the United States is believed to have let its ACSM inventory fall into decline in the aftermath of the Cold War. The two main ACSM missiles used by the United States since the 1990s are the AGM-84 "Harpoon" and AGM-84E "SLAM," which have a maximum range of 240 and 270 kilometers. The state of these weapons is relatively unknown according to a 2015 RAND report. As a result, the United States has begun development of the JASSM-ER, which is a Long-Range Anti-Ship Missile (LRASM). The first LRASMs reportedly entered operational service in December 2018. The advantages that the JASSM-ER have over the AGM-84 and AGM-84E is that it is less susceptible to jamming and other countermeasures. This new missile will update the U.S. ACSM arsenal and bring it into the 21st century.

ACSMs represent a grave threat to U.S. naval forces. In Operation Iraqi Freedom, the U.S. Navy failed to detect five cruise missiles fired by Iraqi forces toward coalition targets. Cruise missiles are affordable and used throughout the globe. There are more than 75,000 cruise missiles in more than seventy-five countries. For example, Iran has exported its ACSMs to the terrorist organization Hezbollah along with unmanned aerial vehicles (UAVs) to use as a launching platform against ships. Using this technology Hezbollah was able to successfully attack an Israeli vessel killing four in 2006.

The development of ACSMs represent a constant threat to navies around the world. Since 1982 they have proven effective at destroying surface vessels in combat. ACSMs remain an affordable option to offset naval power by states such as China and terrorist organizations such as Hezbollah. They will continue to be a constant challenge to naval planners going forward into the future.

Luke Wayne Truxal

See also: DF-26 Anti-Ship Missile; Littoral Combat Ship (LCS)

Further Reading

Goldstein, Lyle. "China's Falklands Lessons." *Survival* 50, no. 3 (2008), 65–82.

Gormely, Dennis. "Missile Contagion." *Survival* 50, no. 4 (2008), 137–154.

Heginbotham, Eric, et al. *The U.S.-China Military Scorecard: Forces, Geography, and the Evolving Balance of Power, 1996–2017.* Santa Monica, CA: RAND, 2015.

Hughes, James. "Chinese Ballistic Missile Developments." *The Journal of Social, Political, and Economic Studies* 32, no. 2 (2007), 153–161.

Meilinger, Phillip. "Range and Persistence: The Keys to Global Strike." *Air and Space Power Journal* 22, no. 1 (Spring 2008), 63–70.

Body Armor, Modern

Wearable materials designed to protect the user against specific physical threats in combat. Metal, ceramic, and especially fiber materials constitute typical modern body armors. Key considerations in design and use of body armor involve the items' weight, their optimized qualities against particular dangers, and ergonomic impact on personnel.

Body armor in ancient and medieval warfare had been primarily effective in protecting against edged weapons, and the rise of firearms led to the gradual elimination of body armor from the 16th century onward. Artillery in World War I led to massive numbers of fatal shrapnel-induced head injuries, and this ushered in the return of body armor in the form of steel helmets in the 20th century. Bullet resistant Kevlar vests first saw combat use by U.S. personnel during its fighting in the Vietnam War during the 1960s.

The Kevlar Personnel Armor System for Ground Troops (PASGT) helmet was first introduced in 1983, and it was subsequently replaced by a Modular Integrated Communications Helmet (MICH), which covers slightly less of the head but allows the use of communication equipment such as tactical headsets. The advantage of the PASGT and MICH over steel helmets is shown by ergonomic improvements and improved protection. Notably, the PASGT helmet and vest together weigh between 3 and 4 pounds, only a few ounces more than the Vietnam-era M1 helmet previously used by U.S. military personnel. These improvements set the stage for further advances in the 21st century.

The United States has taken a notable lead in the development of armor materials and also in its deployment in combat, although counterpart efforts have frequently appeared in several other countries. Perhaps surprisingly, although the People's Republic of China (PRC) is the global leader in the manufacture and export of body armor products, its People's Liberation Army (PLA) has been reported as lagging in the issuance and use of body armor. Since armor exported by PRC carries a reputation for combining good quality with low price, some analysis suggests that the PLA's apparent decision to minimize body armor is based on ergonomic and weight considerations overtaking perceived advantages.

Whereas metals had been the mainstay of body armor in Western countries from the medieval period onward, its role has diminished significantly. In modern body armor, metal plates appear as components within wearable fabric systems. The metal plates typically offer protection against stab and cutting wounds that could be delivered with edged weapons. Other armor plates such as ceramic inserts, and also the specially designed fabric systems themselves, are the components of modern body armor that are intended for protecting the wearer against gunfire or explosive fragments.

Ceramic plates are commonly used as inserts within ballistic vests. Ceramic armor materials have demonstrated the strength to stop a rifle bullet. However, ceramic plates are prone to cracking in the process of stopping a projectile, eliminating the armor's further usefulness. This is particularly disadvantageous for monolithic ceramic armors that are meant to protect a large area of the human form. One response is to design ceramic armor arrays that involve hexagonal or circular armor components. This retains effectiveness of other components in an array even after one is hit, and it also offers greater flexibility that may help mitigate the limited range of movement that armor systems impose.

One especially notable example, called Dragon Skin, used an array of small circular ceramic plates and was used by civilian contractors in Iraq and SWAT members of law enforcement in the United States. However, Dragon Skin was involved in major controversy in 2006. This included U.S. Army and Air Force claims that the armor was not effective—a recall of the armor was conducted by the U.S. Navy—and controversy continued as the company producing Dragon Skin challenged the results of U.S. military testing of their product. The Army ultimately upheld its rejection of the product for its own personnel, and these events coincided with an Army policy in March 2006 banning its personnel wearing privately purchased body armor. This was also during a high point in the controversy about the Second Iraq War (2003–2011) and in the wake of allegations that military units and personnel were being deployed without adequate equipment, including vehicular and body armor. Months of allegations and controversy on Capitol Hill and in the media followed.

Synthetic fiber materials are extremely common in modern body armor designs. The introduction of Kevlar by DuPont in 1965, and the commercial production of Twaron by Akzo in the 1980s and Dyneema by DSM in the 1990s, constitute milestones in the development of fiber materials for body armor.

Fiber materials are organized in successive frameworks that cumulatively form fiber armor. The fibers themselves are made of polymer materials with stiffness and strength as well as thermal and chemical resistance qualities. In production, the fibers are stretched until they are a fraction as thick as a human hair; this stretching enforces an alignment of the polymer's molecular chains and thereby further strengthens the individual fiber's tensile strength.

Multiple fibers are bundled together, either linearly or in a multithread helix, to produce a yarn. The width of these are typically measured in millimeters. These yarns are then interlaced to form a fabric layer, where warp yarns run parallel to the overall fabric edge and weft yarns run perpendicular. Multiple layers of such fabrics are often stacked and stitched together to form ballistic armor packs. In contrast to earlier soft body armors that were entirely constituted of woven armor fabrics, soft body armors now frequently include nonwoven laminate materials among the layers of armor fabric.

Different types of armor are most effective against different kinds of threats. Tightly woven fiber armor offers optimized protection against shards from fragmentation weapons. This is because the force of such a projectile hitting a tightly woven fiber armor layer is dissipated across several different yarns that are each composed of numerous fibers. The same pattern of weaving, however, would not provide similar effectiveness against a bullet, which may carry much more kinetic energy in a deformable lead projectile than existed in a nondeformable steel fragment.

Different materials can be used in concert with one another. For example, fiber vests are frequently equipped with compartments for the installation of metal or ceramic plates in order to provide more complete protection against powerful rifle bullets. Combining armor types optimized against different threats, such as bullets, stabs, or steel fragments, are frequently called "in-conjunction" armor, but synergies are mitigated by weight and ergonomic factors that mean restricted mobility.

Yarn motion within the weave of a fabric armor is an integral factor in the material's effectiveness, because this motion is necessary for the absorption of a projectile's ballistic energy. Designers must keep in mind that armors, including soft body armors, can successfully prevent a penetration by a projectile such as a bullet but still fail in the task of protecting the wearer if the absorbed impact causes indentations in the armor that injure or kill the wearer. For this reason, stopping the projectile's penetration is complemented by the need to dissipate the energy the projectile carries.

The U.S. Army Research Laboratory, BAE Systems, and other institutions are also researching "liquid armor," in which soft body armor fabric materials such as Kevlar are soaked in non-Newtonian fluids, which possess some of the properties of a solid under certain conditions while they otherwise behave similar to a liquid.

Rose E. Sambaluk

See also: Exoskeleton, Powered; Helmet Technologies; Human Universal Load Carrier (HULC); Improvised Explosive Device (IED)

Further Reading

Bhatnagar, Ashok, ed. *Lightweight Ballistic Composites: Military and Law-Enforcement Applications,* 2nd ed. Cambridge, MA: Woodhead, 2016.

Cadarette, Bruce S., et al. "Heat Stress When Wearing Body Armor." Report. Natick, MA: U.S. Army Research Institute of Environmental Medicine, 2001.

Cavallaro, Paul V. *Soft Body Armor: An Overview of Materials, Manufacturing, Testing, and Ballistic Impact Dynamics.* Newport, RI: Naval Undersea Warfare Center Division, 2011.

Franks, Lisa Prokurat, Tatsuki Ohiji, and Andrew Wereszczak, eds. *Advances in Ceramic Armor IV: Ceramic Engineering and Science Proceedings* 29, no. 6. Hoboken, NJ: Wiley, 2008.

Hazell, Paul J. *Ceramic Armour: Design, and Defeat Mechanisms.* Yarralumla, Australia: Argos, 2006.

Horn, Kenneth, et al. *Lightening Body Armor: Arroyo Support to the Army Response to Section 125 of the National Defense Authorization Act for Fiscal Year 2011.* Santa Monica, CA: RAND, 2012.

Jacobs, W. T. "Task Equipping Body Armor." Report. Quantico, VA: Command Staff College Marine Corps, 2009.

Tobin, Laurence B., and Michael J. Iremonger. *Modern Body Armour and Helmets: An Introduction.* Yarralumla, Australia: Argos, 2006.

Boston Dynamics BigDog

A quadruped robot meant to serve as a pack mule for U.S. military personnel in combat. Funded by the Defense Advanced Research Projects Agency, it was initiated in 2005 by Boston Dynamics, Foster-Miller, the National Aeronautics and Space Administration Jet Propulsion Laboratory, and Harvard University.

BigDog aimed to demonstrate the capability of a robot to carry burdens over complicated terrain in a combat environment, thus easing the combat loads carried by human combatants. BigDog was not intended or designed with any armed capability, and in size and shape resembled a headless mule, weighing 240 pounds and being 2.5 feet tall and 3 feet long. A 15-horsepower two-stroke engine drove a hydraulic pump that turned the vehicle's sixteen leg actuators. Each of the vehicle's four legs was equipped with a pair of actuators at the hip and one each in the positions corresponding to the knee and ankle. An onboard Pentium 4 computer and an array of fifty sensors helped the vehicle maintain its balance over complex and uneven terrain.

Demonstrations with an improved BigDog in 2008 also showed that the vehicle could maintain its balance even when pushed or kicked and that it could walk on icy surfaces. Further development in 2012 and 2013 aimed to further increase the vehicle's sensory capabilities, maximum allowable load, and to equip it with an arm. As a robotic porter, BigDog could carry a load up to 340 pounds and could move up to 3.5 miles per hour.

Although a respected technological achievement, it was not deemed practical for use in combat. The noise produced by the gasoline engine was viewed as a major drawback in a tactical environment. The search for quieter alternatives included work on an electrically powered robot called Spot, but this was reported to be able to carry barely one-tenth as much equipment.

Another follow-on project was the Legged Squad Support System (LS3). In tests starting in 2012, LS3 surpassed BigDog in several ways, including a top speed of 7 miles per hour and an engine that emitted a fraction as much sound as the original predecessor. Although LS3 was primarily to be controlled through typed or joystick commands, it was additionally able to recognize and respond to ten verbal commands, including to turn its engine off and on, follow closely, and follow along a passageway. Exercises during 2014 displayed its capabilities but also confirmed that even the much quieter LS3 was too loud to be useful for troops in the field. The Marine Corps retired LS3 into storage in 2015. It, like the BigDog that inspired it, did offer significant learning about the potential for autonomous technology.

Nicholas W. Sambaluk

See also: Boston Dynamics Handle Robot; Defense Advanced Research Projects Agency (DARPA); Unmanned Ground Vehicle (UGV)

Further Reading

Buehler, Martin, et al., eds. *The 2005 DARPA Grand Challenge: The Great Robot Race.* Cambridge, MA: Springer, 2007.

Parker, Philip M. *The 2018–2023 World Outlook for Military Ground Robots.* Las Vegas, NV: Icon Group International, 2017.

Springer, Paul J. *Outsourcing War to Machines: The Military Robotics Revolution.* Santa Barbara, CA: Praeger, 2018.

Technology Development for Army Unmanned Group Vehicles. Washington, D.C.: National Academic Press, 2003.

Boston Dynamics Handle Robot

A robot equipped with a combination of legs and wheels permitting unique maneuverability and implicitly foreshadowing future directions of robotics technologies. Described by some as "moving like an Olympic athlete," others have disparaged it as "nightmare inducing." When on flat terrain, Handle can roll up to 15 kilometers per hour. Its maximum vertical jumping capacity exceeds 1 meter—more than double the average human jump. Handle can easily carry 45 kilograms over 25 kilometers on a single battery charge, and it navigates narrow or cumbersome environments with an ease unseen in earlier robot technology. Videos by its designer showcase Handle leaping over tables and making hairpin turns.

The system's impressive range of motion is achieved through a mere ten articulated joints. This design simplicity lends it resilience in dusty, dirty, or wet environments by limiting maintenance requirements. Unlike BigDog, commissioned by Defense Advanced Research Projects Agency through the U.S. Department of Defense, Handle's military capacity remains heretofore unknown. Its speed and maneuvering capacity make it a natural delivery vehicle for goods across difficult terrain.

Handle's designer, Boston Dynamics, has fended off misgivings about the technologies it develops, including Handle. Founded in 1992 in Waltham, Massachusetts, Boston Dynamics is well known for its combination of cutting-edge technological innovation and quirky marketing, such as publishing YouTube videos of staff kicking its semianthropomorphic BigDog robotic pack mule to demonstrate its stability. Although BigDog was a research project connected with the U.S. military, other Boston Dynamics systems have been pointed at civilian use, such as Spot and SpotMini, whose locomotion resembles domesticated animals, and a 6-foot bipedal robot called Atlas that can spin and jump. Boston Dynamics itself was acquired by Google's Alphabet in December 2013, put up for sale less than two years later, and ultimately sold to Japan's SoftBank Group in June 2017.

Despite military investment, none of the current offerings at Boston Dynamics, including Handle, have been weaponized. Nevertheless, observably faster and stronger than their human makers, Handle and its "brothers" invoke a future where the science-fiction notion of a "robot soldier" does not seem so far flung. Perhaps this is why thousands of Facebook users have gathered to encourage Boston Dynamics to "stop kicking the robots!" Invoking the future while appealing to our prehistoric fears of predators, Handle represents a new frontier in technology. Its applications will determine if it is a new frontier in war.

Lisa Jane de Gara

See also: Boston Dynamics BigDog; Defense Advanced Research Projects Agency (DARPA); Unmanned Ground Vehicle (UGV)

Further Reading

Kneram, Mark S. *Enabling Soldiers with Robots*. Thesis. Carlisle, PA: Army War College, 2012.

Kuindersma, Scott, et al. "Optimization-Based Locomotion Planning, Estimation, and Control Design for the Atlas Humanoid Robot." dbc95c03fb0784dc07f02d9a5ba-44f568acb.pdf.

Shkolnik, Alexander, et al. "Bounding on Rough Terrain with the LittleDog Robot." *The International Journal of Robotics Research* (December 2010), 1–24.

Springer, Paul J. *Outsourcing War to Machines: The Military Robotics Revolution*. Santa Barbara, CA: Praeger, 2018.

BrahMos-II

An Indian hypersonic cruise missile currently under development. Designed by India's National Research and Development Organisation with Russian assistance, the BrahMos-II is powered by a scramjet engine that will provide a speed estimated between Mach 4 and Mach 7. Such speeds allow cruise missile strikes against mobile high-value targets, and Bramos Aerospace specifically contrasts its hypersonic missile speeds to U.S. Tomahawk cruise missiles whose slower speeds meant that a 1998 strike against Al Qaeda leader Osama bin Laden missed him by 1 hour of his movement.

Reports suggest that it will be ready to begin testing in 2020 and that with India's accession to the Missile Technology Control Regime the system will be reworked to double the weapon's range from its currently set 290 kilometers (180 miles). The missile is predicted to be effective against seaborne and ground targets. Overall, the BrahMos-II is expected to dwarf the capability of India's currently fielded ramjet-powered supersonic BrahMos (developed in 2003) and the follow-on BrahMos Mark II.

Nicholas Michael Sambaluk

See also: Anti-Ship Cruise Missile (ASCM); DF-26 Anti-Ship Missile; X-51 Waverider

Further Reading

Joshi, Yogesh, Frank O'Donnell, and Harsh V. Pant. *India's Evolving Nuclear Force and Its Implications for U.S. Strategy in the Asia-Pacific*. Carlisle, PA: Strategic Studies Institute, 2016.

Segal, Corin. *The Scramjet Engine: Processes and Characteristics*. Cambridge: Cambridge University Press, 2009.

Speier, Richard H., et al. *Hypersonic Missile Nonproliferation: Hindering the Spread of a New Class of Weapons*. Santa Monica, CA: RAND, 2017.

The Cloud

A term related to dispersed models for activities such as data storage and computing. The cloud is frequently presented as being a tool facilitating data security, and evidence suggests that this is true regarding some data threats more than others. Although the term deals specifically with computing, its relationship to cyberwar is more conditional. A facet considered to be a principal advantage is the opportunity to conduct computing with distributed resources, allowing a user to rely on already existing resources and thereby reducing up-front costs. In place of these costs, users interested in cloud computing pay for computing power as they consume it.

Cloud-computing advocates project that this model produces an environment in which computing can be done with greater speed and cost efficiency and with reduced levels of maintenance. The origins of cloud computing can be found in experimentation funded by the Defense Advanced Research Projects Agency (DARPA), which eventually gave rise to the ARPANet during the 1960s and 1970s and subsequently developed into the modern Internet. Interest in maximizing the usability of limited computing power formed one of the initial drivers to the research that later yielded the interconnected environment of the early 21st century, but these efforts did not eliminate competition over still limited computing power, and the implications may be applicable to cloud computing in the 21st century as well. Thus major advances in computer technology and in interconnectedness have come about through work meant to distribute access to computer resources.

Although some evidence points to greater financial and energy efficiency in this model of pooled computing resources, predictions of these efficiencies rest on the presumption that demand would remain distributed enough to allow efficient use of computing services without leading to delays due to oversaturation of demand. Advocates of cloud computing have asserted that using the cloud for computing and storage can permit a radically more rapid and responsive approach to updating the technologies that protect software from intrusion and data from compromise.

Cloud storage utilizes cloud technology in order to preserve a user's ability to access data despite the virtualization of infrastructure and the consequence that systems physically storing data are located elsewhere in the world. This system ensures the security of data against a range of physical threats, such as natural disasters, which might eradicate computer and other infrastructure in a locality. Cloud storage would preserve the data, saved also on other distant servers, even in the event of a situation in which other servers have been damaged.

However, critics point to several areas in which cloud storage may actually provide less security, and these threats tend to encompass malicious activities in the digital realm. The copying and distribution of data means that the same data exists at more locations that might be intentionally compromised, and the sharing that stands at the core of cloud storage also increases the number of people potentially able to access data and the number of networks that might be targeted for intrusion. In short, cloud storage massively increases what is known as the attack surface area. Potential accessibility issues exist as well, since interruptions to connectivity would hinder access to remotely stored data in a way that locally stored data would be immune. Copyright concerns, including governmental surveillance as well as espionage from unauthorized entities, pose more complicated challenges when the servers that make the cloud usable are located in other legal jurisdictions.

The first decades of the 21st century have witnessed interest by governments and other large entities toward using the cloud, and corporate giants in the computer industry have energetically developed and promoted their cloud services. One of the primary caveats of being network dependent, even with the most updated security software and accessed-only codes, is that potential cyber threats can threaten an entire nation. Dispersing computational and storage capabilities onto the cloud therefore offers opportunities and potential risks.

Kathy Nguyen and Nicholas Michael Sambaluk

See also: Computer Network Defense (CND); Encryption; Vulnerability

Further Reading

Buennemeer, Timothy K. *A Strategic Approach to Network Defense: Framing the Cloud.* Carlisle, PA: U.S. Army War College Press, 2011.

Gillette, Stefan E. *Cloud Computing and Virtual Desktop Infrastructure in Afloat Environments.* Monterey, CA: Naval Postgraduate School, 2012.

Owen, Tim, et al. *Joint Sensor: Security Test and Evaluation Embedded in a Production Network Sensor Cloud.* Lorton, VA: Defense Research and Engineering Network, 2010.

Powell, Dallas A. *The Military Applications of Cloud Computing Technologies.* Ft. Leavenworth, KS: School of Advanced Military School, 2013.

Wang, Lidong, and Cheryl Ann Alexander. "Big Data Distributed Analytics, Cybersecurity, Cyber Warfare and Digital Forensics." *Digital Technologies* 1, no. 1 (2015), 22–27.

Counter Unmanned Aerial Systems (C-UAS)

Systems designed as a defense against unmanned aerial vehicles (UAVs, commonly known as drones). Throughout the 2010s, hobbyist drones designed for photography became very popular. With their quadcopter design—asymmetrically distributing power to four rotor assemblies for movement—they could physically operate anywhere. As amateur drone pilots took to the skies, near misses with manned aircraft and crashes into people on the ground began to occur.

Police in the Netherlands trained eagles to snatch quadcopters out of the air in response to this new public safety issue. The U.S. Army unveiled a more practical, technological solution at an Army Cyber Institute "CyberTalk" in September 2015, known as the cyber capability rifle (CCR). The weapon consists of a directional antenna replacing the standard barrel of an M-4 carbine, a battery pack located in the magazine well, and a programmable electronic board mounted on the side of the gun. By March of the following year, the CCR had demonstrated its ability to take down a Parrot brand quadcopter by exploiting a known flaw in the drone's code. At a cost of only $150, the CCR is much less expensive than the Netherlands' eagle program, which was terminated in December 2017.

With the increasing employment of fixed-wing (not equipped with rotors) UAVs in the world's militaries, new strategies and equipment are required to counter the potential threats. In October 2016, the U.S. Army released a plan for C-UAS. The plan outlines operational doctrines—both old and modified—to effectively detect, identify, and defeat drones. Defeat involves three main concepts: Left-of-Launch, Standoff, and Close Contact. The strategy of Left-of-Launch is to eliminate UAVs before they can take off, either by preventing their sale to hostile organizations or destroying control facilities. Standoff consists of limiting the operational capabilities of enemy drones once they are already in the air. This includes the elimination of control centers, the creation of more sophisticated electronic jamming systems to disrupt control, and the refinement of current air defense networks to identify and target small UAVs. Close Contact is the "in the field" defense against drones. This involves modification of current unit operating procedures—the use of camouflage against position identification by the UAV, properly reporting enemy UAVs to superior officers, and assessing the risk to the unit from engaging the UAV with weapons.

The U.S. Department of Defense (DoD) announced its desire in 2016 for a weapon system that would physically eliminate a drone, instead of only disrupting it electronically. The so-called "C-UAS Hard Kill Challenge" stemmed from the wastefulness of using multi-million-dollar missiles against drones costing no more than a few hundred dollars.

Meanwhile, some companies have continued to develop electronic countermeasures. SkyDroner, for example, has developed a system that can identify and take control over various types of drones, issuing the electronic commands for them to land or return to their operator. The need for C-UAS technology was proven on August 6, 2018, when two drones loaded with C-4 explosive attempted to assassinate the president of Venezuela.

Elia G. Lichtenstein

See also: Malware; Quadcopter; Swarm Robots; Vehicle-Borne Improvised Explosive Device (VBIED)

Further Reading

Goppert, J. M., et al. "Realization of an Autonomous, Air-to-Air Counter Unmanned Aerial System (CUAS)." *2017 First IEEE International Conference on Robotic Computing (IRC).* Taichung, 2017, 235–240.

United States Army. *Counter-Unmanned Aircraft System (C-UAS) Strategy Extract,* 2015.

Defense Advanced Research Projects Agency (DARPA)

A U.S. Defense Department research entity tasked with developing particularly innovative and revolutionary concepts, especially regarding technology, in order to reduce developmental uncertainty. First established in 1958 as the Advanced Research Projects Agency (ARPA) in the wake of high-profile Soviet missile and space achievements, the word "Defense" was added in 1972, and the organization's focus has adapted since its initial focus on space technology to include various other fields, notably including computer and information technology.

With respect to computer and information technology, DARPA has often been on the leading edge of development. This includes contracting for the fabrication of the first routers, pioneering the use of packet switching to share blocks of information across a network, and the development in 1969 of the Advanced Research Projects Agency Network (ARPANET), which formed the precursor to the Internet. ARPA funded the oN-Line System (NLS) collaboration that included the first hypertext linking, and by 1972 the organization facilitated the development of new communication protocols that would yield the Transmission Control Protocol (TCP) and the Internet Protocol (TCP/IP).

These protocols remain a mainstay of the Internet's underlying technical foundation. Further, ARPA is also credited for inventing the computer mouse as early as 1964.

One of the first major efforts of the agency's Information Processing Techniques Office (IPTO) to support an external civilian institution was the 1964 project on Mathematics and Computation (Project MAC). Project MAC was the world's first large-scale experiment in personal computing at the Massachusetts Institute of Technology (MIT). A major thrust of Project MAC was to develop general-purpose time-sharing capabilities, which later influenced the design of computer systems for commercial and defense uses. Within years of its start, Project MAC would evolve into the world's first online community, complete with online bulletin boards, e-mail, virtual friendships, and open-source software exchange and hackers.

In 2002, through its Personal Assistant that Learns (PAL) program, DARPA created cognitive computing systems with the goals to (1) make military decision making more efficient and effective at multiple levels of command; (2) reduce the need for large command staffs; and (3) enable smaller, more mobile, and less vulnerable command centers. Advances stemming from the agency's PAL program were applied not only to military users but also to enable voice-based interaction with civilian handheld devices. This led to the 2007 launch of Siri Inc., later acquired by Apple Inc., which further advanced and then integrated the Siri/PAL technology into the Apple mobile operating system.

In 2004, as part of the then three-year-old Quantum Information Science and Technology Program (QuIST), DARPA-funded researchers established the first so-called quantum key distribution network, a data-encryption framework for protecting a fiber optic loop that connects facilities at Harvard University, Boston University, and the office of BBN Technologies in Cambridge, Massachusetts.

In 2008, with the goal of developing analysis techniques for massive data sets, DARPA rolled out the Topological Data Analysis (TDA) program, which is still currently in use.

In 2014, DARPA launched the Cyber Grand Challenge, a competition to create automatic defensive systems capable of reasoning about flaws, formulating patches, and deploying them on a network in real time. By acting at machine speed and scale, it is hoped that these technologies could someday overturn today's attacker-dominated status quo. This vision could be realized with breakthrough approaches in a variety of disciplines, including applied computer security, program analysis, and data visualization.

DARPA has also initiated robotics challenges that have spurred on the development of projects such as the BigDog (a quadruped logistical robotic experiment) and the Handle (equipped with a pair of lower wheeled legs and a pair of upper strut legs). DARPA has also been involved in other experimental and research projects dealing with areas including hypersonic flight.

Deonna D. Neal

See also: Boston Dynamics BigDog; Boston Dynamics Handle Robot; X-51 Waverider

Further Reading

Abbate, Janet. *Inventing the Internet.* Cambridge, MA: The MIT Press, 1999.

Jacobson, Annie. *The Pentagon's Brain: An Uncensored History of DARPA, Amercia's Top Military Research Agency.* New York: Little, Brown, 2015.

Roland, Alex, and Philip Shiman. *Strategic Computing: DARPA and the Quest for Machine Intelligence, 1983–1993.* Cambridge, MA: The MIT Press, 2002.

Van Atta, Richard H. *Transformation and Transition: DARPA's Role in Fostering an Emerging Revolution in Military Affairs. Vol. 1—Overall Assessment.* Alexandria, VA: Institute for Defense Analysis, 2003.

Defense Innovation Unit Experimental (DIUx)

A U.S. Department of Defense initiative meant to expedite the development and use of new commercially available technologies in defense applications. DIUx personnel include civilian, active duty military, and reserve components. This combination is in keeping with a philosophy that including the different groups can combine diverse talents and perspectives valuable in rapid problem solving.

The rapid development of experimental technologies requires agile funding processes, and DIUx has touted an ability to typically provide project funding to contractors within ninety days and to present a path to translate a contract for a successful experimental study into a contract regarding a more robust follow-on project. Federal coinvestment provides financial support and is provided in a way that preserves intellectual property rights of the developer. The rapid pace of contract and development is a result of DIUx standing outside the standard federal acquisition regulations that concern most contracting; frequent frustrations with the slower progress under standard acquisition procedures sparked the momentum for establishing DIUx.

Defense Department statements have described DIUx as one of the "key efforts that align with" the Defense Innovation Initiative (DII) that was announced by Defense Secretary Chuck Hagel in November 2014 (U.S. Department of Defense). DII is closely associated with government advocacy of a "Third Offset Strategy," in which rapid development of futuristic technology is asserted to be crucial in addressing future challenges by adversaries. Advocates deem the advent of nuclear weapons to have been the "First Offset" and the combination of stealth aircraft and precision guided munitions to have constituted the "Second Offset." DIUx was created by Hagel's successor Ash Carter ten months after Hagel had unveiled DII, and ten months after that, DIUx was restructured, given new leadership, and was directed to report immediately to the defense secretary rather than through the undersecretary responsible for acquisitions. DIUx has been characterized as a little sibling to the Defense Advanced Research Projects Agency that was established in the late 1950s and tasked with a range of cutting-edge research and development challenges.

The reconstituted DIUx was charged in 2016 with innovating through the application of a philosophy mirroring the "fail fast" mantra of contemporary innovation leaders outside government. This concept prioritizes rapid learning and application of derived lessons. A legislative expansion of the "other transaction authority" in 2016 provided an important procedural vehicle enabling rapid transition of DIUx projects from prototyped models to acquisition of products. One of DIUx's reported successes is a highly miniaturized two-way communication device contracted and fielded by deployed U.S. Air National Guard units within a space of three months. Pentagon figures have also reportedly shown interest in DIUx work to support development of civilian satellites that might be orbited over North Korea. Advocates of DIUx worried that the Barack Obama–era program might be unceremoniously eliminated by the incoming Donald Trump administration, but despite the departure of the DIUx Silicon Valley head in February 2018, the office has continued to operate.

Nicholas Michael Sambaluk

See also: Defense Advanced Research Projects Agency (DARPA)

Further Reading

Ilachinski, Andrew. *AI, Robots, and Swarms: Issues, Questions, and Recommended Studies.* Arlington, VA: CNA Analysis and Solutions, 2017.

Segal, Adam. "Bridging the Cyberspace Gap: Washington and Silicon Valley." *Prism* 7, no. 2 (2017), 67–77.

U.S. Department of Defense. *Defense Innovation Marketplace: Connecting Industry and DOD.* http://www.defenseinnovationmarketplace.mil/DII_Defense_Innovation_Initiative.html.

Wong, Carolyn. *Enhancing ACC Collaboration with DIUx.* Santa Monica, CA: RAND, 2017.

Defense Technical Information Center (DTIC)

A U.S. Defense Department office dedicated to essential, technical research, development, testing, and evaluation (RDT&E) of information rapidly, accurately, and reliably to support the needs of DoD customers. Called the Defense Technical Information Center (DTIC), its first predecessor was called the Air Documents Research Center (ARDC) in 1945. Repeatedly redesignated as the Air Documents Division (ADD), the Central Air Documents Office (CADO), the Armed Service Technical Information Agency (ASTIA), and the Defense Documentation Center (DDC) before becoming DTIC, the organization now exists under the auspices of the Assistant Secretary of Defense for Research and Engineering (ASD(R&E)).

In 2008 DTIC Online was launched to offer customers one comprehensive website to search and access DoD scientific and technical information. In 2018 DTIC published its inaugural issue of the *Journal of DoD Research and Engineering* (JDR&E), a peer-reviewed publication providing an avenue for DoD scientists and engineers to publish their classified and controlled-unclassified research.

DTIC responsibilities to ASD (R&E) are to (1) preserve and disseminate the research that led to the technologies used by today's warfighters; (2) deliver the tools and collections that empower the R&E enterprise to accelerate the development of technologies that will help maintain U.S. technical superiority; (3) stimulate innovation with public and industry access to DoD-funded research and digital data;

(4) maximize the value of each dollar the DoD spends through the analysis of funding data, work in progress and independent R&D (IR&D) to identify gaps, challenges, and the way forward. These areas of responsibility support ASD (R&D) in their efforts to mitigate new and emerging threat capabilities, enable affordable new or extended capabilities in existing military systems, and develop technology surprise through science and engineering.

DTIC also manages Information Analysis Centers (IACs), which provide essential technical analysis and data support to a diverse customer base, to include Combatant Commands (CCMDs), the Office of the Secretary of Defense, Defense Agencies, and the Military Services. IACs actively partner and collaborate with defense research and engineering focus groups and communities of interest in cyber, homeland defense, and defense systems. They are staffed with scientists, engineers, and information specialists to provide research and analysis to customers with diverse, complex, and challenging requirements. DTICs' affiliated organizations include CENDI (Federal Scientific and Technical Information Group) and FEDLINK (Federal Library and Information Network).

DTIC has a number of associated programs across government and with academia to facilitate the dissemination of research information. DTIC also provides valuable technical information to small technology companies engaged in R&D projects for the DoD Small Business Innovation Research (SBIR) and Small Business Technology Transfer (STTR) programs. DTIC databases include unclassified studies on U.S. systems ranging from the military body armor to additive manufacturing to the F-35 Lightning II multirole jet.

In addition to its public access search capabilities, DTIC maintains a Corporate Source Authority System (CSAS) listing all organizations that have contributed information to any of DTICs' three major databases: the Technical Reports Bibliographic Database (TR), the Research Summaries Database (RS), and the Independent Research and Development Database (IR&D). DTIC also maintains the Defense Technology Transfer Information System (DTTIS).

Deonna D. Neal

See also: Defense Advanced Research Projects Agency (DARPA); Defense Innovation Unit Experimental (DIUx)

Further Reading

Molholm, Kurt N. "The Defense Technical Information Center: Expanding Its Horizons." *Government Information Quarterly* 12, no. 3 (1995), 331–344.

Plan to Establish Public Access to the Results of Federally Funded Research. Washington, D.C.: Department of Defense, 2015.

Wallace, Lane E. *The Story of the Defense Technical Information Center, 1945–1995.* Washington, D.C.: Department of Defense, 2009.

DF-26 Anti-Ship Missile

An anti-ship missile designed by the Chinese People's Liberation Army (PLA) and deployed since at least 2015. Along with weapons systems in other categories, such as the stealthy multirole J31 Gyrfalcon fighter plane and the Liaoning aircraft carrier, the DF-26 reflects PLA development of technologies that support a more robust armed capability across the western Pacific Ocean.

A ballistic rather than cruise-type missile—it is deployed on a mobile erector-launcher vehicle—the missile is estimated to have a range of up to 2,500 miles. The weapon joins an array of existing missile types with considerably shorter ranges that nonetheless are believed to possess anti-ship potential. The DF-26 is believed to be capable of carrying either a conventional or a nuclear warhead to its target. As a result, it has sometimes been referred to outside China as a "Guam Killer," since the prominent U.S. Air Force base there is less than 2,000 miles from the Chinese coastline and since the DF-26 appears to be the first PLA Rocket Force weapon with the range to reach the island.

Nicholas W. Sambaluk

See also: Anti-Ship Cruise Missile (ASCM); J31 Gyrfalcon; Liaoning Aircraft Carrier

Further Reading

Bonds, Timothy M., et al. *What Role Can Land-Based, Multi-Domain Anti-Access/Area-Denial Forces Play in Deterring or Defeating Aggression?* Santa Monica, CA: RAND, 2017.

Erickson, Andrew S. *Chinese Anti-Ship Ballistic Missile: (ASBM) Development: Drivers, Trajectories and Strategic Implications.* Washington, D.C.: The Jamestown Foundation, 2016.

Kelly, Terrence K., et al. *Employing Land-Based Anti-Ship Missiles in the Western Pacific.* Santa Monica, CA: RAND, 2013.

Directed-Energy Weapon

Weapons that damage targets through the emission of highly focused energy, rather than by the kinetic force of a projected material. Although several countries, including India, the People's Republic of China (PRC), Russia, and the United Kingdom have each conducted work toward directed-energy weapons, the United States as of 2018 appeared to be the leader in research and development of directed-energy weapons projects and expected the first of these to become operational in numbers as early as the mid-2020s.

Directed-energy weapons offer several important advantages compared with chemical-propellant weapons. The minimal impact of gravity on light means that directed-energy weapons like lasers have a flat trajectory and offer greater precision, speed, and range. The cost per shot, in terms of money and storage space, is far less with a directed-energy system. Calibration of weapon power brings the ability to engage different targets and to engage in potentially lethal or nonlethal ways. Directed-energy weapons also offer fast engagement times that assist in the defeat of several successive targets.

Directed-energy weapons are not a panacea. Although minimally impacted by gravity, they are line-of-sight weapons incapable of indirect fire. Furthermore, atmospheric absorption and adverse weather patterns can hamper their usefulness. Extended laser use can create local atmospheric changes that result in "thermal blooming," which defocuses laser power and erodes its weaponized effectiveness.

Targets can be physically hardened, as PRC researchers appear to be doing with the use of rare earths and carbon fiber materials meant to negate the combat effectiveness of lasers. Other countermeasures could include rotating a target so that a defending laser beam could not concentrate and deform one part of an attacking cruise missile.

Several classes of directed-energy weapons types exist either in theory or in development. These include microwave weapons, sonic weapons projects, and weaponized lasers. Particle-beam weapons and plasma weapons are a theoretical possibility, and reportedly research is underway in Russia, but to date no public demonstration of such a weapon has occurred.

Microwave weapons include the Active Denial System meant as a riot-control project to heat water within a human target's skin resulting in intense pain but nonpermanent injury. Certain types of radar are potentially adaptable to microwave emission purposes that would ruin a targeted computer's memory or electronic components.

The Long Range Acoustic Device (LRAD) is a sonic technology with nonlethal effects meant for riot control. Continuous exposure to certain low frequency tones inflicts damage on brain tissue. In 2017, nearly two dozen U.S. diplomats in Cuba experienced sonic-induced pain including cases of brain injury, leading to speculation that Cuba deployed sonic weapons against U.S. diplomatic personnel. Lab studies have demonstrated that ultrasound exposure can damage organs and tissue.

These developments notwithstanding, the most salient progress in directed-energy weapons has involved weaponized lasers. These range considerably in terms of power and capability. High-energy lasers are usually considered to constitute those with powered beams of 10 kilowatts (kW) or more.

The U.S. Navy's Laser Weapons System (LaWS) met particular success, being developed and tested between 2009 and 2014 at a cost of $40 million and quickly being found so successful as to be authorized as an operational ship-defense system on board its testing platform, the USS *Ponce*, whose service life was extended by five years so that it could serve as a test bed for a helicopter-borne mine-clearance research project. The ship's detailing to this task created the opportunity for the ship to also host the trial deployment of LaWS starting in August 2014. Success with LaWS, a weapon in the range of 15–50 kW, encouraged U.S. Navy interest in the development of far larger and more powerful lasers of up to 300 kW.

Although LaWS was designed primarily to combat enemy fast-attack craft and unmanned aerial vehicles, more powerful systems might be deployed against more robust or challenging targets. A Northrup Grumman research project resulted in a 100-kW electric laser in early 2009. Separately, in mid-2009, the Office of Naval Research awarded Raytheon with a contract to develop a 100-kW free electron laser (FEL). A FEL, which uses high-speed electrons moving within a magnetic structure, is a tunable laser with an especially wide-ranging capability in terms of frequency and therefore weaponized applicability. This opens the possibility of engaging targets such as enemy aircraft or even anti-ship cruise missiles. A full-power FEL prototype was scheduled for delivery in 2018.

In contrast to the electromagnetic railgun (EMRG), the U.S. Navy has asserted that it does not foresee use of lasers against land targets. Demonstrated systems such as LaWS address ship-defense needs, and larger laser weapons have been described as focusing on ship-defense against different kinds of targets. Given the line-of-sight nature of directed-energy weapons (including lasers), this is entirely logical.

At the other end of the spectrum are much smaller systems such as the personnel halting and stimulation response rifle (PHASR) developed by the U.S. Air Force Research Laboratory. Due to the 1995 United Nations Protocol on blinding laser weapons, to which the United States agreed in January 2009, the PHASR prototype was a low-intensity laser weapon so that its blinding effect would be temporary and thus would not defy the protocol. A Chinese design, the ZM-87, was designed at the end of the 20th century and was canceled only in 2000 as a result of the ban, although examples may have been distributed to Russia and North Korea and used during the 21st century. Unlike PHASR, ZM-87 was purposed for permanently blinding humans; its capability to damage technologies like rangefinders was incidental.

Nicholas Michael Sambaluk

See also: Electromagnetic Railgun (EMRG); Laser Weapon System (LaWS); Solid State Laser (SSL)

Further Reading

Dubinskii, Mark, et al. *Laser Technology for Defense and Security XII 19–20 April 2016.* Washington, D.C.: Bellingham, 2016.

Geis, John P. *Directed Energy Weapons on the Battlefield: A New Vision for 2025.* BiblioGov, 2012.

Jenkins, William C. *Navy Shipboard Lasers: Background, Advances, and Considerations.* New York: Nova, 2015.

McAulay, Alaster. *Military Laser Technology for Defense: Technology for Revolutionizing 21st Century Warfare.* Hoboken: Wiley, 2011.

Nielsen, Philip. *Effects of Directed Energy Weapons.* CreateSpace, 2012.

O'Rourke, Ronald. "Navy Lasers, Railgun, and Hypervelocity Projectile: Background and Issues for Congress." Congressional Research Service, May 27, 2016.

Perram, Glen. *An Introduction to Laser Weapon Systems.* Directed Energy Professional Society, 2009.

Titterton, D. H. *Military Laser Technology and Systems.* Boston: Artech, 2015.

Zohuri, Bahman. *Directed Energy Weapons: Physics of High Energy Lasers (HEL).* Cham, Switzerland: Springer International Publishing, 2016.

Division Multiple Access (DMA)

Systems supporting the simultaneous operation of myriad devices in a network, leveraging a limited number of frequencies. As such, the development of reliable and effective multiple access methods can be considered a prerequisite to the creation of modern cellular networks and a significant factor in the expanded availability and use of connected devices in the early 21st century. Several important kinds of multiple access frameworks have been developed, notably systems establishing channel access through space and time divisions, via frequency subdivisions, and through coding.

Space-division multiple access (SDMA) systems leverage awareness of the special location of devices to yield more efficient communication. Whereas earlier forms of cellular networks had involved the radiation of signal in keeping with the tradition of other kinds of radio signal, space division makes use of smart antenna technology that distinguishes the special location of the mobile devices to achieve two important advantages. One is that knowledge of a device's location enables a much more efficient signal to be sent to the proper location rather than across a cell's area in hope of including the proper location; this more focused approach to the sending of signal helps save energy. A complementary benefit is that the saved energy relative to a signal-radiating approach also precludes the signal interference that would have occurred when adjacent "cochannel" cells use the same frequency. Cochannel interference had been experienced not only with early cellular telephone systems but can also occur for FM and AM radio signals, and it can be exacerbated by flawed planning and mapping of frequencies for broadcasters, by saturation of a radio spectrum as in some urban areas, and temporally through the influence of certain adverse weather conditions.

Time-division multiple access (TDMA) systems use a different approach for deconflicting channel access by multiple users. Whereas SDMA differentiates mobile devices by their contemporary locations, TDMA systems alternate users of a common frequency within a rapid succession. In the latter system, each user operates within a specified intermittent time slot, and since these slots are both brief and close together, the effect is to simulate a constant operation by multiple users on different devices, despite their sharing the same frequency channel.

Whereas spatially defined systems must cope with the changing location of mobile devices and the impact on systems maintaining track of devices and sending focused signal to the appropriate position, time-defined systems face a different kind of challenge. As users carrying TDMA devices move, their changing location can minutely impact the time required for uplink signal from the device to a base station. The precision involved in recurrently slicing the time slots within a frequency channel used by a TDMA system means that even a very small change in the timing advance needed for such transmissions can complicate or even hinder effective transmission. Helping to compensate for this, time-division systems are able to detect transmitters and conduct other measurements between the intervals in which the phone is assigned its recurrent time slots. This feature facilitates the transition from one frequency to another in "interfrequency handovers," which can be required when a mobile device user's position shifts so that different cells or different frequencies may be needed in support of an ongoing communication.

TDMA concepts dominated the emergence of 2G cellular systems which spread, particularly in the United States, in the early 21st century. The rise of 2G networks made possible the rapid expansion in the availability and popularity of cellular devices, and this arguably helped fuel the increasing proliferation of increasingly sophisticated mobile devices.

In addition to space- and time-defined multiple access systems, frequency-division multiple access (FDMA) and code-division multiplexing were developed as a means to differentiate signals. Frequency division entails the subdivision of a frequency into subchannels, and it currently appears in satellite communication systems and also in landline telegraphy, as FDMA systems are used most often within coaxial cable systems or in signals sent via microwave beams. Although FDMA can be applied to digital signal, it is much more frequently used with analog signals. Predictably, frequency division does not carry the timing vulnerabilities that are inherent in TDMA systems, although crosstalk can pose challenges in FDMA.

Work on code-division multiple access (CDMA) systems for radio communication date to the mid-1930s in the former Soviet Union. By the late 1950s, different code-division radio telegraphy systems were in limited use in the Soviet Union, with different devices weighing between 1 and 24 pounds (0.5 to 11 kg), depending on the application. Although first envisioned as a technology solely for use by emergency and government authorities within the Soviet Union, by the mid-1960s the system was expanded into the ultra-high-frequency and very-high-frequency mobile telephone Altai. The radiotelephone technology represented by Altai was generally displaced by later cellular technologies such as those 2G systems based on TDMA, although radiotelephones remain in some use in isolated areas underserved by cellular infrastructure. Later applications for code-division multiplexing systems included global positioning satellites and development into 3G cellular technologies, where code division systems dominate.

Nicholas Michael Sambaluk

See also: Global Positioning System (GPS); International Mobile Subscriber Identity-Catcher (IMSI-Catcher)

Further Reading

Amouris, K. "Space-Time Division Multiple Access (STDMA) and Coordinated, Power-Aware MACA for Mobile Ad Hoc Networks." *IEEE Xplore* (August 2002).

Biswas, Ankita. *Robust Data Center Network Design Using Space Division Multiplexing.* Windsor, Canada: University of Windsor, 2018.

Lee, Junhee, et al. "Multi-Channel Time Division Multiple Access Timeslot Scheduling with Link Recovery for Multi-Hop Wireless Sensor Networks." *International Journal of Distributed Sensor Networks* (August 2017).

Rao, Raghuveer, and Sohail Dianat. *Basics of Code Division Multiple Access (CDMA).* Bellingham, WA: International Society for Optics and Photonics, 2005.

Zigangirov, K. *Theory of Code Division Multiple Access Communication.* New York: Wiley, 2004.

Electromagnetic Pulse (EMP)

A source of severe damage to most forms of electrical equipment, potentially caused by either a solar flare or a high-altitude nuclear blast. A solar flare in August 28, 1859, called the Carrington Event, caused sparks, fires, and failure of overheated electrical equipment at telegraph stations. These solar explosions produce a pulse of ionization in the atmosphere, as well as dramatic effects on the earth's magnetic field. Between 1872 and November 2003 five solar storms ranked close to this event for intensity. Tsurutani and colleagues (2003) document twelve strong magnetic storms between 1859 and 1989, concluding that one big flare per eleven-year solar cycle has the potential to create a storm of similar intensity. The size of electrical blackout produced, and the damage to electric power systems, from solar flares on July 13, 1977, and on March 13, 1989, lead experts to fear that a solar flare like the Carrington Event occurring today would cause widespread disruptions and damage to electric grids. NASA estimates the likelihood of such a huge solar storm at 12 percent per decade.

Nuclear detonations can also result in EMPs, as when the July 9, 1962, U.S. test of a 1.4-megaton device at an altitude of 250 miles (400 kilometers) over the Pacific Ocean's Johnson Island created an unplanned EMP that damaged electrical systems in the Hawaiian islands nearly 1,000 away. The same year, the Soviets reported significant damage to electrical systems more than 350 miles from their tests of 300-kiloton weapons at altitudes of 37, 93, and 186 miles (60, 150, and 300 kilometers, respectively).

A nuclear EMP produces a set of three different energy pulse waves. First, gamma rays interact with the atmosphere and the earth's magnetic field to produce high-energy free electrons inducing an oscillating electric current that destroys electronic equipment. The first electromagnetic shock lasts a few billionths of a second over a very large area, which damages anything containing electrical components, including the systems used to protect equipment from lightning strikes. The effect of the second component resembles lighting, but lower in amplitude and more widespread. Because it comes a small fraction of a second after the first pulse has damaged protective equipment, it can damage a great deal more electrical infrastructure. The third pulse rises slower and can last a minute, resembling the effects of a solar flare magnetic storm. It distorts the earth's magnetic field and ionizes the atmosphere below the blast. It disrupts long electric distribution and transmission lines, causing even more extensive damage.

Between 1995 and 2000, the U.S. Congress conducted a number of open hearings to educate other members and the public about the nature of the threat from either a nuclear EMP attack or the EMP from a large solar storm. This testimony established that either event could produce a protracted nationwide blackout. In 2001, Congress established the Commission to Assess the Threat to the United States from Electromagnetic Pulse (EMP) Attack (EMP Commission), which was authorized until 2008. The commission tested modern electronics in EMP simulations.

On July 22, 2004, members of the EMP Commission testified before the House Armed Services Committee and presented a five-volume report, of which three volumes were classified, with an Executive Summary and Assessment of U.S. Critical Infrastructure left unclassified. Critics disputed the value of the report, claiming a small likelihood of a large-scale EMP attack and that critical infrastructure would survive. Commission members respond that the electric power industry had relied on flawed and incomplete testing, leading to false confidence in the survivability of electrical equipment. Meanwhile, increasing dependence on electronic devices employing increasing miniaturization has raised the vulnerability. Furthermore, they argued that failure to make a visible effort to protect computer systems and critical infrastructure invited an EMP attack.

The report of its conclusions in 2008, based on additional testing of electrical components, the commission estimated that the electric grid and vulnerable infrastructure could be protected for $2 billion. This report made one hundred recommendations. Critics dispute the claims of the commission regarding the feasibility for a nuclear EMP weapon. The commission responded with documents from the military doctrine of Russia, China, North Korea, and Iran that call for conducting EMP attacks on the United States.

The EMP Commission complained that the Defense Department withheld funding and security clearances to which the commission was legally entitled, and that the electric power industry's private Electric Power Research Institute (EPRI) had produced junk science to resist recommendations to add shielding and other protective measures to transformers and power substations. Members of the commission continue to complain of industry resistance, through the North American Electric Reliability Corporation, while the U.S. Federal Energy Regulatory Commission (FERC) lacks necessary regulation enforcement power and fails to use the power it has.

In 2015, the U.S. Government Accountability Office testified to Congress that the government had failed to implement even one major recommendation of the EMP Commission. Therefore, in 2015 Congress reestablished the EMP commission for two years to reexamine the threat and make new recommendations. Members of both major political parties have sponsored legislation requiring protection of the electric grid, without success until in 2016 Congress passed the Critical Infrastructure Protection Act by including it in the National Defense Authorization Act. In late 2015, the Department of Energy directed the development of an EMP resilience strategy, which it published in January 2017. This "Electromagnetic Pulse Resilience Action Plan" addresses at least partially eleven of fifteen recommendations related directly to the electric power system in the 2008 EMP Commission report. In January 2017, President Trump signed an executive order designating improving the U.S. electric grid as a "High Priority Infrastructure Project."

Jonathan K. Zartman

See also: Directed-Energy Weapon; Solid State Laser (SSL)

Further Reading

Electromagnetic Pulse Resilience Action Plan. Washington, D.C.: U. S. Department of Energy, January 10, 2017.

Graham, William R., et al. *Report of the Commission to Assess the Threat to the United States from Electromagnetic Pulse Attack. Volume 1: Executive Report.* McLean, VA: Electromagnetic Pulse (EMP) Commission, 2004.

Pry, Peter Vincent. *EMP Manhattan Project: Organizing Survival against an Electromagnetic Pulse (EMP) Catastrophe.* CreateSpace Independent Publishing Platform: EMP Task Force on National and Homeland Security, 2018.

Tsurutani, B. T., W. D. Gonzalez, G. S. Lakhina, and S. Alex. "The Extreme Magnetic Storm of 1–2 September 1859." *Journal of Geophysical Research* 108, no. A7 (2003), 1268.

Wilson, Clay. *High Altitude Electromagnetic Pulse (HEMP) and High Power Microwave (HPM) Devices: Threat Assessments.* CRS Report No. RL32544. Washington, D.C.: Congressional Research Service, July 21, 2008.

Electromagnetic Railgun (EMRG)

A weapon system that uses the discharge of high-voltage electrical energy (instead of traditional chemical or explosive energy) and a series of electromagnets to rapidly accelerate a metallic projectile to hypersonic velocity providing increased range, speed, and kinetic lethality exceeding that of traditional weaponry. EMRG projectiles often rely solely on speed and kinetic impact to cause damage to a target and thus usually do not require inclusion of an explosive warhead. High speed and extended range may also allow an EMRG with a flatter firing trajectory, decreasing projectile detection and target reaction time while increasing penetration and damage. At a higher trajectory, the railgun capability extends engagement ranges into the hundreds of miles. However, due to the forces and electrical discharge involved, EMRG barrels experience wear over time and must be replaced, often at significant expense due to their complexity. Additionally, the electrical power requirements of an EMRG are considerable, requiring extensive design and accompanying hardware to produce and store the power required to fire the projectile. However, if the technical and logistical hurdles are overcome, an EMRG offers significant military advantages over conventional military munitions and artillery systems.

An electromagnetic railgun consists of a power generator or generating mechanism, a bank of capacitors to store electrical energy, and two parallel electromagnetic rails, one with a positive polarity and the other with a negative polarity. Firing the weapon requires a pulse of power in the tens of megajoules (MJ), which is released in a pulse from the capacitors. This pulse creates a strong magnetic field between the oppositely charged rails. A carrier called an "armature" rides between the rails, conducting the pulse of electricity passing between the polarized rails, and is pushed outward (down the "barrel") by the net electromagnetic force acting between the rails. The armature transports and accelerates a given projectile to hypersonic velocity (Mach >5, or >3,800 mph), after which the projectile exits the barrel and flies on to the target, impacting with high speed and kinetic energy. Different types of armatures have been tested, some of which are solid conductive metal, and some which are vaporized conductive gasses. The key capabilities of either type of armature are that it functions to conduct electricity between the coils and that it accelerates the projectile. The projectile itself is a solid streamlined impactor, and requires no warhead, but may incorporate a guidance system and small control surfaces for increased accuracy in longer range engagements. EMRG research and design has suggested different types of potential projectiles, which can be tailored for the size and type of target, ranging from large hardened structures through "softer" targets such as vehicles and personnel.

The U.S. Armed Forces, primarily the U.S. Army and U.S. Navy, have researched and experimented with design of electromagnetic railguns since the mid-1980s. Due to the technology available at the start of research and high electrical power requirements, the first experimental prototypes required a large room full of supporting systems and capacitors. Continued Research and Development (R&D) with aerospace contractors through the 1990s and 2000s shrunk the supporting systems while increasing railgun capability and performance. As of 2018, electromagnetic railguns are nearing operational size and performance requirements, allowing the potential for them to be deployed on naval surface vessels. Supporting power peripheral equipment and capacitors could be stored internally within a naval vessel or housed within a larger gun space depending on deployment configuration. Railgun power demands and competing vessel requirements are important design considerations, and drive requirements for production as well as potentially limiting the railgun firing rate. In 2008, the U.S. Navy envisioned an EMRG-enabled platform engaging targets into the hundreds of miles of range, multiple times a minute. However, this creates a significant demand on resources, and questions remain for weapon employment tactics, given the energy generation, storage, and release required to operate a weapon.

As new weapon technology, EMRG systems have advantages and disadvantages when compared to existing artillery and missile systems. Size and safety of ammunition are significant benefits of an EMRG system over conventional shells and munitions fired by traditional artillery and naval guns. The fact that solid EMRG projectiles do not contain explosive warheads reduces operational risk. This reduces the hazard of storage and relieves delicate handling required for conventional ammunition. ERMG projectiles also are smaller than conventional artillery rounds. This means that a greater amount of ammunition may be safely shipped to the battlespace, transported, handled, and stored onboard the vessel until required—benefiting the entire supply and logistics chain. The high-speed projectile has the advantage of speed and range, increasing artillery and naval engagement capability from tens to hundreds of miles, and may allow flatter trajectories with increased accuracy over shorter range engagements. The EMRG may also be fired in elevated or ballistic trajectories to increase range and overcome distant obstacles. However, current EMRG prototypes experience significant wear in the rail and armature system over successive firings. This may require extensive rail maintenance or the need to carry spare rails as replacements at sea or afield. The massive power requirement per shot and time between firings depends greatly on the power generation and storage capacity of the carrying vessel or vehicle. Even with reduction in size and complexity, EMRGs are still large in size. Depending on the types of materials involved, the armature mechanism can malfunction, catastrophically destroying the rails. Lastly, as a new weapon system, EMRGs remain untried in combat and are expensive and technically complex. Adaptation and operational testing will require significant continued investment and development to realize final system fielding.

Daniel J. Schempp

See also: Directed-Energy Weapon; Green Ammunition

Further Reading

Adams, Eric. "Is this What War Will Come To?" *Popular Science* 264, no. 6 (June 2004), 64–66.

Heppenheimer, T. A. "Electromagnetic Guns." *Popular Science* 231, no. 2 (August 1987), 55–58.

Kwon, Y. W., N. Pratikakis, and M. R. Shellock. "Multiphysics Modeling of a Rail Gun Launcher." *International Journal of Multiphysics* 2, no. 4 (2016), 421–436.

O'Rourke, Ronald. *Navy Lasers, Railgun, and Gun-Launched Guided Projectile: Background and Issues for Congress.* Washington, D.C.: Congressional Research Service, 2018.

Wong, Wilson. *Emerging Military Technologies: A Guide to the Issues.* Recent Titles in Contemporary Military, Strategic, and Security Issues. Santa Barbara, CA: Praeger, 2013.

Exoskeleton, Powered

Wearable systems that permit increased strength and endurance for the user. Military applications drove the first significant research efforts in the 1960s, and although other likely 21st-century applications include various civilian sectors such as medicine, military uses remain prominent. Military research has noted that "an exoskeleton is not a panacea and will likely still require parallel development with other robotic alternatives" (Reese, 2010, 15), such as load-carrier unmanned ground vehicles. The Naval Research Advisory Committee in the early 2000s concluded that a U.S. Marine squad's equipment weighed almost 1,000 pounds more than the optimal carrying loads of its thirteen soldiers. This is a particularly pressing issue as advocates of distributed operations envision small units at the company, platoon, and squad level undertaking actions that traditionally required a whole battalion or company to accomplish. Increasingly large tasks require bringing larger amounts of equipment and ammunition.

Advocates anticipate that in military usage, exoskeletons could permit soldiers to carry more equipment, ammunition, and body armor, while also being able to traverse greater distances. The combat load carried by military personnel, and particularly by infantrymen, has been a chronic challenge for modern militaries. Efforts to significantly diminish combat loads have been stymied, and the invention of new and important pieces of equipment threatens constantly to exacerbate the problem. Exoskeleton designs have frequently been powered by batteries, and future systems could be powered with batteries or with internal combustion engines or fuel cells.

Notable improvements occurred between the 1960s and the early 2000s. General Electric's Hardiman suit, a hydraulic and electric system codeveloped with the U.S. military in the 1960s, would amplify the wearer's strength by a factor of 25; however, the suit itself weighed over 1,000 pounds and the system's design built delays into the suit's response time, which contributed to a tendency for violent motions that precluded any worn testing of the complete suit. By the 1970s, an exoskeleton developed in Yugoslavia focused on medical rehabilitation with greater success. Los Alamos Laboratories' LIFESUIT program began in the 1960s, and its successor projects in the 2000s delivered the ability to simulate a human walking gait, walk for a mile on a single full electrical charge, and lift 200 pounds for the wearer.

Exoskeleton research transitioned from the Defense Advanced Research Projects Agency to the U.S. Army's Natick Labs in the early 2000s. Research on the Human Universal Load Carrier (HULC) began in 2000, and by 2012 Lockheed Martin refined it into an exoskeleton system able to function in a load-carrying marching mode for 8 hours at a time. The objective was to produce a wearable system able to carry 200 pounds at a sustainable speed of 10 miles per hour. Whereas systems like the earlier Hardiman had meant to be a full-body suit, HULC focused on magnifying the capabilities of the legs. The battery-powered HULC distributed the load weight onto the front and rear of the system in order to minimize the burden felt by the wearer.

SARCOS Labs and Raytheon collaborated to produce the XOS full-body system. Although its power amplifies the wearer's strength by a factor of 25, the system draws power through a tethered cable. XOS therefore offers promise in areas such as vehicle maintenance and logistics, whereas systems like HULC point toward use by personnel on the battlefield itself.

While systems like the HULC, XOS, and earlier Hardiman are active exoskeletons, passive exoskeletons such as the U.S. Marine Corps Mojo work to reduce shock and vibration felt by personnel riding in swift vehicles. Researchers in the military have also noted that "snap on" armor components might be of interest for protecting the wearer and the system itself when exoskeletons are deployed in the future. "The exoskeleton is a radically new technology that offers ground forces the potential to influence the fight persistently, over greater range, and more diverse terrain than ever imagined" (Reese, 2010, 15).

Nicholas W. Sambaluk

See also: Boston Dynamics BigDog; Boston Dynamics Handle Robot; Human Universal Load Carrier (HULC)

Further Reading

Alleyne, Andrew. *Optimized Power Generation and Distribution Unit for Mobile Applications.* Research report. Champaign, IL: University of Illinois, 2006.

Bonato, Paolo. *Skeletal and Clinical Effects of Exoskeletal-Assisted Gait.* Charlestown, MA: Spaulding Rehabilitation Hospital, 2016.

Hansen, Jack, et al. *Human-Robot Interaction to Address Critical Needs to the Present and Future.* Arlington, VA: Office of Naval Research, 2006.

Mooney, Luke Matthewson. *Autonomous Powered Exoskeleton to Improve the Efficiency of Human Walking.* Cambridge, MA: Massachusetts Institute of Technology, 2016.

Reese, Travis C. *Exoskeleton Enhancements for Marines: Tactical-Level Technology for an Operational Consequence.* MA thesis. Quantico, VA: School of Advanced Warfighting, 2010.

Wearable Robotics: Challenges and Trends: Proceedings of the 2nd International Symposium on Wearable Robotics, WeRob2016, October 18–21, 2016, Segovia, Spain. Houston: Springer, 2016.

F-22 Raptor

An advanced stealth fighter jet developed by Lockheed Martin, Boeing, and General Dynamics. Fighter jets have been categorized into "generations," with each generation representing an improvement in speed, sensor, and system capabilities. The F-22 is a fifth-generation fighter, meaning that its airframe has been designed with stealth as a major factor—weapons are stored in internal bays to maintain external radar appearance—and the flight computers are powerful enough to control the extreme maneuverability of the aircraft. Most of the United States' current air fleet, including the F/A-18 and the F-15, are fourth-generation fighters.

A mixture of advanced systems and performance but lack of stealth is what defines the "4.5" generation of fighters, which includes the Eurofighter Typhoon.

The F-22 Raptor demonstrated its abilities in 2014 during operations conducted against the Islamic State of Iraq and Syria (ISIS). The United States—accompanied by its allies—was fighting the militant group in Syria, whose government had not approved the operations. Discretion by the coalition was required in order to avoid setting off Syria's air defense network. The Raptor's stealth made it the ideal candidate for the job. The aircraft further proved itself when, on one mission, an airborne warning and control system (AWACS) aircraft suffered radar issues. The Raptor's sensors give it a view of the entire battle area, indicating enemy and friendly aircraft and hostile facilities on the ground. An F-22 was able to take over the AWACS role and direct all involved forces to a successful mission conclusion.

Despite its impressive performance, the F-22 has seen very little usage in combat. The aircraft's first action was in Syria in 2014, but it had been designated combat ready since 2005. As of 2018, the U.S. Air Force (USAF) inventory counted only 187 Raptors due to the high cost per aircraft—$412 million. This has led to a reduction in the number of aircraft per squadron to between eighteen and twenty-one planes from the traditional twenty-four, and the number of squadrons per fighter wing from three to two. When only part of a squadron is deployed, the higher maintenance and operational demands of the aircraft result in a disproportionate deployment of aircraft and support staff, severely crippling the aircraft remaining at home base. Additionally, the purpose for which the F-22 was designed in 1986 has faded significantly. The potential enemies envisioned in the 1980s, China and the then-Soviet Union, still stand as major powers and are designing fifth-generation fighters of their own, but conflicts throughout the 1990s, 2000s, and 2010s have predominately pitted U.S. forces against insurgent groups and smaller nation-states. These organizations do not possess intricate air-defense networks or modern fighter aircraft, so the demand from ground troops has been for close air support and some threat elimination. Work is being done to develop "light attack aircraft" for this purpose, which have so far consisted of armed versions of both crop dusters and military trainer aircraft.

Unlike its younger brother the F-35 Lightning II, another fifth-generation fighter designed by Lockheed Martin, the F-22 Raptor is prohibited from being exported. Because the F-22 represents the pinnacle of stealth technology and is the top fighter in the USAF, the U.S. government wanted to prevent other nations from seeing its full capabilities. Would-be F-22 buyers have been purchasing alternatives such as the F-35 and the Eurofighter Typhoon. The United Kingdom, Germany, Italy, and Spain did not prioritize stealth and were able to collectively create a very successful multirole warplane in the Eurofighter Typhoon. It proved its ground attack capabilities during campaigns in Libya conducted by the United States and its allies in 2011—in which the F-22 did not participate. It is also successful in intercepting Russian military aircraft that fly too close to European airspace, which is the extent of the Typhoon's aerial combat experience. Despite having both the F-35 and the Typhoon in its air fleet, in 2018 the British government unveiled a project with BAE Systems to create a stealth replacement for the Typhoon, named the Tempest.

Lockheed Martin has proposed an F-22/F-35 hybrid as part of a modernization program for the Japanese Self-Defense Forces (JSDF). The new aircraft would have the larger body and more powerful engines of the F-22, and the more advanced computer systems of the F-35. This, however, may face congressional scrutiny due to the F-22 export restrictions. Stateside, the USAF will hopefully reorganize its Raptor force and finally make use of its air superiority fighter.

Elia G. Lichtenstein

See also: F-35 Lightning II; J31 Gyrfalcon; Su-57

Further Reading

Dorr, Robert F. *Air Power Abandoned: Robert Gates, the F-22 Raptor and the Betrayal of America's Air Force.* Self-published, 2015.

Hirschberg, Michael J., Albert C. Piccirillo, and David C. Aronstein. *Advanced Tactical Fighter to F-22 Raptor: Origins of the 21st Century Air Dominance Fighter.* Reston, VA: American Institute for Aeronautics and Astronautics, 1998.

Niemi, Christopher J. "The F-22 Acquisition Program: Consequences for the US Air Force's Fighter Fleet." *Air and Space Power Journal* (November–December 2012), 53–82.

F-35 Lightning II

Also known as the Lightning II and the Joint Strike Fighter (JSF), a fifth-generation fighter produced by Lockheed Martin. A single-seat, single-engine aircraft designed to have all-weather capabilities intended for multiple roles, it nonetheless reflects emphasis on an air-to-ground striking power. More notable even than its speed, maneuverability, and stealth characteristics are its sensors and significant data collection ability that are enabled, in part, by the system's twenty-four million lines of code. Some have described the plane as a flying computer encased in an airplane frame.

The Department of Defense (DoD) first intended the F-35 to be an affordable fighter in a way that replicated the Air Force's marriage of the expensive twin-engine F-15 air superiority fighter with the more affordable, smaller, and single-engine F-16. The F-35 also is intended to replace the A-10 and the AV-8. The first major portion of updating these fourth-generation aircraft was completed, in part, with the arrival of the F-15's fifth-generation successor, the F-22. That project was cut short by then-Defense Secretary Robert Gates, leaving the Air Force as sole operator of the country's stable of 187 F-22 air superiority fighters.

The Air Force and the Navy partnered to develop the F-35, and Congress subsequently mandated that the Marine Corps participate as well. The underlying assumption was that the partnership would result in cost savings, but, in keeping with other historic attempts, different service requirements drove the cost substantially upward. The F-35 became the most costly defense project in world history, drawing criticism as even the F-35's helmet attracted significant mockery for its $400,000 per unit price tag. Project reorganization in 2011 led to some cost reductions, and in 2017 the newly inaugurated President Donald Trump announced that he had convinced Lockheed Martin to lower its per unit price to under $100 million.

The F-35 exists in three variants. The F-35A offers conventional takeoff and landing (CTOL) for the Air Force. By contrast, the F-35B is short takeoff and vertical-landing (STOVL) capable. In taking off and landing like a helicopter, it best services the Marine Corps' responsibilities for amphibious warfare. Finally, the Navy has the F-35C, which is the Catapult Assisted Take-Off But Arrested Recovery model for use on carriers.

The United States planned to purchase almost 2,500 F-35s. Partner nations have purchased or plan to buy the F-35 as well. These countries include the United Kingdom, Australia, Canada Norway, Italy, the Netherlands, and Turkey. The aircraft is manufactured in three separate sections, by Northrup Grumman in California, by Lockheed Martin in Texas, and by BAE Systems in the United Kingdom, before components are then assembled in Texas. Analysts have noted the diversification of component origin, and because various parts of the aircraft are manufactured in almost every state across the United States, the program remains popular with politicians because of the jobs it provides in many constituencies. First flying in 2006, the F-35C first became operational with the Marine Corps in 2015 and, subsequently, in the Air Force's A model only in 2016. Orders are expected to continue through 2037, and the plane is anticipated to fly until 2070.

Cost marks one controversy regarding the F-35 program. Experts believe that the early designs for the F-35 program were prominent targets in an extensive digital information exfiltration conducted between 2008 and 2010. Reports that defense contractors had been hacked reappeared in 2017 and were answered with assertions that no classified data regarding the F-35 was lost.

Some analysts also question whether the F-35 represents an excessive U.S. reliance on stealth, while others point to a January 2015 fly-off event in which the already forty-year-old F-16 fighter was reported to have outclassed the F-35 in a dogfight. The F-35's reliance on advanced technology, particularly cyber capabilities, also raises the question of if it can be hacked, especially through its autonomic logistics information system (ALIS), which is designed to streamline maintenance and to assess aircraft safety. In the future, some anticipate a number of unmanned aerial vehicles (UAVs) working in tandem with an F-35 in manned-unmanned teaming. Others believe that the F-35 will be the last manned fighter the United States ever builds, with it being replaced next by a UAV capable of air-to-air combat.

Heather Pace Venable

See also: F-22 Raptor; J31 Gyrfalcon; Su-57

Further Reading

Gertler, Jeremiah. *F-35 Joint Strike Fighter (JSF) Program.* Washington, D.C.: Congressional Research Service, 2018.

Gillette, S. E. *Joint Strike Fighter Tactics Development—Truly Joint?* Master's thesis. Quantico, VA: Marine Corps Combat Development Command, 2010.

Harmon, Bruce R. *The Role of Inflation and Price Escalation Adjustments in Properly Estimating Program Costs: F-35 Case Study.* Alexandria, VA: Institute for Defense Analyses, 2016.

Keijsper, Gerard. *Joint Strike Fighter: Design and Development of the International Aircraft.* Barnsley, UK: Pen and Sword, 2008.

Gas Centrifuge

A tool used for separating the isotopes of radioactive uranium via high-speed spinning centrifuges (sometimes arrayed in a cascade) in order to produce enriched nuclear fuel for use in science, energy (nuclear reactors), and military weapons. An isotope of an element, although being the same element as the primary source, differs in atomic mass and properties. In the case of uranium, most naturally occurring uranium is of the isotope uranium-238 (U-238), but nuclear applications require enrichment to a higher concentration of the more radioactive isotope uranium-235. In this context, "enrichment" refers to the percentage of a uranium product containing U-235.

A centrifuge is a scientific, technical, medical, and industrial tool used to separate solids, liquids, and/or gases by spinning a sample in a high-speed revolution. The centripetal and centrifugal forces created by the high-speed rotation separate different masses and densities of a sample. Parts of a sample with higher mass are pulled by centrifugal force to the outside of a container (the furthest from the axis of rotation, and moving at highest speed). In a gas centrifuge, a gaseous mix of uranium is spun at high speed by a centrifuge in a vacuum to separate the lighter radioactive U-235 isotope from the more massive and less radioactive U-238 isotope. The process can be repeated with the product (somewhat enriched U-235 gaseous mixture) to further refine the percentage of radioactive U-235 in the product. To expedite this process, several gas centrifuges may be arranged in a "cascade," which allows efficient removal of the U-235 and automatic repetition of the process. Using this process, an enriched product of radioactive uranium material may be produced at whatever percentage is required for the intended use.

Nuclear power plants require low enriched uranium as fuel, while nuclear weapons require highly enriched uranium to sustain a destructive nuclear chain reaction. Naturally mined uranium must undergo significant processing and refinement as described in order to separate the most highly radioactive content. Historically, enrichment of uranium was a technically complex, time-consuming, dangerous, and expensive process. However, technologists in the United States and the Soviet Union began experimentation with gas centrifuges as a lower cost and technologically more accessible path to nuclear enrichment, and development efforts yielded further progress starting in the 1960s. Gas centrifuges remain a proliferation concern due to the relative simplicity and low cost of the technology.

Daniel J. Schempp

See also: Operation Olympic Games; Stuxnet

Further Reading

Albright, David. *Peddling Peril: How the Secret Nuclear Trade Arms America's Enemies.* 1st Free Press Hardcover ed. New York: Free Press, 2010.

Kemp, R. "The End of Manhattan: How the Gas Centrifuge Changed the Quest for Nuclear Weapons." *Technology and Culture* 53, no. 2 (2012), 272–305.

Krige, John. *Sharing Knowledge, Shaping Europe: U.S. Technological Collaboration and Nonproliferation.* Transformations. Cambridge, MA: The MIT Press, 2016.

Obeidi, Mahdi, and Kurt Pitzer. *The Bomb in My Garden: The Secret of Saddam's Nuclear Mastermind.* Hoboken, NJ: Wiley, 2004.

Global Positioning System (GPS)

An integrated system of space-based satellites, ground control elements, and individual equipment that enhances navigation, transmits precise time to receivers worldwide, and helps users pinpoint locations. GPS is available to both civilian and military users, and while civilians value the system, military leaders robustly incorporate GPS into the planning and conduct of military operations.

GPS, sometimes referred to by the legacy term "Navstar" or "Navstar GPS," offers users a near-perfect assessment of position, velocity, altitude, and time practically worldwide and in all weather conditions. Space-based assets include a constellation of about thirty satellites that orbit the earth and transmit signals. Space professionals command and control this system through ground-based master control stations, monitoring facilities, and antennas. The ground control segment incorporates redundancy, backup systems, and protections. Millions of users equipped with many types of receivers use GPS in thousands of ways. For example, civilians use GPS while flying private and commercial aircraft. Sportsmen use GPS to traverse the outdoors. Motorists use GPS as they go about daily routines. Military leaders use GPS to achieve significant military advantage.

Military leaders have always pursued technology to accurately determine geographic positions, direction and distance to desired positions, altitude if appropriate, and precise time. For example, travelers have used compasses to navigate since the 11th century. Basic map reading, an early means of land and aerospace navigation, was improved by flight instruments, dead reckoning, radio aids, radar, and celestial navigation. GPS not only complements these early efforts but has significantly affected military activities since achieving legitimacy and fame during the 1991 Gulf War.

In 1964, about twenty-five years before the 1990 buildup prior to the Gulf War known as Desert Shield, and the 1991 commencement of hostilities known as Desert Storm, leaders and visionaries conceptualized the idea of GPS. The project, branded by the term "Navstar," was a joint Air Force–Aerospace Corporation study. Stakeholders struggled for decades with both congressional and Air Force funding shortfalls. Despite these and other challenges, they launched the first research and development satellite on February 22, 1978, and then placed sixteen satellites in orbit by the onset of the 1991 Gulf War.

Precision, typically enhanced by GPS, became a key theme in the 1991 Gulf War. U.S. Air Force MH-53J Pave Low III helicopters, with unique capabilities to include GPS, acted as pathfinders for U.S. Army attack helicopters on the first night of the war. Stealth bombers relied on GPS as they penetrated perilous air defenses. B-52 bombers leveraged GPS as they flew long combat missions. Special operations forces used GPS to navigate featureless terrain. Fighter and bomber aircrew members, such as F-16 pilots armed with the Low-Altitude Navigation and Targeting Infrared for Night (LANTIRN) system, found GPS effective in enhancing the accuracy of their attacks. Many military personnel purchased their own handheld GPS units. Some aircrew members, assigned to aircraft without an integrated GPS system, attached civilian handheld GPS receivers with Velcro to cockpit areas. Finally, Gulf War leaders introduced the public to Precision-Guided Munitions (PGMs).

Millions of viewers worldwide witnessed PGMs going through windows and down air shafts while destroying key targets. The practical application of GPS precision in 1991, along with the public marketing of capability, significantly affected future warfighting tactics, techniques, operations, and strategy.

The ongoing validation of GPS technology continues to unveil enormous military benefits. Military leaders in the decades after the 1991 Gulf War leveraged GPS during rescue and humanitarian missions worldwide. Airmen employed many PGMs during the 1999 air war over Serbia. Military leaders continue to leverage GPS within the U.S. Central Command area of responsibility. GPS satellites transmit near-perfect time transfer signals to synchronize digital communications to permit frequency hopping. Finally, GPS complements other space-centric systems such as meteorological and communication assets.

GPS today assists military users by enhancing navigation, transmitting precise time, and helping pinpoint locations. GPS is a unique and significant technology that is deeply integrated into military planning and operations.

John Blumentritt

See also: Anti-Satellite Weapon (ASAT); Precision-Guided Munitions (PGMs)

Further Reading

Boyne, Walter J. *Beyond the Blue: A History of the U.S. Air Force*. New York: St. Martin Griffin, 1997.

Hallion, Richard P. *Storm over Iraq: Air Power and the Gulf War*. Washington and London: Smithsonian Institution Press, 1992.

Krolikowski, USAF Maj. Jennifer. "Navstar Global Positioning System." In *AU-18 Space Primer*. Prepared by Air Command and Staff College Space Research Elective Seminars. Maxwell AFB, AL: Air University Press, 2010, 217–226.

Green Ammunition

A U.S. military program aiming to find and deploy effective replacement components to some of the hazardous materials that constitute small arms ammunition. An Army office was first formed for the purpose in 1995 with representation from the national laboratories at Oak Ridge, Tennessee, and at Los Alamos, New Mexico.

An important category for research has been the primer and related materials. Research into "smokeless" propellants represented part of a revolution in chemistry in the late 1800s, allowing vast increases in shooting range and accuracy relative to the black powder propellants universally used until the 1880s. Concurrently, metallic cartridges facilitated rapid-fire weapon mechanisms. Sealant and primer materials were required for the new cartridges, and these were typically developed and selected based on their ballistic implications and on economic considerations. Often, however, these volatile chemical compounds present health hazards. Examples include barium nitrate and methyl chloroform used in tracer ammunition and for incendiary ammunition, as well as toluene that frequently appeared as a primer material and in explosives. Research into green ammunition has therefore included work to eliminate environmentally harmful materials, whose effects are evident at sites of ammunition manufacture as well as sites of use such as gun training ranges.

Another area of concern is the projectile itself. Lead and antimony are heavy metals posing known dangers to humans and animals. The displacement of lead bullets with copper is an example of shifting toward more green ammunition. These efforts can prove complicated, since sometimes the replacement material is itself subsequently judged to raise environmental and health concerns. The use of tungsten-alloy materials and of tungsten-nylon combinations for bullets was attempted in the early 2000s, but by 2008 the Army identified enough health hazards from the new tungsten ammunition to discontinue manufacture.

These efforts have led to the development of 5.56 millimeter ammunition compatible with U.S. military small arms, which has been reported as comparably effective relative to other non-green ammunition types used in the early 21st century. Although the goal of green ammunition is to retain effectiveness and improve environmental impact, the research is contemporary to research and development of alternatives to battlefield firearms.

Nicholas Michael Sambaluk

See also: Directed-Energy Weapon; Electromagnetic Railgun (ERMG); Laser Weapon System (LaWS); Solid State Laser (SSL)

Further Reading

Arnemo, Jon M. "Health and Environmental Risks from Lead-Based Ammunition: Science versus Socio-Politics." *EcoHealth* (2016), 618–622.

Bellinger, David C., et al. "Health Risks from Lead-Based Ammunition in the Environment: A Consensus Statement of Scientists." Santa Cruz: University of California, 2013.

Branco, Pelagio Castelo, et al., eds. *Defense Industries: Science and Technology Related to Security: Impact of Conventional Munitions on Environment and Population.* Dordrecht, Germany: Kluwer, 2004.

Guided Small Arms

Precision-sighted arms and/or accompanying ammunition that may have the ability to receive and execute corrections to increase accuracy and ensure a higher probability of the projectile hitting a target. The term "small arms" refers to weapons and associated ammunition which are man- or team-portable and do not require heavy weapons vehicle mounts (i.e., are not cannons, mortars, or large caliber weapon systems like tanks). The term typically refers to rifle and pistol caliber weapons carried by individuals or used in support of armed personnel, although it can extend to light machine guns or crew-served weapons, typically .50 caliber (12.7 mm) or below. In this context, "caliber" refers to the size of a projectile fired by a given diameter weapon barrel. As of 2018, much of the effort is still theoretical and dominated by research and development (R&D) work. There are currently two general approaches to creating guided small arms: the first emphasizes the weapon, while the second concentrates on a guided, correcting projectile. Future systems may incorporate elements of both approaches.

The weapon approach focuses on precision firearms equipped with sensitive sensors and electronic computer sighting mechanisms, which will fire only at the correct conditions to ensure a perfect shot. Only a small number of companies have pursued this route, although registered U.S. Patents describe the process. Designed to eliminate uncertainty in the shot from human and environmental factors, the weapon sight is used to designate and optically track an aim point on a given target. Given inputs from a suite of other sensors onboard the weapon, the sighting computer calculates the correct shot and, when conditions are right, fires the weapon. Equipped with the sighting computer, sensors, and designation mechanism, the weapon is described as a precision-guided firearm (PGF), although the projectile itself is not guided.

Alternatively, the projectile approach focuses on in-flight corrections to a fired bullet or munition in order to guide it to a designated target. The 20th century saw the development of precision-guided missiles and munitions as advances in technology and warfare enabled development of advanced weapon systems. Guided systems require methods to control and execute in-flight corrections in order to reach or intercept their target. This may be done onboard the weapon by sensors, guidance system calculations, and steerable control surfaces; by corrections sent by communication with the launch platform; or both methods combined. Precision sensors and their associated guidance systems are complex and costly to develop and acquire, and must be small enough to fit inside a given projectile housing. Additionally, they must be "flight tested" to ensure they reliably perform under the physical stresses of launch and target intercept. Advancement in the complexity, capability, and computing power of advanced microsystems have allowed miniaturization, although developing guided small arms caliber projectiles remains a difficult engineering problem. The ability to guide a projectile requires a method of control or steering (most often via fins or small control surfaces), a system to calculate (or receive) necessary corrections and send commands to those control surfaces, and finally the ability to execute corrections quickly and accurately in the short time of projectile flight. All of those processing, communication, and control systems must fit in a small diameter space within a bullet-sized projectile. Additionally, those systems must reliably perform after undergoing the explosive forces, heating, and acceleration of being fired from a weapon.

Guided projectiles could provide an unprecedented tactical advantage, allowing fewer rounds to be fired at longer ranges in an engagement and find their target. The U.S. Air Force and Auburn University in Alabama collaborated in the 1990s to examine barrel-launched adaptive munitions (BLAMs) for use by aircraft. This concept uses a sensor, electronics, and electrically driven movable surfaces on the projectile itself to change the airflow around the bullet and influence direction. The target would need to be externally tracked by either radar or laser energy to illuminate the target for the BLAM sensor.

Recent efforts by both Defense Advanced Research Projects Agency (DARPA) and Sandia National Laboratories have focused on guided .50 caliber projectiles. The DARPA Extreme Accuracy Task Ordnance (EXACTO) program focuses on a guided bullet firing through a conventional sniper rifle; the Sandia National Laboratories project envisions a smooth bore rifle firing a finned guided projectile. Both approaches require the projectile to optically track a laser illuminated target (using the laser for guidance corrections) and contain actuators or fin controls to change airflow around the moving bullet. Other potential developers have examined designing a bullet with micro electromechanical strips on the leading portion, which require very little power and can flex on command to influence airflow around the bullet and guide its flight.

Daniel J. Schempp

See also: Defense Advanced Research Projects Agency (DARPA); Green Ammunition

Further Reading

Boffard, R. "Target Acquired: The Future of Bullets." *Engineering & Technology* 11, no. 1 (2016), 36–39.

Gibson, Neil. "Sandia Creates Self-Guided Small Calibre Bullet." *Jane's International Defense Review* 45 (2012), 8–9.

Jones, James F., Brian Kast, Marc Kniskern, Scott Rose, Brandon Rohrer, James Woods, Ronald Greene, and Sandia National Labs. *Small Caliber Guided Projectile.* U.S. Patent 7,781,709 B1, 2010.

Lipeles, Jay, and R. Glenn Brosch. *Guided Bullet.* U.S. Patent 6,474,593, 2002.

Lupher, John Hancock, Stefanie Kwan, Douglas Scott, and Michael Toal. *Precision Guided Firearm Including an Optical Scope Configured to Determine Timing of Discharge.* U.S. Patent 9,127,907, 2015.

Helmet Technologies

Materials, shapes, and attachments appearing in the designs of 21st-century military combat headgear in order to optimize their usefulness in protecting personnel and facilitating their activities in the combat zone.

The United States has been a leading country in the development and fielding of helmet technology since the middle of the 20th century. Proactive helmets, which had been a common part of body armor in antiquity and the medieval era, had fallen out of favor during the 18th and 19th centuries because of their weight and their inability to protect the head from small arms bullets or artillery. The widespread use of new shrapnel shells during World War I led to a massive increase in the number of fatal head wounds. This forced the return of the steel helmet, and steel continued to be the preferred helmet material during most of the 20th century. A new Personnel Armor System for Ground Troops (PASGT), including both a ballistic vest and a new combat helmet, was introduced to the U.S. military in 1983. The PASGT helmet was superseded by a succession of follow-on types, but the PASGT marked the transition away from steel (which had dominated helmet technology in the 20th century as well as the helmets of the earlier medieval and Renaissance eras as well) and instead toward a range of other materials for ballistic protection.

U.S. helmets of the 21st century include the Modular Integrated Communications Helmet (MICH), the Lightweight Helmet (LWH), the Advanced Combat Helmet (ACH), and the U.S. Enhanced Combat Helmet (ECH). MICH, introduced in 2001, borrowed directly from PASGT, its hallmark characteristic being a reworked helmet shape in order to facilitate the use of tactical headsets for communication. LWH was adopted by the Navy and Marines in 2003 as a successor to PASGT.

LWH, colloquially called the "Fritz helmet" because of its resemblance to the steel 20th-century German Stahlhelm, weighs slightly less than PASGT and offers better protection, also incorporating a protection system for the nape of the user's neck.

ACH, currently in service with the U.S. Army, was introduced in 2002, and it too added an attachable "nape pad" by 2007. The following year, in view of the military operations in occupying Iraq and the proliferation of improvised explosive device (IED) attacks and personnel injuries, an Army study began emplacing sensors on some ACH helmets in order to better study IED-induced head injuries. This new information would be incorporated into considerations during the design of future helmet types. U.S. soldiers in Afghanistan were first issued a MultiCam cover for their ACH types during 2009.

A joint Marine-Army effort to further refine the helmet type and to ultimately consolidate on a single pattern for ground services worked toward the creation of ECH. Although ECH was to have a profile similar to that of ACH, the helmet was designed to be thicker walled, so that its ultra-high-molecular-weight polyethylene materials could protect against rifle bullets. Reportedly, tests of ECH designs have successfully exceeded the 35 percent protective improvement (relative to current ACH helmets), which had been set as a precondition to acceptance. ECH development began in 2007, and by mid-2009 a total of five competing designs had been drawn up. Refinements in selection of materials and further testing ultimately led to the letting of a contract with Ceradyne in March 2012 for the construction of the first helmets. Their issuance a year and a half later would mark the beginning of a shift toward ECH, and older types such as MICH and LWH would gradually be shifted to rear echelon units and for training activities.

Many other nations have also followed in the footsteps of the U.S. PASGT helmet. For example, Australia adopted what it called its Enhanced Combat Helmet in 2004; this is an Israeli-produced version of the U.S. MICH, but it is lighter in weight, comes in four head sizes, and carries headband and suspension similarities to PASGT. New Zealand followed suit in 2009.

The French SPECTRA helmet is named for the Spectra fiber used in its construction. Design of the SPECTRA was prompted by its experience during its participation in United Nations activities in the Balkans during the 1990s. Contemporary French steel helmets had proven ineffective against adversary snipers in Sarajevo, and the French military concluded that it needed a new helmet with improved effectiveness against rifle bullets. SPECTRA was developed throughout most of the 1990s and 2000s and was introduced in 2016.

Israel's OR-201, developed in the 1960s, was one of the first Kevlar helmets in the world, and it is still used in sixteen countries including Ecuador, Egypt, Mexico, India, Portugal, and commandos in South Africa. Beginning in the mid-1990s, Israeli Defense Forces have also issued "mitznefet" helmet coverings in order to obscure the outline of the issued helmet and therefore assist in camouflaging the wearer when prone in the countryside.

The People's Republic of China is reported to be in the midst of transitioning from its GK80 steel helmet to a new model called QCF 02/03. The GK80 was a 1980 refinement of an earlier model of 1960s vintage. In keeping with the pattern set by many other countries in the late 20th century and early 21st century, QCF 02/03 is constructed of ballistic-rated heat-resistant synthetic fibers rather than of steel.

The United Kingdom's Mk 7 helmet, introduced in 2009, is an ergonomic improvement on its 1980s-era Mk 6. Both are made of ballistic nylon, and the newer model has been reshaped so that it is easier to use when fighting prone. The United Kingdom has exported thousands of Mk 6 and Mk 7 helmets to Ukraine following the invasion of the eastern half of Ukraine by paramilitary forces in 2014.

Different helmets appear in other settings, including for flight crew in aircraft, incorporating communications and other equipment but not designed around ballistic protection crucial for helmets in ground combat. The F-35 Joint Strike Fighter was conceptualized to include sensors on the plane that would transmit information to the pilot's head-up display (HUD) so that the pilot's perspective would simulate an ability to see in all directions and through the pilot's own aircraft.

Nicholas W. Sambaluk

See also: Body Armor, Modern; Improvised Explosive Device (IED)

Further Reading

Bhatnagar, Ashok, ed. *Lightweight Ballistic Composites: Military and Law-Enforcement Applications.* Cambridge, UK: Woodhead, 2006.

Chen, Xiaogang, ed. *Advanced Fibrous Composite Materials for Ballistic Protection.* Cambridge, UK: Woodhead, 2016.

Crouch, Ian G., ed. *The Science of Armour Materials.* Cambridge, UK: Woodhead, 2017.

Franklyn, Melanie, and Peter Vee Sin Lee. *Military Injury Biomechanics: The Cause and Prevention of Impact Injuries.* Boca Raton, FL: CRC Press, 2017.

Inspector General, Department of Defense. *Advanced Combat Helmet Technical Assessment.* Washington, D.C.: Department of Defense, 2013.

Sparks, E., ed. *Advances in Military Textiles and Personal Equipment.* Oxford, UK: Woodhead, 2012.

Human Universal Load Carrier (HULC)

An active exoskeleton system built in the early 2000s to assist the U.S. military to address the problem of combat load. Although not deployed in combat by the U.S. military, Lockheed Martin deemed the technology promising enough to enter into a licensing agreement with Ekso Bionics, who had initiated work in 2000.

The equipment carried by military personnel, and especially by infantry in combat, regularly exceeds optimal burdens defined by military researchers, and attempts to reduce the amount and weight of needed equipment, ammunition, and supplies have repeatedly proven unequal to the challenge. A wearable active exoskeleton system is able to reduce the burden felt by the user.

Unlike some other exoskeleton concepts, HULC aimed solely to amplify the carrying capacity of the wearer. HULC's 10:1 ratio of support means that a soldier using the system can carry 200 pounds and feel the burden as if the weight is only 20. The weight is placed on the front and rear of the frame to accomplish this. Although the 10:1 ratio is a far more modest amplification than that of the contemporary SARCOS Labs–Raytheon tethered XOS system, HULC is not dependent on external power and is therefore an inherently more mobile device.

This was of particular importance because the military need involve not only enabling wearers to carry more weight but also that the users could actively move in the field as infantry. This represents a significant challenge for the battery-operated exoskeleton. HULC's designers aimed for a sustainable 10 miles per hour capability. Early prototypes would manage only 4 hours of movement at a speed of 3 miles per hour, but by 2012 refinements of the system were able to operate in a marching mode for twice as long.

Increasing the duration of use on a single battery drove the need to contemplate a larger battery, which increased the power required for the suit, and ultimately the U.S. military opted not to adopt HULC. Nonetheless, an unmotorized derivation called iHAS and a follow-on Mantis attracted some civilian interest. Work on HULC advanced the state of the art in exoskeleton technology.

Rose E. Sambaluk

See also: Boston Dynamics BigDog; Boston Dynamics Handle Robot; Exoskeleton, Powered

Further Reading

Bonato, Paolo. *Skeletal and Clinical Effects of Exoskeletal-Assisted Gait.* Charlestown, MA: Spaulding Rehabilitation Hospital, 2016.

Hansen, Jack, et al. *Human-Robot Interaction to Address Critical Needs to the Present and Future.* Arlington, VA: Office of Naval Research, 2006.

Mooney, Luke Matthewson. *Autonomous Powered Exoskeleton to Improve the Efficiency of Human Walking.* Cambridge, MA: Massachusetts Institute of Technology, 2016.

Reese, Travis C. *Exoskeleton Enhancements for Marines: Tactical-Level Technology for an Operational Consequence.* MA thesis. Quantico, VA: School of Advanced Warfighting, 2010.

Hypersonic Missiles

A powered or unpowered maneuverable projectile that is capable of sustained, guided flight at speeds exceeding Mach 5 in the sensible atmosphere (in the atmosphere at <100-km altitude). Hypersonic missiles may be generally separated into two types: unpowered hypersonic glide vehicles (HGVs) and powered hypersonic cruise missiles (HCMs).

Flight speed within the sensible atmosphere may be broken down into three regimes:

- Subsonic: 0–760 mph, Mach 0–0.99, most aircraft, some small caliber ammunition
- Supersonic: 761–3,805 mph, Mach 1–4.99, missiles, supersonic aircraft, ammunition
- Hypersonic: 3,806–19,030 mph, Mach 5–25, spacecraft, test aircraft, hypersonic missiles

The desire is for HGVs and HCMs to operate at the highest achievable hypersonic speed at which control is still possible, in order to increase range and flexibility while reducing target response time. Current hypersonic vehicles achieve

Mach 5 to Mach 7. Flight at higher hypersonic speeds causes extreme environmental effects and technical difficulties for propulsion, control, guidance, and vehicle structure, particularly with respect to heat that must be absorbed or mitigated by the vehicle.

A hypersonic glide vehicle is the simplest hypersonic missile, typically launched on a ballistic trajectory from a ground-based ballistic missile into the high atmosphere (~100 km in altitude). In its initial launch and boost phases an HGV behaves in a similar manner to a conventional ballistic payload. However, the HGV takes a shallow trajectory "gliding" or riding on the top levels of the atmosphere over an extended range to the selected target, in contrast to a traditional ballistic arc and nearly vertical reentry. An HGV contains control surfaces and guidance and may be maneuvered repeatedly, maximizing targeting flexibility. Control surfaces or winglets provide lift in the minimal atmosphere. The near-vertical launch profile and high atmospheric flight imparts the HGV with sufficiently high speed and kinetic energy that it does not require additional propulsion. The HGV's long reentry profile extends the total flight time in the atmosphere and the heating across vehicle surfaces relative to that encountered by a ballistic missile warhead. An HGV may contain a warhead (explosive, chemical/biological/nuclear, or other) or may simply use its extreme kinetic energy to destroy a target.

A hypersonic cruise missile, by contrast, resembles a traditional cruise missile, may be either ground or air-launched, and uses a combined series of engines to accelerate to supersonic and then hypersonic velocities. This presents a significant technical design challenge, as single type of engine cannot accelerate the HCM in both the supersonic and hypersonic realms, and the vehicle must be able to maintain aerodynamic control at both speeds. In most current designs, a rocket engine is used to accelerate the HCM to and through supersonic flight, with a scramjet engine activating for sustained hypersonic flight. A scramjet engine uses direct injection of fuel into the supersonic engine air stream, which combusts and sustains high-speed flight. HCMs may use a hydrocarbon fuel or rely on liquid hydrogen as dual-use fuel and coolant. HCMs tend to travel at lower altitudes than HGVs (<30 km in altitude), are maneuverable, and may contain a warhead.

The United States, China, and Russia reportedly lead the world in research and development of hypersonic missiles, and partner in research with many of their allied nations. The United States pursued the promise of hypersonic technology since the 1930s, but maturation of materials and propulsion system technologies in the late 1990s and 2000s have only recently made operational hypersonic missiles a viable possibility. Hypersonic missiles present a deterrence concern due to their potentially destabilizing influence. Their range, unpredictability, ability to evade detection/defenses, and speed all shorten the decision-making time for a targeted nation. This could potentially escalate a conflict. Additionally, hypersonic missiles may present a proliferation concern if entire systems or key technology are irresponsibly shared. As hypersonic missiles develop, it is likely nations will look to nontraditional countermeasures, such as directed-energy weapons to counter the threat.

Daniel J. Schempp

See also: BrahMos-II; X-51 Waverider

Further Reading

Besser, Hans-Ludwig, et al. "Hypersonic Vehicles: Game Changers for Future Warfare?" *Joint Air Power Competence Centre* 24 (2017), 11–27.

Kemburi, Kalyan M. *Diffusion of High-Speed Cruise Missiles in Asia: Strategic and Operational Implications*. Singapore: Institute of Defence and Strategic Studies, 2014.

Sibu, C. M., et al. "Advanced Development in Hypersonic Cruise Missile—Inertial Guidance System, Universal Launch Platform & Ramjet Engine." *Journal of Basic and Applied Engineering Research* 1, no. 10 (October 2014), 15–29.

Speier, Richard H., George Nacouzi, Carrie Lee, and Richard M. Moore. *Hypersonic Missile Nonproliferation: Hindering the Spread of a New Class of Weapons*. Santa Monica, CA: RAND, 2017.

IAI Heron

A reconnaissance unmanned aerial vehicle (UAV, commonly known as drone) developed by Israeli Aerospace Industries (IAI). The Heron is officially described by its designers as a "multi-role medium altitude long-endurance (MALE)" UAV. The aircraft has a unique design that differs from many other UAVs that have conventional configurations. A conventional configuration involves one main wing for lift, and smaller wings—called the horizontal and vertical stabilizers—at the rear of the aircraft for directing it up/down and left/right, respectively.

The Heron has a central fuselage containing surveillance, guidance, and control systems. On each side of the central fuselage, there is a small portion of wing that connects to a long, narrow fuselage known as a boom. The horizontal stabilizer is located between the end of the two booms, and a vertical stabilizer is placed perpendicular to this on each boom. The main wing extends outward from each boom, a continuation of the wing connecting the boom to the central fuselage. A pusher propeller is located on the back of the central fuselage, using a turboprop engine for propulsion. This configuration closely resembles the P-38 Lightning built by Lockheed during World War II. The advantage of this design is that the tail can exert the same amount of force on the aircraft without using a longer central fuselage, which would increase the weight of the aircraft.

There are two models of the Heron: the standard or "Machatz-1" model, and the model known as the "TP" or "Eitan" for higher altitude and larger payload. Both models have many autonomous features: they can automatically take off and land, and they can follow a predesignated flight path without external input from its ground controllers. The controllers can, if need be, pilot the Heron manually.

The Heron is in use by several militaries, not just the Israeli armed forces. In support of the United Nations Multidimensional Integrated Stabilization Mission in Mali (MINUSMA) operations, the German armed forces use the Heron for intelligence gathering and enforcing security perimeters. The aircraft had proven itself previously to the Bundeswehr during operations in Afghanistan beginning in 2010. The French Air Force also uses the Heron in Mali. IAI has finally produced something that has been on India's wish list for years: armed drones. These are export versions of the Heron TP—designated Heron TP-XP—that are able to carry missiles. Maritime versions of the Heron are in use by countries including El Salvador and Australia. In July 2018, IAI tested the aircraft with new sonar buoy ("sonobuoy") systems for detecting submarines.

In the future, the Heron is expected to be acquired by more countries for various purposes. The Georgian military used Israeli-made drones during the Russian military's attempted invasion in 2008. The effectiveness of the UAVs in artillery targeting so impressed the Russians that the invading country eventually bought its own drones from IAI in 2010. Since 2015, more purchases have been made as Russia has increased its involvement in the Syrian Civil War.

Elia G. Lichtenstein

See also: MQ-1 Predator; MQ-9 Reaper; Unmanned Aerial Vehicle (UAV)

Further Reading

Majumdar, Sayan. "IAI's Heron—the Unmanned Sentinel." *Vayu Aerospace and Defence Review*, Issue 3 (May/June 2014), 85.

Tsach, Shlomo, Jacque Chemla, and D. Penn. "UAV Systems Development in IAI—Past, Present & Future." 2nd AIAA "Unmanned Unlimited" Conf. and Workshop & Exhibit, San Diego, 2003.

Improvised Explosive Device (IED)

A homemade bomb, potentially fabricated from military or nonstandard parts. Although not a new tool of warfare, these weapons burst onto the global stage in the mid-2000s as Iraqi and Afghani insurgents utilized them against conventional American and coalition forces. These devices generally consist of some form of initiation mechanism connected to a detonator on an explosive charge and projectile. IEDs can be used in a variety of roles, ranging from antipersonnel to roadside bombs intended to target large vehicles. Ease of construction, rapid emplacement, and relative effectiveness make them a common weapon of unconventional forces.

Although IEDs have become much more popular in the present day, insurgent groups throughout history have used such weapons. The Russian "People's Will" rebel group *Narodnaya Volya* used IEDs throughout the 1870s and 1880s in an attempt to challenge the Tsarist rule of Alexander II and assassinate him in 1881, and the Irish nationalist group *Clan na Gael* used variations of IEDs against their British occupiers in the 1880s. Anarchists across Europe continued to use IEDs against monarchial rule at the turn of the century, and insurgent groups in Southeast Asia, Africa, and Latin America utilized IEDs in anticolonial wars and rebellions.

Triggering systems vary, but they include command-initiated designs activated by humans as well as autonomously initiated types. Modern command-initiated IEDs are often constructed by mechanisms that send a signal to another component, often attached to the detonator; examples of this construction include cell phones, pagers, remote-controlled toys, and other such technologies. Due to the human interaction necessary to initiate IEDs of this design, insurgents must be able to observe the target as it approaches the device, thus placing the attacker at greater risk of detection. Autonomously initiated IEDs use designs such as pressure plates and trip wires. Autonomous designs allow insurgents to evade detection and also reduce the chances that an IED's existence is suspected by adversaries who identify an emplaced insurgent.

Construction varies as well. Artillery and mortar rounds are often favored materials, since their explosives tend to be reliable, and they are often used to target vehicles or larger objectives. Homemade projectiles, such as pipe bombs filled with fragmentary material, are often used in an antipersonnel role. As IEDs have continued to demonstrate their effectiveness on the modern battlefield, additional modifications have been made to improve the lethality of such devices. In Iraq, Shiite militia groups developed explosively formed projectiles—shape-charge IEDs with a copper liner—to target and penetrate heavily armored vehicles.

Although often rudimentary in design and construction, IEDs have proven to be effective. Easily camouflaged and emplaced, defeating such weapons has proved problematic. Examples of such locations include being buried along avenues of approach, hidden in debris, and booby-trapping houses. To counter these emplacement techniques, conventional forces developed heavily armored engineering vehicles to conduct reconnaissance and destroy identified IEDs along major roads. Likewise, explosive-sniffing dogs are also utilized to find these hidden devices on the modern battlefield. Many militaries, including the United States, have applied additional armor to their existing vehicles in an effort to mitigate the effects of these weapons. The U.S. military developed a new form of vehicle, the Mine-Resistant, Ambush-Protected vehicle, to counter the blast characteristics of most IEDs. Additional changes have been made by conventional forces to attempt to defeat the detonation systems utilized by insurgents, such as wireless jamming. Although effective to a certain degree, insurgent forces continue to adapt the designs of their IEDs and their initiating systems to counteract the technological and tactical adaptations of their foes. Due to their varied construction, emplacement, and initiation, IEDs will continue to be a mainstay of the modern battlefield and difficult to defeat. In order to truly attempt to defeat such weapons, conventional forces must effectively target the individuals and networks that support and finance the building and emplacement of such weapons.

Christian Garner

See also: Mine-Resistant Ambush-Protected (MRAP) Vehicles; Vehicle-Borne Improvised Explosive Device (VBIED); Vehicle-Mounted Active Denial System (VMADS)

Further Reading

Good, Matt T. *The IED: Tactical Solutions for a Tactical Problem.* Quantico, VA: Marine Command and Staff College, 2010.

North Atlantic Treaty Organization Glossary of Terms and Definitions (AAP-06). Brussels, Belgium: NATO Standardization Office, 2018.

Revill, James. *Improvised Explosive Devices: The Paradigmatic Weapon of New Wars.* Cham, Switzerland: Palgrave Macmillan, 2016.

Smith, Andrew. *Improvised Explosive Devices in Iraq, 2003–09: A Case of Operational Surprise and Institutional Response.* Carlisle, PA: Strategic Studies Institute, 2011.

Infrared-Guided Weapons

A passive-guidance system that identifies and tracks targets based on their infrared signature, which is a function of their heat. Humans and operating vehicles register prominent infrared signatures and thus are both relevant and vulnerable targets. The passive nature of infrared guidance also denies warning to an enemy that it is being targeted. Infrared guidance systems are believed to have caused 90 percent of all combat losses of U.S. Air Force planes in the post–Cold War era.

Interest in infrared detection led to research efforts during World War II in hopes of use by armor and aircraft, but the first infrared-guided weapons included the U.S. AIM-4 Falcon and AIM-9 Sidewinder air-to-air missiles in the mid-1950s. Early Sidewinder variants outclassed the Falcon and other contemporary infrared-guided systems such as the Soviet reverse-engineered copy of the Sidewinder, but through the 1960s none of these types demonstrated kill probabilities above 20 percent. Research efforts in the United States, the Soviet Union, and Britain between the 1950s and 1970s also sought to devise infrared-guidance systems for man-portable air defense (MANPAD) capabilities that could reduce the vulnerability of infantry to air attack. Although the Soviet 9K32 Strela represented the best of the early efforts, it was impressive by comparison only.

Early infrared sensors proved relatively successful in detecting jet exhaust, and as a result enemy aircraft were usually targeted from the rear, where their exhaust would be prominent and would lead sensors to the plane. This method, however, meant that early MANPADs could be effectively used only by a soldier who had already been attacked and had survived an overpass by an enemy jet. This did not lead to popularity of the first generations of MANPAD designs.

Major technical improvements occurred in the 1970s and 1980s as infrared guidance could be thought of reaching greater maturity. For aircraft use, technological advance occurred in combination with revised training for pilots using infrared-guided weapons. The Falklands War in 1982 highlighted the technological leap, as British pilots deploying new L model Sidewinders achieved kill probabilities exceeding 80 percent while their Argentine opponents struggled with Vietnam-era B model Sidewinders only a fraction as effective. In 1984, the Soviet Union unveiled the R-73 infrared-guided missile, which could be fired beyond line of sight and initially oriented via a sight mounted on the pilot's helmet. It posed a significant enough challenge to adversaries that it remains in use with almost two dozen air forces across Asia, northern and eastern Africa, eastern Europe, and South America. Sidewinder variants are in service in forty-four countries spanning all six inhabited continents.

Further research into man-portable air defense (MANPAD) systems by the late 1970s resulted in the U.S. FIM-92 Stinger. Stingers were covertly smuggled into Afghanistan for use by mujahidin combatants to shoot down Soviet helicopters during the country's decade-long counterinsurgency effort. Before the arrival of Stinger MANPADs, only the country's terrain and altitudes had hindered helicopter deployments, and the impact of Stingers coincided with policy changes in the mid-1980s that publicized the previously quiet U.S. arms export to mujahidin forces. Nevertheless, the supposed 79 percent kill probability of Stingers against helicopters has been the subject of debate since the claim was first made.

Improved infrared-guidance technologies have involved refinements in tracking missiles as well as for more effective scanning by the missile's guidance system itself. As these systems advanced and missiles became more effective in combat, research into countermeasures proceeded. Flares were quickly adopted as a practical way to misdirect many infrared-guidance systems, prompting the guidance to harmlessly follow the flare instead of the intended target. Dual frequency seekers in turn negated the impact of flares and enabled use of infrared guidance from different angles, instead of necessarily chasing the heat in a jet's exhaust. Different jammer technologies have emerged. Early variants were meant to dazzle the sensor with different reflected light until the missile flew aimlessly, but follow-on improvements in guidance systems made this approach counterproductive. In contrast, laser jammers direct light at a seeker with the intention of blinding rather than confusing the guidance system.

Nicholas Michael Sambaluk

See also: Precision-Guided Munitions (PGMs)

Further Reading

Hept, George B. *Infrared Systems for Tactical Aviation: An Evolution in Military Affairs?* Maxwell AFB, AL: Air War College, 2002.

Kuperman, Alan P. "The Stinger Missile and US Intervention in Afghanistan." *Political Science Quarterly* 114, no. 2 (Summer 1999), 219–263.

Lenfle, Sylvain. "Toward a Genealogy of Project Management: Sidewinder and the Management of Exploratory Projects." *International Journal of Project Management* 32, no. 6 (August 2014), 921–931.

Mosavi, M. R., et al. *Design and Simulation of an Infrared Jammer Source for an Infrared Seeker.* Rasht, Iran: Amir Kabir University, 2007.

Iron Dome

A system for intercepting rockets and artillery shells with ranges up to 70 kilometers, developed in Israel by Rafael Advanced Defense Systems, in cooperation with Elta Systems.

The system consists of a radar to detect incoming threats, linked to a sophisticated command and control computer to estimate the impact points, and three missile launchers that each hold twenty Tamir interceptor missiles. The system software only seeks to intercept those missiles projected to hit populated areas. This system has reduced the damages inflicted by relatively cheap missiles with a simple design fired by Hamas in the Gaza Strip toward the closest Israeli towns. System development began in 2005. After successful final tests in late 2010, on March 28, 2011, the Israeli Air Force deployed the first battery to protect the Beersheba area, and the second the next week to protect Ashkelon. During Operation Cast Lead in August 2011, the Israeli public observed successful interceptions and demanded an accelerated deployment schedule. By November 2012, the fifth battery entered service during Operation Pillar of Defense. Between November 14 and 21, 2012, Hamas fired 1,506 rockets at Israel. The Israel Defense Forces (IDF) claimed 58 rockets fell in populated areas killing 5 people and injuring 240, while the Iron

Dome system intercepted 421 rockets (the rest either fell in open areas or failed in flight). Although the IDF claimed an 84 percent success rate, a number of critics have argued for lower effectiveness.

Critics offer several arguments regarding limitations of the Iron Dome system. The response time to identify the threat and estimate its point of impact prevents the system from responding to rockets or shells fired closer than 5 to 7 kilometers, or to rockets with flat trajectories up to 16–18 kilometers. Furthermore, the cost of $40,000–$50,000 per interceptor rocket—some threats requiring two interceptors—creates a high cost per interception compared to the relatively low cost of the rockets fired by Hamas. Because this system provides defense for a relatively small area, protecting any large territory in full would prove prohibitively expensive. Finally, critics expect that a sufficiently large and rapid onslaught of the more sophisticated rockets held by Hezbollah would oversaturate the system's response capability. Modest rates of effectiveness in the early versions of the system gave way to significantly improved performance by the time of Operation Protective Edge in 2014. The number of casualties and fatalities fell even faster than the property damage claims. Analysis attributes this to the robust civil defense system for warning the population, and reinforced shelters. Total tonnage of warheads sent by rocket also fell, and researchers claim that Hamas responded to the drop in rocket effectiveness with exploration of alternative strategies for offensive actions.

In addition to technical and financial limitations, analysts also note some political and strategic consequences. The protection of civilians and property offered by this system has enabled Israeli political and military leaders to forego a ground invasion to destroy rockets on the ground, thus saving lives on both sides and avoiding property destruction in the Gaza Strip. It also makes the current stalemate tolerable and reduces the Israeli incentive to negotiate with Hamas. The United States contributed significant funds to develop this system: researchers claim the United States spent $1.3 billion, as of 2018, to develop Iron Dome and $1.7 billion for Israel's other missile defense efforts, plus $500 million annually for the next decade. Other experts have noted the difficulty in calculating the cost due to the necessity for sensors, satellites, communication, and logistical infrastructure necessary for these systems to function.

Despite all the criticisms—economic, strategic, and technical—the tangible and psychological benefits of protection offered to the Israeli public by the Iron Dome system generates a strong political determination in Israeli political leaders to expand and perfect this system and similar strategies of active defense.

Jonathan K. Zartman

See also: Counter Unmanned Aerial Systems (C-UAS); Laser Weapon System (LaWS); Terminal High Altitude Area Defense (THAAD)

Further Reading

Armstrong, Michael J. "The Effectiveness of Rocket Attacks and Defenses in Israel." *Journal of Global Security Studies* 3, no. 2 (2018), 113–132.

Dombrowski, Peter, Catherine Kelleher, and Eric Auner. "Demystifying Iron Dome." *The National Interest* (July/August 2013).

Kurz, Anat, and Shlomo Brom, eds. *Lessons of Operation Protective Edge.* Tel Aviv, Israel: Institute for National Security Studies, November 2014.

Shapir, Yiftah S. "Lessons from the Iron Dome." *Military and Strategic Affairs* 5, no. 1 (May 2013).

J31 Gyrfalcon

A People's Liberation Army (PLA) Air Force fifth-generation multirole stealth fighter also known as the "J-21 Snowy Owl." In the absence of consensus about the definition of a "fifth-generation fighter," it can be thought of as a jet fighter with stealth, advanced avionics, and networking capabilities for linking with other friendly combat platforms. The J31 has been steeped in controversy due to evidence that its designers borrowed heavily from technological research digitally exfiltrated from aerospace companies in the United States.

Conflicting accounts from Chinese sources have somewhat obscured the intended purposes of the airplane. Some statements indicate that the J31 was meant as a carrier aircraft for use by the PLA's naval arm, and features in the plane's landing gear have supported this suggestion. Other indicators have pointed to the possibility that the J31 might be meant as a military export to countries desiring affordable yet stealthy combat aircraft. Since only the United States and the People's Republic of China had more than one stealth aircraft design in existence, the J31 would logically then be a competitor to the F-35 Lightning II as a military export.

Since the unveiling of the prototype, observers have widely noted the extremely close resemblance of the J31 to U.S. stealth technology, the F-35 in particular. The F-35 began development first, and both planes have similar overall shape, twin engines, twin internal weapon bays, as well as four heavy and two smaller hardpoints for externally carried weapons. This, combined with PLA cyberactivities and exfiltration of data in actions such as Operation Aurora and Shady RAT, has led to the J31 being called "the biggest free ride in the history of national security" (Moss, 2012).

Despite the obvious parallels between the J31 and the F-35, much still remains a mystery. The plane's surface is believed to be coated in stealth materials that prevent detection by some military radar technologies and impede identification by others. Early flight demonstrations with a "slick" prototype without weapons on its hardpoints have brought observers to different conclusions about the plane's flight characteristics and therefore also to different ideas about the plane's combat potential against other fifth-generation fighter planes such as the U.S.-produced F-22 and F-35, or designs under way in India, Japan, and Turkey.

Nicholas Michael Sambaluk

See also: DF-26 Anti-Ship Missile; F-22 Raptor; F-35 Lightning II; Liaoning Aircraft Carrier; Operation Aurora; Operation Shady RAT

Further Reading

Cliff, Roger, et al. *Shaking the Heaven & Splitting the Earth: Chinese Air Force Employment Concepts in the 21st Century.* Santa Monica, CA: RAND, 2011.

Garafola, Cristina L., and Timothy R. Heath. *The Chinese Air Force's First Steps toward Becoming an Expeditionary Air Force.* Santa Monica, CA: RAND, 2017.

Hallion, Richard P., Roger Cliff, and Phillip C. Saunders. *The Chinese Air Force Evolving Concepts, Roles, and Capabilities.* Washington, D.C.: National Defense University Press, 2015.

Moss, Trefor. "China's Stealth Attack on the F-35." *The Diplomat,* September 27, 2012. https://thediplomat.com/2012/09/the-fake-35-chinas-new-stealth-fighter.

Rupprecht, Andreas. *Modern Chinese Warplanes: Combat Aircraft and Units of the Chinese Air Force and Naval Aviation.* Philadelphia: Casemate, 2013.

Joint Direct Attack Munitions (JDAMs)

Munitions resulting from the pairing of global positioning system (GPS) kits and ordinance. Originally developed by McDonnel Douglas and produced by Boeing, JDAM kits can be attached to unguided munitions or "dumb bombs." Once the JDAM is attached the weapon becomes a precision-guided munition (PGM). JDAMs have historically been used in close air support (CAS) and air interdiction roles. Typical munitions that use JDAMs are the 2,000-pound Mk 84. The largest weapon to use a JDAM is the 21,000-pound Massive Ordnance Air Blast (MOAB), which was first tested in 2003.

To improve its accuracy, the JDAM uses both GPS and Inertial Navigation Systems (INS) to operate in poor weather conditions. JDAMs can still operate even with the loss of GPS. The INS is dependent on information received up until the bomb is released. As a result, its accuracy decreases dependent on the time it takes from the moment the bomb is released to the time it hits the target. For instance, in the span of 100 seconds, the accuracy range increases from 5 meters to 30 meters. For context, JDAMs used by the U.S. Air Force and U.S. Navy have a range of anywhere between 500 and 5,000 kilometers. Current American artillery has a range of approximately 40 kilometers and can miss the target by as much as 150 meters. JDAMs using both the INS and GPS guided munitions have an accuracy range of 5 meters.

JDAMs are considered relatively cost effective because of the reliability and effectiveness derived from their precision capabilities. As early as 2001, the cost of a JDAM kit was $21,000. In April 2002, the U.S. Air Force purchased 236,000 JDAM kits. During the same month, the Chairman of the Joint Chiefs of Staff, General Richard Myers, stated that the United States could purchase several thousand JDAMs for the same amount of money that would be needed to purchase two Eurofighters.

The first JDAM saw combat in 1999. In Operation Allied Force, the NATO bombing of Yugoslavia, American planes dropped 652 JDAMs, which hit over 80 percent of their targets. Early JDAMs had difficulty operating in cloudy conditions and therefore could not be used unless the air was clear. The JDAM improved on laser-guided munitions developed during the Vietnam War by increasing the accuracy of munitions during poor weather conditions. Older laser-guided weapons require clear weather conditions and cannot be used during cloudy or smoky conditions.

In Operation Iraqi Freedom the United States attached JDAM kits to BLU-109 munitions, or bunker buster bombs, to carry out precision strikes against Suddam Hussein's underground bunkers. These attacks were effective in destroying a number of Iraqi underground bunkers during the invasion of Iraq. During the war B-1B bombers frequently carried a twenty-four load of JDAMs during the conflict. During the war a B-2 could carry as many as eighty 500-pound JDAMs. The development of 500-pound JDAMs allowed the United States to convert more of its aircraft into an air support role.

The fighting in Afghanistan has led to a number of changes in the use of JDAMs. First, the U.S. B-52 Stratofortress, a strategic bomber developed to deliver nuclear weapons during the Cold War, switched from being primarily a strategic bomber to an effective CAS aircraft. The B-52 was first used in a CAS role in Vietnam, when it carried out carpet bombing missions to support U.S. troops. Those tactics changed during the war against the Taliban. Due to the accuracy of the JDAM, carpet bombing missions are no longer seen as a viable military option. Instead, B-52s can now loiter over the battlefield delivering PGMs as needed. Another change resulting from the use of the JDAM in Afghanistan revolves around bomb technology. Due to the nature of the fighting in Afghanistan, the United States has recommended using smaller explosives to increase accuracy and reduce collateral damage. The Small Diameter Bomb (SDB) program aims to achieve these goals by producing 250-pound weapons. These weapons are significantly smaller than the smallest JDAM, which is the 500-pound GBU-38/B. The smaller munitions allow aircraft to carry larger payloads and can be used in a CAS role.

The fighting in Afghanistan has also shown that improvements in accuracy still can be made. There have been a number of cases of friendly fire or collateral damage. In three instances during the initial invasion of Afghanistan, U.S. and coalition forces were either struck by friendly JDAMs or the bomb landed far from its intended target. On October 12, 2001, a U.S. F-18 Hornet dropped a bomb that struck the house of a civilian. The original target, a Taliban airfield, was a mile away from where the bomb landed. This case of collateral damage was due to human error when the coordinates were incorrectly input. On November 25, 2001, Afghan and U.S. forces were struck when a JDAM landed too close. This resulted in the death of a number of Afghan military personnel and five wounded Americans. On December 5, 2001, a JDAM dropped by a B-52 killed three U.S. personnel and wounded nineteen U.S. and Afghan troops.

Recent developments in technology have allowed the JDAMs to be used in naval mining operations. On September 23, 2014, a U.S. Air Force B-52 dropped a GBU-62B Quickstrike-ER using a JDAM tail kit to guide the weapon 40 miles into shallow water. This was the first successful testing of an aerial mine using the JDAM. Using the JDAM to guide the mine successfully significantly increased the range of aerial mine delivery. Since 1943 aerial mines had to be dropped at lower altitudes for greater accuracy, due to the drift of the parachutes used to deliver the mines. The addition of the JDAM to the Quickstrike-ER now gives U.S. aircraft the capability to drop mines as far away as 40 miles at an altitude of 35,000 feet.

The JDAM has transformed aerial warfare. By relying on the GPS/INS system to guide the bomb to the target, bombing accuracy is less restricted by weather conditions. Thanks to the relatively inexpensive costs of JDAMs, the weapon has made PGMs more abundant in the U.S. military. Finally, the United States has started to expand the use of JDAMs to use in mine-laying operations.

Luke Wayne Truxal

See also: Global Positioning System (GPS); Precision Guided Munitions (PGMs)

Further Reading

Boyne, Walter. *Operation Iraqi Freedom: What Went Right, What Went Wrong, and Why.* New York: Tom Doherty Associates, 2003.

Hasik, James. *Arms and Innovation: Entrepreneurship and Alliances in the Twenty-First Century Defense Industry.* Chicago: University of Chicago Press, 2008.

Pietrucha, Michael W. "Twenty-First-Century Aerial Mining." *Air & Space Power Journal* 29, no. 2 (2015), 129–150.

Wong, Wilson, and James Gordon Ferguson. *Military Space Power: A Guide to the Issues.* Santa Barbara, CA: ABC-CLIO, 2010.

Joint Land Attack Cruise Missile Defense Elevated Netting Sensor System (JLENS)

An unmanned and unarmed aerostat-based air defense monitoring system identifiable by its prominent lighter-than-air platforms. Initiated in 1996, the resulting system was a U.S. Army–led joint effort with the Navy and Air Force. JLENS followed in the tradition of most 20th-century military applications for lighter-than-air craft in their surveillance roles, although 20th-century uses also included bombardment platforms during World War I and air defense obstructions during World War II. At first produced through a partnership between Hughes Aircraft and Raytheon, the latter company soon acquired Hughes and consequently became the sole contractor. Trials and testing eventually led to the program's cancellation in 2015.

JLENS was intended to complement conventional air assets by constituting a long-endurance platform for sensing enemy missile attack. Tethered aerostats 77 feet long would be carried aloft by nearly 600,000 cubic feet of helium. As a lighter-than-air platform filled with a nonflammable gas, designers intended JLENS to be resistant to a range of interception technologies that might be effective against other types of aircraft, such as fixed-wing planes or balloons with flammable gases. Comparatively impervious to interception, the system was intended to operate on-station, providing sensory and warning information for up to thirty days at a time.

Among JLENS's array of instrumentation were VHF-band surveillance radar and X-band fire-control radar meant to facilitate its identification of enemy missiles and its collecting and communication information to defense systems such as the ground-based Patriot antimissile missile system and the air-to-air AIM-20 AMRAAM. Its designers wanted it to provide 360 degree coverage of an area, protecting a circle of terrain over 300 miles out from the system. Potential defended areas were thought to include U.S. cities or recurring hot spots such as the Korean Peninsula. If meeting its expected performance, it would match the air defense capabilities of several aircraft whose operating costs would exceed that of JLENS by a factor of five.

A prototype aerostat was lost during a storm in 2010, but the program began to encounter significant trouble from 2012 onward. This included a scaling back of unit procurement, a consequent rise in per unit cost, and a Nunn-McCurdy cost breach because the program's costs would now be covered by a smaller number of units. The same year, testing by the Pentagon raised questions about the system's overall reliability, and this became a topic of concern for several retired military officials as well. Many of these critics suggested deploying air defense technologies in a number of existing and proven systems rather than concentrating them into a single platform which was hoped to be economical and survivable but which implicitly constituted a lucrative target for an enemy interested in using missile attacks.

Technical analysis by 2013 undercut earlier assumptions that JLENS would survive in the hostile environments for which it had been intended. When introduced to the ongoing Operation Noble Eagle air defense tests conducted by the United States and Canada since September 2001, JLENS is reported to have successfully identified simulated adversaries at a distance of 140 miles, although this appears to have fallen short of the more ambitious intended capability.

Leak-induced instability in a JLENS aerostat caused the severing of its tether in October 2015, and the system reported drifted through Pennsylvania damaging power lines and interrupting service for 20,000 people. Despite assurances that JLENS carried radar but not cameras, the program also encountered continuing criticism from the civil liberties community on the suspicion that JLENS was or could easily be adapted into a ubiquitous surveillance platform. The tactical value of video feed from JLENS brought aloft by aerostat was examined in the context of combat zones, such as in efforts to detect and neutralize the impact of improvised explosive devices (IEDs) by enemy forces. JLENS was canceled in early 2016.

Nicholas W. Sambaluk

See also: Improvised Explosive Device (IED); RQ-4 Global Hawk; Terminal High Altitude Area Defense (THAAD)

Further Reading

Gormley, Dennis M. *Cruise Missiles and NATO Missile Defense: Under the Radar?* Paris: IFRI Security Studies Center, 2012.

Government Accountability Office. *Defense Acquisitions: Future Aerostat and Airship Investment Decisions Drive Oversight and Coordination Needs.* Washington, D.C.: GAO, 2012.

Inspector General of the Department of Defense. *Joint Land Attack Cruise Missile Defense Elevated Netted Sensor System Not Ready for Production Decision* (declassified). Alexandria, VA: Department of Defense Office of Inspector General, 2012.

Whiteman, Shannon J. *Improving Situational Awareness in the Counter-IED Fight with the Utilization of Unmanned Sensor Systems.* Monterey, CA: Naval Postgraduate School, 2009.

Laser Weapon System (LaWS)

Also called AN/SEQ-3 Laser Weapon System, a U.S. Navy ship-defense system meant to target incoming vehicles and projectiles. Development starting in 2009 led to prototype testing in 2012 and to installation on the USS *Ponce* destroyer in August 2014 to begin a twelve-month trial deployment. Just a month into this trial, LaWS was declared an operational asset.

U.S. Navy interest in lasers dates to the Cold War, and although safety concerns foreclosed the practicality of 1980s chemical lasers, the LaWS system, which borrowed from private sector developments in laser technology, proved successful during tests. An array of six solid state lasers, powered by a diesel generator that is separate from the ship's other electrical system, combine to form the destructive directed-energy beam. It is designed to be mounted on U.S. destroyers or on Littoral Combat Ships, and naval officials note that LaWS was intended to complement other shipboard defense systems such as guns and missiles, rather than to replace them.

Among the chief advantages of LaWS or other directed-energy weapons is the savings in weight and cost involved in propellants. A missile costs hundreds of thousands or even millions of dollars, whereas firing one shot with LaWS reportedly cost less than 60 cents. Overall, research and development into LaWS totaled $40 million.

As with other directed-energy weapons, LaWS offers several advantages beyond cost alone. These include precision engagement of targets, fast engagement times that can help a ship defend against multiple targets in succession, and the opportunity to graduate power to meet different kinds of threats. However, weather conditions can pose difficulties for LaWS. Although the system performed well in the wake of a dust storm and in humid conditions, it has not been tested during a heavy sandstorm and is not expected to function well in such conditions.

Although its exact range remains classified, LaWS can destroy enemy fast-attack craft, unmanned or manned aerial vehicles, and cruise missiles at a distance of a few miles. Use against human targets, while technologically feasible, is forbidden by the Convention on Certain Conventional Weapons as amended in the 1990s and agreed to by the United States in 2009. Experts believe that LaWS employs power in the 15–50 kilowatt range in order to engage targeted vehicles, and the extreme success enjoyed by LaWS has reportedly encouraged U.S. Navy interest in larger laser weapons of almost ten times as much power.

Nicholas Michael Sambaluk

See also: Directed-Energy Weapon; Littoral Combat Ship (LCS); Solid State Laser (SSL)

Further Reading

Jenkins, William C. *Navy Shipboard Lasers: Background, Advances, and Considerations.* New York: Nova, 2015.

McAulay, Alaster. *Military Laser Technology for Defense: Technology for Revolutionizing 21st Century Warfare.* Hoboken, NJ: Wiley, 2011.

O'Rourke, Ronald. "Navy Lasers, Railgun, and Hypervelocity Projectile: Background and Issues for Congress." Congressional Research Service, May 27, 2016.

Titterton, D. H. *Military Laser Technology and Systems.* Boston: Artech, 2015.

Zohuri, Bahman. *Directed Energy Weapons: Physics of High Energy Lasers (HEL).* Cham, Switzerland: Springer International Publishing, 2016.

Liaoning Aircraft Carrier

Commissioned in 2012, the first aircraft carrier of the People's Republic of China (PRC). Its impact on PRC military power is debated, and the ship is listed as a training vessel. This, combined with evidence of activity toward construction of at least two other comparable ships, may indicate a concerted program toward the People's Liberation Army Navy (PLAN) building a robust maritime airpower power projection capability across the western portion of the Pacific Rim.

Before being acquired by PRC, the hull had been constructed for the Soviet Union during the 1980s and the ship had been intended as a Soviet Navy Kuznetsov-class warship equipped with a combination of anti-ship missiles and a contingent of maritime aircraft. The ship's construction was interrupted by the collapse of the Soviet Union, and its fate appeared murky as its Ukrainian inheritors searched for a way to dispose of the hull. Its engines were effectively mothballed, and it was purchased by a front company based in the PRC and headed by retired PLAN officers, with the dubious cover that it would be towed from the Black Sea to the Chinese specially administered region of Macau to become a floating casino. The hull arrived in PRC in early 2002, following a complex towing procedure that lasted a year and a half. The ship was refitted over the next nine years and underwent sea trials in mid-2011 as PLAN's first aircraft carrier.

In terms of size, Liaoning's estimated 65,000 tons loaded is about comparable to the United Kingdom's Queen Elizabeth, commissioned in December 2017. Liaoning's sistership Kutznetsov remains in Russian service. Liaoning exceeds the French Charles de Gaulle by about 20,000 tons and dwarfs the Russian-built Kiev-class aircraft carriers possessed by Russia and India by a similar degree. Liaoning itself is half the displacement of any of the fully equipped Nimitz- and Ford-class aircraft carriers currently in the U.S. inventory. As a jump-deck carrier in the style of all modern aircraft carriers (with the exception of U.S. supercarriers), Liaoning's aircraft contingent focuses on relatively small and light aircraft types. It reportedly carries two dozen jet fighter aircraft and a dozen helicopters of various makes, with the latter reportedly equipped for antisubmarine warfare, early warning, and air-sea rescue. Its J-15 fighters are thought to slightly outmatch U.S. Navy F-18 carrier aircraft types but are believed to fall short of the capabilities of U.S. F-22 and F-35 stealthy designs.

Analysts debate the power projection impact that Liaoning itself provides to the People's Liberation Army Navy (PLAN). Few countries in the Pacific and Indian Ocean possess any comparable capabilities: India's one carrier is considerably smaller than Liaoning; Australia, Japan, South Korea, and Thailand possess helicopter carriers but no ships deploying fixed-wing aircraft; no other countries in eastern or southern Asia possess an aircraft carrier. PLAN arguably commands a power projection capability unique to countries of the region. Rumor exists of construction efforts for either one or two further carriers of similar size.

However, ownership of a single aircraft carrier or even of two or three such ships falls far short of the eleven supercarriers currently in U.S. service, each of which outclasses Liaoning. Global commitments mean that this U.S. naval presence extends across the Pacific Ocean but also to other regions around the world as well.

Attempts to measure and compare naval power have also contextualized this point with data about the aerial refueling ability of U.S. carrier planes that PLAN currently lacks and additionally with speculation about PRC Anti-Access/Area Denial (A2AD) technologies and the impact of A2AD on power projection. In the context of other defense systems, such as the J31 stealthy jet fighter, anti-ship cruise missile (ASCM) capabilities, and mounting evidence of cyber capabilities emanating from informal hacktivists and from organized People's Liberation Army units such as Unit 61398 and Unit 61486 to form an advanced persistent threat in the digital realm, it seems credible to believe that the nation's carrier program aims to establish regional rather than global projection.

Other technological factors buttress this conclusion in the eyes of some analysts. The absence of long-range antisubmarine aircraft from the Liaoning's aircraft compliment and the lack of carrier-borne in-flight refueling have been identified as factors that implicitly limit the distance to which the ship itself might range from friendly coastlines and necessarily limit the distance to which its aircraft might venture from the carrier itself. Reports deem the ship to pilot PLAN's exploration of carrier-based aviation but that Liaoning itself consolidates PRC's regional ability to project power rather than signal a more global reach.

Nicholas W. Sambaluk

See also: Advanced Persistent Threat; Anti-Ship Cruise Missile (ASCM); F-22 Raptor; F-35 Lightning II; J31 Gyrfalcon; People's Liberation Army Unit 61398; People's Liberation Army Unit 61486/"Putter Panda"

Further Reading

Agnihotri, Kamlesh. *Strategic Direction of the Chinese Navy: Capability and Intent Assessment.* New York: Bloomsbury, 2016.

Cheung, Tai Ming, ed. *Forging China's Military Might: A New Framework for Assessing Innovation.* Baltimore: Johns Hopkins University Press, 2014.

Cole, Bernard D. *China's Quest for Great Power: Ships, Oil, and Foreign Policy.* Annapolis, MD: Naval Institute Press, 2016.

Erickson, Andrew S., and Andrew R. Wilson. "China's Aircraft Carrier Dilemma." *Naval War College Review.* Newport, RI: China Maritime Studies Institute, 2006.

Holmes, James R., and Toshi Yoshihara. *Chinese Naval Strategy in the 21st Century: The Turn to Mahan.* New York: Routledge, 2008.

Nguyen, Binh. "The People's Liberation Army Navy: The Motivations behind Beijing's Naval Modernization." *CUREJ: College Undergraduate Research Electronic Journal* (April 2013).

O'Rourke, Ronald. *China Naval Modernization: Implications for U.S. Navy Capabilities—Background and Issues for Congress.* Washington, D.C.: Congressional Research Service, 2017.

Saunders, Phillip C., et al., eds. *The Chinese Navy: Expanding Capabilities, Evolving Roles.* Washington, D.C.: National Defense University Press, 2011.

Littoral Combat Ship (LCS)

Versatile, fast, and controversial U.S. Navy surface combatant designed to function in the littoral zone. The Littoral Combat Ship (LCS) incorporates a modular design that allows it to perform a variety of missions. Its capabilities include anti-submarine warfare, mine countermeasures, and surface warfare.

The LCS program was part of the U.S. Navy's response to the end of the Cold War. By the 1990s, U.S. Navy planners determined that many emerging threats were likely to arise in the shallow waters known as the littoral zone. These threats included anti-access/area denial (A2/AD) strategies that large surface combatants (LSCs) were not equipped to combat. Studies called for a small surface combatant (SSC) that could operate in littoral waters against small boats engaged in swarming attacks. Planners also envisioned the LCS providing many of the functions performed by frigates, minesweepers, and patrol boats.

The LCS was announced on November 1, 2001, as part of the DD(X) program. Contracts were awarded to General Dynamics and Lockheed Martin to develop two designs. General Dynamic's trimaran hull design known as the Independence Class is constructed by the Austal USA shipyard in Mobile, Alabama. Lockheed Martin's Freedom Class is a single-hulled design built by the Marinette Marine shipyard in Marinette, Wisconsin. Even numbers of both classes have been procured. Freedom Class LCSs are homeported in Mayport, Florida, and Independence Class LCSs are homeported in San Diego, California.

The capabilities of both classes are similar. The LCS's shallow draft allows operation in the littoral zone. Both classes are believed to have a maximum speed in excess of 40 knots through the use of a water-jet propulsion system. The displacement of the LCS is around 3,000 tons. The Freedom and Independence classes are armed with a single 57-mm cannon, four .50 caliber machine guns, Rolling Airframe Missiles (RAMs) for air defense, and anti-ship missile countermeasures.

In support of the antisubmarine mission, LCSs may carry an MH-60 helicopter, as well as equipment for mine countermeasures and surface warfare. Mission modules may be changed to allow the LCS to perform a different primary mission. Other suitable missions for the LCS may include humanitarian relief, homeland security, support of amphibious operations and intelligence gathering, surveillance, and reconnaissance (ISR).

The LCS was designed to be a less costly alternative to larger surface combatants such as frigates and to be operated with a much smaller crew, under one hundred sailors, assisted by substantial automation. Rotation of two crews known as Blue-Gold crews allows the LCS to deploy more frequently than an ordinary single-crew model. The first LCS, the USS *Freedom*, was commissioned in November 2008, and the namesake lead ship of the Independence Class was commissioned in January 2010. The first four ships of each class will operate as testing ships.

Mechanical challenges, including with power and propulsion, confronted both classes on their initial deployment. Concerns that arose about firepower have been partly addressed by plans to add Longbow Hellfire guided missiles to the LCS armament, although the LCS's ability to survive an enemy attack is rated at a lower level than other SSCs such as frigates. Cost growth, from a projected $220 million per seaframe to $646 million per ship in fiscal year 2018, has contributed further controversy.

These concerns have led the number of LCSs procured to drop and has contributed therefore to the rise in per unit cost of the ships to be delivered. The number of planned LCSs has been reduced from fifty-two to thirty-two as of fiscal year 2018. A planned frigate known as the FFG(X) will likely take on some of the tasks originally envisioned for the LCS. The FFG(X) may be based on either the Freedom or Independence Class LCS or be chosen from another competing design.

Andrew Harrison Baker

See also: Anti-Ship Cruise Missile (ASCM); DF-26 Anti-Ship Missile

Further Reading

Murphy, Sean P. *Acquiring the Tools of Grand Strategy: The U.S. Navy's LCS as a Case Study.* Unpublished dissertation. Old Dominion University, December 2017.

O'Rourke, Ronald. *The United States Navy Current Issues and Background.* New York: Nova Science Publishers, 2003.

O'Rourke, Ronald. *Navy Littoral Combat Ship (LCS) Program: Background and Issues in Congress.* Washington, D.C.: Congressional Research Service, 2018.

U.S. Government Accountability Office. *Navy Shipbuilding: Significant Investments in the Littoral Combat Ship Continue Amid Substantial Unknowns about Capabilities, Use, and Costs.* GAO-13-530. Washington, D.C., 2013.

U.S. Government Accountability Office. *Littoral Combat Ship: Knowledge of Survivability and Lethality Capabilities Needed Prior to Making Major Funding Decisions.* GAO 16-201. Washington D.C., 2015.

Mine-Resistant Ambush-Protected (MRAP) Vehicles

A collection of wheeled armored vehicles that offers occupants protection in hostile conditions. Mine-Resistant Ambush-Protected (MRAP) vehicles, frequently referred to as MRAPs, were a customized response and adaptation to enemy Improvised Explosive Device (IED) attacks in Iraq and Afghanistan from 2003 to 2010. The MRAP program illustrates a record-setting effort in rapid acquisition during wartime.

The vulnerability of relatively light High Mobility Multipurpose Wheeled Vehicles (HMMWVs), as challenged by sophisticated IEDs, accounted for the death and injury of thousands of U.S service members early in the Iraq and Afghanistan conflicts. This generated an urgent need for a more survivable alternative. National security decision makers, domestic and international companies, and other stakeholders devised a customized vehicle to address under-vehicle detonations, explosively formed penetrators, and powerful enemy bombs.

The resultant MRAP collection of vehicles incorporated a V-shaped hull to direct blasts away from vehicle occupants. High ground clearance provided better dissipation of blast intensity. Customized armor plating was superior to up-armored HMMWVs.

However, national security decision makers did not unanimously support technically superior MRAPs at the onset. There was opposition. Pragmatism and bureaucratic challenges prompted valid concerns. Some U.S. Department of Defense officials favored light and expeditionary forces over slow and formidable armor. Heavy MRAP vehicles would be inconsistent with this light and expeditionary mind-set. MRAP vehicles were also expensive. They would compete with existing programs. Finally, some believed the wars in Iraq and Afghanistan would end before the existing bureaucratic processes could field such a vehicle. Decision makers, for three years after combatants surfaced a request for a replacement vehicle, delayed critical action. The MRAP concept almost slipped into the dustbin of flat ideas.

Attacks by IEDs escalated in the following years, and mounting casualties led to a decision by Secretary of Defense Robert Gates in May 2007 to force consensus and action. He designated the MRAP program the highest acquisition priority in the Department of Defense. Stakeholders supplanted inaction for an acquisition effort in record-setting time. Suppliers delivered thousands of MRAPs to the battlefield. Casualty rates decreased.

MRAPs reign significant in 21st-century warfare. National security decision makers, domestic and international companies, and other stakeholders provided a customized response and adaptation despite practical concerns and bureaucratic challenges. Combatants received thousands of MRAPs in record time. MRAPs reduced casualty rates. As such, these unique vehicles and their rapid acquisition contribute to the evolution of technology, armor, and warfare.

John Blumentritt

See also: Improvised Explosive Device (IED); Unmanned Ground Vehicle (UGV)

Further Reading

Feickert, Andrews. *Mine-Resistant, Ambush-Protected (MRAP) Vehicles: Background and Issues for Congress.* Washington, D.C.: Congressional Research Office, 2009.

Gansler, Jacques S., William Lucyshyn, and William Varettoni. "Acquisition of Mine-Resistant, Ambush-Protected (MRAP) Vehicles: A Case Study." Presentation at the 7th Annual Acquisition Research Symposium, Monterey, CA, May 12–13, 2010.

Howitz, Colonel Michael C. *The Mine Resistant Ambush Protected Vehicle: A Case Study.* Carlisle Barracks, PA: U.S. Army War College Strategy Research Project, 2008.

Van Atta, Richard H., R. Royce Kneece Jr., and Michael J. Lippitz. *Assessment of Accelerated Acquisition of Defense Programs.* Alexandria, VA: Institute of Defense Analysis, 2016.

MQ-1 Predator

The first medium-altitude unmanned aircraft used by the U.S. Air Force (USAF) and the Central Intelligence Agency (CIA) and was widely used in the Global War on Terror (GWOT). They are remotely piloted aircraft operated by multiple militaries for armed reconnaissance and intelligence collection. Built by General Atomics Aeronautical Systems, the Predator can accomplish several types of missions without risking the lives of the aircrew who operate it.

Approximately the size of a small civilian aircraft, the Predator is powered by a four-cylinder engine capable of 115 horsepower. The engine drives a two-bladed propeller that is mounted on the rear of the aircraft. The body of the aircraft houses several radio antennas while the bulbous nose contains the satellite receiver. Mounted on the bottom of the nose are several cameras and sensors, which provide near real-time surveillance during both day and night. The MQ-1 can also carry two wing-mounted Air-to-Ground Missiles (AGMs) that are guided by a laser that is housed next to the cameras underneath the aircraft nose.

The Predator is operated by a pilot and a sensor operator who sit side-by-side in a cockpit called a Ground Control Station (GCS). Through secure satellite connections, the GCS transmits the actions of the aircrew to the aircraft. Along with the aircrew and the communication systems, teams of intelligence analysts provide support and coordinate the missions.

First developed in the 1990s, the Predator completed its first USAF mission in July 1995. Originally, the Predator was designated the RQ-1 as the "R" denotes reconnaissance, the "Q" denotes that the aircraft is remotely piloted, and the number "1" meaning that it was the first model of that series. In 2002, the Predator designation changed to MQ-1 with the addition of Hellfire AGMs, as the "M" denotes that the aircraft can accomplish multiple missions. Along with providing reconnaissance, the Predator can provide close air support to ground forces, coordinate strikes on targets for other aircraft, and conduct search and rescue operations. The Predator was officially retired by the USAF on March 9, 2018.

Operation Allied Force (OAF) in Kosovo saw the first use of the Predator in combat. In 2001, MQ-1 use increased in conflicts all over the world but was most often associated with the GWOT. USAF and CIA use of the MQ-1, along with the newer MQ-9 Reaper, dramatically increased during the first years of President Barack H. Obama's administration and became the preferred method of attack and reconnaissance against valuable targets. The ability to provide intelligence over a long mission and strike targets, all without risking aircrew, has proven to be an attractive option in modern warfare.

Along with widespread recognition, remotely piloted aircraft have been at the center of a debate about current and future combat. Not only has the MQ-1 and other remote aircraft changed military tactics and ethics, but it has reshaped the experience of warfare. Through remotely piloted aircraft, aircrew can accomplish missions across the globe through satellite connections. This ability has redefined what it means to be a combatant.

Andrew O. Hunstock

See also: IAI Heron; MQ-9 Reaper; Precision-Guided Munitions (PGMs); RQ-4 Global Hawk; RQ-170 Sentinel; Unmanned Aerial Vehicle (UAV)

Further Reading:

Cockburn, Andrew. *Kill Chain: The Rise of the High-Tech Assassins.* New York: Picador, 2015.

Gusterson, Hugh. *Drone: Remote Control Warfare.* Cambridge, MA: The MIT Press, 2016.

Whittle, Richard. *Predator: The Secret Origins of the Drone Revolution.* New York: Henry Holt, 2014.

MQ-9 Reaper

Remotely piloted aircraft operated by multiple militaries for armed reconnaissance and intelligence collection. General Atomics Aeronautical Systems built the Reaper as a replacement for the MQ-1 Predator with upgraded capabilities. The Reaper is the primary unmanned aircraft currently in use by the U.S. Air Force (USAF) and the Central Intelligence Agency (CIA) to observe and strike dynamic targets.

Much larger than its Predator predecessor, the MQ-9 is roughly the same size as a Fairchild Republic A-10 Thunderbolt II. The aircraft is powered by a 900-horsepower turboprop engine that turns a four-bladed propeller located at the rear of the aircraft. The fuselage contains two Ultra High-Frequency (UHF) radio antennas, as well as other communication receivers including the Satellite Communication (SATCOM) equipment. The large nose of the Reaper holds the satellite dish receiver as well as electrical equipment. Underneath the nose of the aircraft is the Multi-Spectral Targeting System (MTS) that is a collection of cameras and sensors, as well as the laser designator for weapon employment.

For armament, the Reaper carries four wing-mounted Hellfire Air-to-Ground Missiles (AGMs) while the Predator only carried two. The AGMs are highly accurate weapons and are used primarily against smaller targets such as enemy combatants and small vehicles. The high accuracy of the missiles helps keep collateral damage to a minimum. Along with the AGMs, the MQ-9 can carry two bombs attached to the underbelly of the fuselage. Each bomb weighs 500 pounds and can be both laser guided to the target, like the AGMs, or guided by an internal Global Positioning System (GPS).

The Reaper is operated in a similar fashion as the Predator, with both a pilot and a sensor operator sitting in a cockpit called a Ground Control Station (GCS). The control inputs from the aircrew are translated digitally and sent from the GCS to a satellite in orbit above the earth. The satellite then sends that information to the aircraft. The aircrew are supported by a team of intelligence analysts who monitor the mission and update relevant information to senior leaders. Also included in the support team is a mission coordinator who observes and manages the mission execution. Although the Reaper is remotely operated, each mission involves a team of several dozen people to be successfully accomplished.

The MQ-9 is used primarily as a hunter-killer aircraft that can find a target, accurately fix the target's position, and finish the engagement with weapon employment. These abilities allow the Reaper to be employed in close-air support for ground forces, combat search and rescue missions, and perform strike coordination for other aircraft, as well as reconnaissance. The increased speed of the MQ-9 also allowed the Reaper to strike moving targets with increased accuracy. Upgraded avionics and anti-icing equipment make the Reaper much more reliable than the Predator.

Development began in February 2001, building on early success with the MQ-1, also built by General Atomics. Initially this project was called "Predator B" as it was considered an upgraded version of the MQ-1 Predator. The USAF bought the aircraft and named it the Reaper in October 2007. Although similar in nature to the Predator, the Reaper is considered a true "multimission" aircraft with dynamic targeting capabilities. The MQ-9 was designed primarily for combat against ground targets, while the combat capabilities of the MQ-1 were developed after its initial use.

The MQ-9 has shifted the roles and responsibilities of what remote aircraft can accomplish in modern warfare. Intelligence gathering is still a valuable mission for unmanned vehicles, but the Reaper has dramatically expanded remote aircraft capabilities. The MQ-9 is considered nearly equal to other strike aircraft in competency, with a much lower operating cost and risk to personnel. The combat record of the Reaper in the Global War on Terror (GWOT) and Operation Inherent Resolve (OIR) show that it is often the weapon of choice against extremist groups and asymmetric threats.

The increased abilities of the Reaper have only furthered the ethical discussion on using unmanned aircraft or "drones" in war. Under the administration of President Barack H. Obama, the MQ-9 was increasingly used as a low-cost, high-efficiency weapon against terror cells in GWOT and OIR. This has helped promote the argument that the MQ-9 makes war "too easy" and therefore unethical. The Reaper has helped redefine how we describe combatants in modern warfare as well as how war is fought and seen by the public.

Andrew O. Hunstock

See also: IAI Heron; MQ-1 Predator; Precision-Guided Munitions (PGMs); Unmanned Aerial Vehicle (UAV)

Further Reading

Cockburn, Andrew. *Kill Chain: The Rise of the High-Tech Assassins.* New York: Picador, 2015.

Gusterson, Hugh. *Drone: Remote Control Warfare.* London: The MIT Press, 2016.

Whittle, Richard. *Predator: The Secret Origins of the Drone Revolution.* New York: Henry Holt, 2014.

Precision-Guided Munitions (PGMs)

Munitions that are guided onto a specific target using an electronic guidance system. There are three types of PGMs that have been developed during the past century. The first PGMs relied on a radio guidance system. The Laser Guided Bomb (LGB) was the second type of guidance system, which was developed during the 1960s and 1970s. Modern PGMs use the Global Positioning System (GPS) to guide the munitions. Currently, the United States attaches Joint Direct Attack Munition (JDAM) kits to its munitions, which relies on GPS to guide the munition to the target.

Since the 1930s the United States has shown a preference for a precision bombing doctrine. During World War II the United States developed the Norden bombsight, which enabled American bombers to attack targets from high altitudes. Bombing accuracy was determined by whether or not a bomb landed within 1,000 feet of the target. Throughout World War II, American bombers struggled to achieve this level of accuracy.

Also during the war, the United States began to experiment with its earliest PGMs. The first PGM used by the United States was the radio-guided azimuth only glide bomb (AZON), which was first used against bridges in the China-Burma Theater. These weapons were bombs equipped with radio-controlled surfaces. The AZON did not see widespread use because the U.S. Army Air Forces shifted to area bombing tactics in late 1944 and 1945 against Germany and Japan. Despite abandoning precision bombing later in the war, the United States demonstrated a preference for a precision bombing strategy and PGMs.

The Korean and Vietnam Wars created a new need for precision bombing. Without the large industrial cities of Germany and Japan to attack in fire bombing missions or nuclear strikes, the United States once again began to develop precision-guided munitions. In Korea the United States developed another radio-controlled bomb known as the Tarzon, which was successfully used against bridges. During the Vietnam War the U.S. Air Force improved its precision-guided munitions technology with the development of laser-guided and television-guided bombs. The first large-scale use of these weapons came during Operation Linebacker I in 1972. During this period laser-guided glide bombs destroyed more than one hundred North Vietnamese bridges. The most famous precision-guided munition raid of the Vietnam War took place on May 10 and May 12, 1972. The U.S. Seventh Air Force destroyed the Paul Doumer and Thanh Hoa bridges in North Vietnam using laser-guided munitions. The raid was followed up on May 26, 1972, when a single flight of F-4 Phantoms dropped laser-guided bombs on the Son Tay warehouse and storage area, destroying the three buildings attacked with only three bombs. Seventh Air Force Director of Intelligence stated that laser-guided munitions used during Linebacker I revolutionized tactical bombing. This significantly reduced the number of sorties needed to destroy a specific target. As the Linebacker I offensive intensified in September and October 1972, the Seventh Air Force flew more sorties using laser-guided munitions. In September 1972 alone the Seventh Air Force flew 111 sorties using laser-guided munitions. During this period pilots started using the Long Range Navigation (LORAN) bomb delivery techniques, which allowed the bombers to operate during poor weather. This represented a transitional point in the U.S. Air Force's munition choice.

After Vietnam the U.S. Air Force began to focus on using PGMs as the U.S. Air Force returned to a precision bombing doctrine that had originally been developed during the 1930s. Colonel John A. Warden III encouraged the use of precision-guided weapons, which had been made more effective with developments in satellite and stealth technology. Warden did make some noticeable changes to the precision bombing doctrine that the United States developed during the 1930s. Instead of destroying key economic nodes, the U.S. Air Force was to target key command and control systems. By using precision attacks against these targets, Warden believed that a nation's military would be paralyzed, laying the groundwork for its destruction by ground and air forces.

The first large-scale employment of laser-guided precision munitions began during Operation Desert Storm, the campaign to remove Iraqi forces from Kuwait. The laser-guided munition used during the conflict, the BLU-109, had an error range of 10 feet. Approximately 80 percent of the 2,065 precision-guided weapons released from stealth aircraft found their designated targets. The new doctrine melded to the PGM's capabilities was able to disable 50 percent of Saddam Hussein's ground forces prior to the employment of the ground forces.

In the 1995 air campaigns over Bosnia and Herzegovina more precision-guided weapons were used over "dumb" bombs. During the 1995 air campaign to weaken the Bosnian Serbs, Operation Deliberate Force, NATO aircraft flew 3,535 sorties and used more than 1,100 bombs at the expense of only one aircraft. Over 75 percent of the munitions used during the campaign were PGMs. The effectiveness of the PGMs can be demonstrated by the low number of Serbian deaths attributed to collateral bomb damage. A little over two dozen civilians were killed during the bombing.

In Operation Allied Force, the NATO air campaign against the forces of Slobodan Milošević in Kosovo, American bombers struggled to employ PGMs successfully. During the campaign only 30 percent of the munitions used were PGMs. Due to the shortage of PGMs used during the campaign, civilian targets were struck, causing a diplomatic crisis for NATO forces. On May 7, 1999, American forces struck an airfield near Niš killing fourteen civilians and wounding another twenty-eight. Also on May 7, the United States struck the Chinese embassy killing another three. These two incidents led to a temporary pause in the bombing campaign.

Despite these setbacks, PGMs improved during the 1990s. The most significant improvement came with the development of the Joint Direct Attack Munition (JDAM). With development starting in 1992, the JDAM is a guidance system kit that can be attached to munitions. Relying on both the Global Positioning System (GPS) and the Inertial Navigation System (INS) to guide the munition to the target, the JDAM made it possible to use PGMs in all weather conditions. This addressed a major flaw with PGMs used during the Gulf War, which were still weather reliant. The JDAM not only increased on the accuracy of previous PGMs, but they also proved cost efficient. JDAM kits could be attached to munitions lacking a guidance system, or "dumb bombs." Once attached the JDAM converted the munition into a PGM. As a result, the JDAM increased both the accuracy and number of PGMs in the U.S. arsenal starting in the late 1990s.

The U.S. wars in Afghanistan and Iraq have seen an increasing number of PGMs used in combat, and it was during these conflicts that the proportion of PGMs to nonguided munitions shifted so that PGMs began to constitute an actual majority of weapons used. Nonetheless, as the conflicts evolved into counterinsurgency operations, assessing the effectiveness of PGMs has become more difficult due to the nature of the targets selected. After 2003 the United States started targeting terrorist and insurgent leaders in drone strikes using PGMs. Leadership targets were more likely to hide in urban or residential areas, making any precision strike difficult. GPS-guided munitions gave American military planners the belief that these targets could be eliminated using PGMs. As a result, there has been an increase in air attacks in close proximity to civilians. This has resulted in greater collateral damage and civilian deaths. Unfortunately, it is hard to assess the effectiveness of PGMs in Iraq and Afghanistan since there is no official count of civilian casualties resulting from coalition air strikes.

Since its earliest inception, precision bombing has remained a part of the American bombing culture. As a result, the development of PGMs continued throughout the development of the U.S. Air Force and American air strategy. PGMs became more popular after the success of the Gulf War in 1991. After the war the United States developed the JDAM. Due to its cost effectiveness and improved accuracy, the JDAM made PGMs the preferred munitions for First World militaries.

Luke Wayne Truxal

See also: Global Positioning System (GPS); Joint Direct Attack Munitions (JDAMs)

Further Reading

Clodfelter, Mark. *The Limits of Air Power: The American Bombing of North Vietnam.* New York: The Free Press, 1989.

Gillespie, Paul G. *Weapons of Choice: The Development of Precision Guided Munitions.* Tuscaloosa, AL: University of Alabama Press, 2006.

Gray, Colin S. *Airpower for Strategic Effect.* Maxwell AFB, AL: Air University Press, 2012.

Hasik, James. *Arms and Innovation: Entrepreneurship and Alliances in the Twenty-First Century Defense Industry.* Chicago: University of Chicago Press, 2008.

Haulman, Daniel L. "Precision Aerial Bombardment of Strategic Targets: Its Rise, Fall, and Resurrection." *Air Power History* 55, no. 4 (2008), 24–33.

McFarland, Stephen. *America's Pursuit of Precision Bombing.* Tuscaloosa: University of Alabama Press, 2008.

Sewall, Sarah. *Chasing Success: Air Efforts to Reduce Civilian Harm.* Maxwell AFB, AL: Air University Press, 2016.

Pukkuksong-1 Submarine-Launched Ballistic Missile (SLBM)

A North Korean submarine-launched ballistic missile (SLBM) type successfully tested in August 2016 and believed to be first deployed sometime between 2017 and 2018. South Korean and U.S. analysts have disagreed in their estimates of the missile's potential range, although the highest estimates suggest a range of up to 1,500 miles.

The traditional purpose of SLBMs is to diversify a nation-state's strategic deterrent, by possessing a credible nuclear force that can be hidden aboard stealthy nuclear-powered launching submarines and thereby protected against an adversary's preventive first strike. North Korea is currently believed to lack any submarine capable of evading U.S. antisubmarine warfare detection, and as a result, conjecture exists about Pukkuksong-1's purpose. Analysts have suggested that variants may be deployed in submerged facilities along North Korea's coastline.

A mobile variant called Pukkuksong-2 was demonstrated in February 2017. As a solid fuel vehicle, the Pukkuksong series offer a significantly diminished time to launch and therefore a more rapid ability to strike in the event of war. Defense analysts have also debated the relevancy of the U.S. Terminal High Altitude Area Defense (THAAD) system in defeating Pukkuksong missiles, due to the SLBM's predicted angle of entry.

Nicholas W. Sambaluk

See also: Terminal High Altitude Area Defense (THAAD)

Further Reading

Nam, Kisung. *Sources of Evolution of the Japan Air Self-Defense Force's Strategy.* Monterey, CA: Naval Postgraduate School, 2016.

Schiller, Markus. *Characterizing the North Korean Nuclear Missile Threat.* Arlington, VA: RAND, 2012.

Steinitz, Chris, et al. *Transregional Threats and Maritime Security Cooperation.* Washington, D.C.: Office of the Chief of Naval Operations, 2017.

Quadcopter

A remotely piloted aircraft consisting of four helicopter-like rotors mounted on arms connected to a central body, which contains the guidance/control systems and a camera for flight tracking and photography. With the advent of smaller, high-quality digital cameras in the 2010s, interest mounted in unmanned aerial vehicles (UAVs, commonly known as drones) that could be used by amateur and professional photographers. By 2015, interest in quadcopters peaked, and multiple companies began producing these unique UAVs for just such a purpose. The design of quadcopters is simplistic—no wings, no tail, no control surfaces. The direction of flight is obtained by asymmetrically distributing power to each rotor assembly. In comparison, a typical remote-controlled aircraft either resembles a full-size helicopter or airplane.

Companies such as Amazon have begun research and development for large-scale derivatives of the hobbyist quadcopter. Due to the design's ability to carry a load and take off and land vertically, it is seen as a viable platform for deliveries of packages to a customer's front door. This would allow the seller to ensure delivery and safe handling of a product, possibly even charging a premium for the service. Ideally, the delivery UAVs would be able to fly themselves from company distribution centers to their intended destinations and back again.

The Federal Aviation Administration (FAA)—the organization that oversees the skyways of the United States—has been slow to develop policy and regulation for commercial-usage UAVs. Originally, the FAA thought that integration of these craft into U.S. airspace might be straightforward, in keeping with evidence of successful integration of military UAVs into high-trafficked skies during American military operations in the Middle East. The introduction of hobbyist quadcopters complicated the issue. With such tiny aircraft able to operate anywhere, citizens have been pressing to know how much of the airspace they own over their homes. The FAA has conducted safety campaigns to educate amateur pilots and establish some rules, such as requiring "line-of-sight" operation. The agency has implemented other restrictions for commercial usage, which is mostly limited to professional photography and power line inspection. Even with these stop-gap measures, quadcopters have already caused problems. A number of incidents have involved UAVs operating within approach paths to airports, bringing them less than 100 feet away from passenger aircraft. People on the ground have also become endangered, such as in 2015 when a quadcopter lost power and fell onto a crowd below in Seattle.

Despite these civilian/commercial airspace issues, quadcopters have great benefits in the military realm. Much work has been done to create intercommunication systems between self-flying quadcopters, allowing multiple units to work together on a mission. Delivery of supplies to troops on the ground is currently being worked on, merely requiring quadcopter technology to be scaled up. Even larger quadcopters have been proposed for the retrieval of injured troops, allowing the rest of a military unit to move forward without expending resources and time to go back to base.

Elia G. Lichtenstein

See also: Counter Unmanned Aerial Systems (C-UAS); Unmanned Aerial Vehicle (UAV)

Further Reading

Rupprecht, Jonathan. *Drones: Their Many Civilian Uses and the U.S. Laws Surrounding Them.* Self-published, CreateSpace, 2015.

Wallace, Ryan J., and Jon M. Loffi. "Examining Unmanned Aerial System Threats & Defenses: A Conceptual Analysis." *International Journal of Aviation, Aeronautics, and Aerospace* 2, no. 4 (October 2015), 1–33.

RQ-4 Global Hawk

A jet-powered unmanned aerial vehicle (UAV) developed by Northrop Grumman. Military surveillance satellites are invaluable assets for planning operations and keeping eyes on the world from above. However, the nature of orbital mechanics keep satellites from being easily repositioned to observe a target area, and the Earth's atmosphere degrades picture quality. Reconnaissance and spy aircraft compensate for these deficiencies, flying within the Earth's atmosphere and being deployable wherever it is logistically possible.

The current spy plane in use by the American military is the Lockheed U-2 Dragon Lady, designed and put into service in 1956 in the early period of the Cold War. The increased demand for surveillance, however, added major strain on the aircraft and its pilots. Long-duration missions conducted in rapid succession has led to U-2 pilots suffering from decompression sickness, known to ocean divers as "the bends." The change in pressure from the altitude of a U-2's home base to its cruising altitude of 70,000 feet combined with its partially pressurized cockpit are ideal conditions for the bends to occur.

The RQ-4 Global Hawk—with its cruising altitude of 60,000 feet—flies nearly as high as the U-2, without posing the same physical risks to its pilot. The RQ-4 is controlled from the ground by three pilots—one for landing and takeoff, one for the duration of the mission, and one to monitor the sensor data being sent back by the aircraft. The data transfer and operator commands are accomplished by satellite link. The RQ-4 has a large wingspan—130 feet—providing it with enough lift for the high altitude. It has a "V-tail" configuration that combines the functions of the horizontal and vertical stabilizer into one, subsequently reducing weight. The single turbofan engine is located on top of the main fuselage just in front of the tail. The unique design of the Global Hawk results in an endurance of over 34 hours and a range of 12,300 nautical miles without refueling. The U-2 could retire from its long career knowing that it has a more than capable replacement.

Work is already under way to further increase aerial surveillance capabilities. Although UAVs such as the RQ-4 can refuel in mid-air, the procedure currently requires a manned tanker aircraft to fly into dangerous operating areas. Plans to create tanker versions of existing UAV models would eliminate this issue and allow the RQ-4 to remain on site for even longer than 34 hours. Northrop Grumman—for the U.S. Navy—has been creating the MQ-4C Triton, a version of the RQ-4 that has been beefed up structurally and equipped with new systems, which allow it to operate at lower altitudes in bad weather. This new aircraft will minimize the manpower required to conduct reconnaissance missions over a larger area of ocean.

Elia G. Lichtenstein

See also: MQ-1 Predator; RQ-170 Sentinel; Unmanned Aerial Vehicle (UAV)

Further Reading

Headquarters, Department of the Army. *Army Unmanned Aircraft System Operations* (Field Manual Interim No. 3-04.155).

Thomas, Ricky. *Global Hawk: The Story and Shadow of America's Controversial Drone.* Self-published, 2015.

RQ-170 Sentinel

A potentially stealthy, jet-powered, flying-wing unmanned aerial vehicle (UAV, commonly known as a drone) developed by Lockheed Martin. A conventional aircraft has a main wing and a tail, which consists of both a horizontal and vertical stabilizer. The horizontal stabilizer is the location of the elevator, the control surface that helps the plane point its nose up or down. The vertical stabilizer is the location of the rudder, which points the aircraft's nose left or right. The main wing possesses the flaps—which increase lift at lower airspeeds—and ailerons, which roll the plane left or right. A flying-wing aircraft has all of these control surfaces incorporated into the main wing. This is done by more complex control surfaces such as elevons—elevator/aileron—and split flaps, which replace the rudder. A flying wing is much less detectable on radar due to this lack of external components.

To further reduce its radar signature and also its aerodynamic drag, the engine of the RQ-170 is located inside a hump blended into the main wing. Lockheed Martin, the company that designed the Sentinel, has been a pioneer in the field of stealth aircraft since the 1970s. Three aircraft built by the company—the F117, F-22, and F-35—are constructed from materials that absorb radar waves. Many experts believe this to be the case with the RQ-170 as well.

Although spy satellites have been and are incredibly useful to militaries at war, there will always be the need for spy planes and reconnaissance UAVs. The resolution of images from satellites is reduced because of interference from the Earth's atmosphere and from weather conditions. Planes and UAVs fly high enough to get the big picture of a target area and avoid weather, but low enough to avoid atmospheric interference. The RQ-170 flies even lower due to its mission and sensor payload. The Sentinel is said to "see, hear, and sniff," which consists of imagery intelligence (IMINT), signals intelligence (SIGINT), and measurement and signature intelligence (MASINT). IMINT involves taking photos and videos across the visible and ultraviolet spectrum. SIGINT is the collection/interception of microwaves and radio waves, such as cell phone usage and Wifi. MASINT includes the detection of nuclear radiation and chemical compounds. The Sentinel can pick up the presence of minute nuclear radiation in the air using its sensors. All of this data is sent back to intelligence officers for analysis.

The RQ-170 demonstrated its IMINT and SIGINT abilities in 2010 during the hunt for Osama bin Laden. After his theorized whereabouts at a compound in Pakistan, Sentinels reportedly babysat the compound collecting communication signals. This was instrumental in confirming bin Laden's presence and enabling his eventual elimination in 2011. The RQ-170 is extremely well suited to this kind of investigative work. It can also be used to enforce nuclear treaties in hostile countries because of its stealth-like capabilities and its MASINT sensors.

Elia G. Lichtenstein

See also: MQ-1 Predator; RQ-4 Global Hawk; Unmanned Aerial Vehicle (UAV)

Further Reading

Headquarters, Department of the Army. *Visual Aircraft Recognition* (Training Circular No. 3-01.80).

Scharre, Paul, Ben FitzGerald, and Kelley Sayler. *A World of Proliferated Drones: A Technology Primer.* Washington, D.C.: Center for a New American Security, 2015.

Singer, Peter

Author and scholar widely considered one of the leading world experts in the changing nature of war in the 21st century. Educated at Princeton's Woodrow Wilson School of International Affairs and Harvard, his first book, *Corporate Warriors* (2003), examined both the benefits as well as ethical tensions that ensue when warfare becomes privatized, while his second (*Children at War*, 2005) was among the first to examine the global use of children as soldiers. His third book, *Wired for War* (2009), was one of the first books to explore the implications of robotics for the future of war, and *Cyberwar and Cybersecurity* (2014) examined the security implications of an increasingly networked world.

Most recently, Singer cowrote the novel *Ghost Fleet* (2015), blending research and imagination to portray what a future war may look like in the new global and technological age. Singer has worked for the Office of the Secretary of Defense, was the founding Director of the Center for 21st Century Security and Intelligence at the Brookings Institution, and is currently a Strategist for New America.

Deonna D. Neal

See also: Rid, Thomas; Stiennon, Richard

Further Reading

Singer, P. W. *Wired for War: The Robotics Revolution and Conflict in the 21st Century.* New York: Penguin, 2010.

Singer, P. W. *Cybersecurity and Cyberwar: What Everyone Needs to Know.* Oxford: Oxford University Press, 2014.

Singer, P. W. *Ghost Fleet: A Novel of the Next World War.* Boston: Houghton Mifflin Harcourt, 2015.

Solid State Laser (SSL)

A laser using a gain medium that is in a solid form rather than a liquid or gas. Typically, this involves a crystalline material infused with a dopant material such as a rare earth. As is the case with other directed-energy weapons, the potential for scalability has attracted interest in the potential for SSL weapon technologies. Although the semiconductor laser dates to 1962, its notable applicability to weaponization is much more current. The U.S. Army was reported in early 2017 to be interested in testing a truck-mounted SSL slightly more powerful than the Navy's recently tested ship-defense Laser Weapon System (LaWS).

The U.S. Navy itself is known to have tested an SSL in 2011 as part of its Maritime Laser Demonstration (MLD) project. MLD focused on the ability to destroy a small vessel representing an enemy target. Reports as early as 2002 promised incorrectly that an SSL weapon for the F-35 Joint Strike Fighter was nearly operational, and newer reports from 2015 have again predicted modular laser weapons to be introduced to the F-35 within the following years.

Nicholas Michael Sambaluk

See also: Directed-Energy Weapon; Electromagnetic Railgun (EMRG); F-35 Lightning II; Laser Weapon System (LaWS)

Further Reading

Jenkins, William C. *Navy Shipboard Lasers: Background, Advances, and Considerations.* New York: Nova, 2015.

O'Rourke, Ronald. "Navy Lasers, Railgun, and Hypervelocity Projectile: Background and Issues for Congress." Washington, D.C.: Congressional Research Service, May 27, 2016.

Titterton, D. H. *Military Laser Technology and Systems.* Boston: Artech, 2015.

Zohuri, Bahman. *Directed Energy Weapons: Physics of High Energy Lasers (HEL).* Cham, Switzerland: Springer Publishing International, 2016.

Su-57

Russia's first stealthy, fifth-generation combat plane. Although much of the program remains cloaked in some secrecy, it has been identified as an effort to answer new combat plane technology such as the F-22 Raptor and F-35 Lightning II Joint Strike Fighter produced by the United States. Reportedly, one defense analyst has been quoted saying "'performance-wise it looks to compete with the [F-22] Raptor'" (Majumdar, 2018). Despite its apparent promise, the Su-57 faced significant budgetary challenges and delays, and its scale production was canceled in July 2018.

The plane is reported to have a supercruise speed above Mach 1.5 and a maximum range of approximately 2,000 miles, although as is the case with high-performance aircraft the range would be reduced when flying at maximum speed. Its armament reportedly includes an integral 30-millimeter cannon as well as up to a dozen hardpoints for ordinance such as bombs and missiles. Six of these appear on the exterior of the plane, while an internal bay is located on the underside of the plane between the engines, and a smaller weapons bay has been noted at each wing root.

Stealth is an important component of a fifth-generation fighter, and this characteristic imposes some strictures in terms of armament. Internal bays are highly preferred over external hardpoints with respect to stealth, since the carefully designed surfaces of the airframe become cluttered with external ordinance and stealthy characteristics are thereby lost. The mix of internal and external hardpoints arguably provides some flexibility to the design, so that it could either fight with six weapons in an environment demanding stealth or with twelve weapons in a more permissive environment.

Stealth also shapes some design aspects of the plane itself. Analysts observe an emphasis on frontal stealth rather than on all-around stealthiness. The plane's overall flat appearance from the side is in keeping with most of the stealth designs from the end of the 20th century and early 21st. Radar-absorbent materials are used extensively in the aircraft's fabrication, as are composite materials that appear on seven-tenths of Su-57's exterior and reportedly comprise a quarter of the plane's weight. The resulting blended wing body is barely 3 percent as observable on radar as the Su-37 airframe it is meant to replace.

Although the program originated at the end of the 1970s, economic stress in the Soviet Union and political turmoil with the Soviet collapse contributed to successive delays. Further budgetary hurdles slowed the program during the 2000s, although this did also mean that Russian designers watched planes like the F-22 and Eurofighter Typhoon emerge, and the design reflects awareness of contemporary warplanes' performance. Certain features, such as thrust vectoring and a notable weapons bay between the plane's engines, are thought to have been built into the Su-57 in part from a belief among Russian designers that their absence detracts from the performance of the U.S. F-22.

As with areas such as supersonic and hypersonic missile research and development, Russia moved into partnership with India on fifth-generation fighter aircraft (FGFA) development, essentially constituting an Su-57 variant. Russian-Indian partnership was announced in 2007, and by 2010 officials at Sukhoi predicted that up to 1,000 Su-57 types might be produced, with 200 each for Russia and India and 600 others exported worldwide. The first Su-57 flight took place in January 2010. Russian think tanks predicted that Su-57 variants would be available for export in 2025, and in 2013 Russia reportedly invited Brazil to become a partner in the program. These moves appear in retrospect to have been overly optimistic.

Similar to U.S. experience with the F-35, cost issues have precipitated a scaling back of intended purchases. Russian plans to acquire 150 of the jets were adjusted, and by 2015 the country's defense minister outlined the retention of existing fourth-generation aircraft to be bolstered by a dozen Su-57s. Two Su-57 aircraft were reported in Syria during February and March 2018 in connection with Russia's air support program assisting Syrian dictator Bashar Assad during that country's civil war.

The decision to deploy prototype technology in combat, even when the adversary lacked much air defense capability, was the subject of some controversy and criticism. India backed out of the program in April 2018 after repeatedly voicing concern over program reliability, and efforts to prepare for mass production ended in July of that year. Nonetheless, media reports asserted that the pair deployed in Syria were equipped with less advanced engines than others might possess even though "the actual number of aircraft delivered by the end of 2020 . . . could be just two" (Gady, 2018).

Nicholas Michael Sambaluk

See also: F-22 Raptor; F-35 Lightning II; J31 Gyrfalcon

Further Reading

Gady, Franze-Stefan. "Russia's 5th Generation Stealth Fighter Jet to Be Delivered to Russian Air Force in 2019." *The Diplomat,* October 17, 2018.

Majumdar, Dave. "Air War: Russia's Deadly Su 57 vs. China's Stealth J-20 Fighter (Who Wins?)." *The National Interest,* May 23, 2018.

Russia, Military Power: Building a Military to Support Great Power Aspirations. Washington, D.C.: Defense Intelligence Agency, 2017.

Swarm Robots

Large numbers of generally simple robotic systems. Applications for swarm robots proceed from the belief that a desired collective behavior will emerge not only from the interaction among the robots themselves but also from the interaction between the robots and the environment.

The term refers to the group of locally interacting individuals sharing common goals, and swarm robotics takes inspiration from swarms in nature, such as social insects (ants, termites, bees, etc.) as well as fish and certain mammals. Swarms can range in size from a few individuals living in small areas to highly organized colonies that could occupy large territories and consist of millions of individuals. Individuals often have poor abilities on their own, but complex group behaviors can emerge from the whole, which no individual member might have independently performed.

A single robot capable of undertaking a sophisticated task is likely to require a complicated and costly structure and set of control modules and to be vulnerable as a system to breakage of a particular component. In contrast, multiagent systems focus on behaviors of multiple static agents in known environments, and robots in the multirobot system (which require external control) are small. Redundancy, decentralized coordination, design simplicity, and the multiplicity of sensing across different members of a swarm mean that a swarm is robust relative to a particular robot. As a result, a swarm can continue to operate, though at a lower performance, despite failures in the individual or disturbances in the environment. Other important characteristics of swarm robotics is that they are autonomous, decentralized, use local sensing and communications, are homogeneous, and flexible.

Swarm robots must be able to perform certain sets of behaviors and tasks in order to be suitable for real-world applications. The following list of tasks are offered in ascending order of complexity. First, they must be able to *aggregate*, that is, gather in order to perform other tasks. They must also be able to *disperse*, which is also known as area coverage. The goal is to distribute robots in space to cover as much area as possible without losing the connectivity between them. *Pattern formation* is also another key task, where a global shape is created by changing the position of the individual robots. A robot must also be able to perform *collective movement.* There are two types of collective movement: formations and flocking. In formations, robots must maintain predetermined positions and orientations among them. In flocking, relative positions can vary and are not strictly enforced. Robots must be able to perform *task allocation*, also known as division of labor. *Source search* is a task where robots collectively search for an object, smell, or sound. Swarm robotics is also very promising for the *collective transportation of objects.* Finally, *collective mapping*, that is, the ability to disperse over a wide area, map an assigned area, and then regroup to combine the information is a task that may be well suited to swarm robots as well. In swarm robotics, the main challenge is to develop methods and strategies that guide the collective execution of tasks by robots, by designing simpler behaviors such as subtasks.

The future applications of swarm robotics are uncertain, but it is hoped that the use of swarm robotics might range from surveillance operations to mine disarming in hostile environments. Current challenges for swarm robotics are developing algorithms for individual behaviors that would emerge into a collective group behavior. Some security challenges for developers of swarm robotic systems are physical security, identity and authentication protocols among the robots, and thwarting communication attacks and intercepts from agents outside the robotic system.

Deonna D. Neal

See also: Counter Unmanned Aerial Systems (C-UAS); Unmanned Aerial Vehicle (UAV); Unmanned Ground Vehicle (UGV)

Further Reading

Harmann, Heiko. *Swarm Robotics: A Formal Approach.* Cham, Switzerland: Springer International Publishing, 2018.

Sahin, Erol, and William M. Spears. *Swarm Robotics: SAB 2004 International Workshop (Revised Selected Papers).* Santa Monica, CA, July 17, 2004.

Tan, Ying, and Zhong-Yang Zheng. "Research Advance in Swarm Robotics." *Defense Technology* 9 (2013), 18–39.

Trianni, Vito. *Evolutionary Swarm Robotics: Evolving, Self-Organizing Behaviors in Groups of Autonomous Robots.* Studies in Computational Intelligence, Vol. 108. Spinger-Verlag Berlin Heidelberg, 2008.

Tactical Air Delivery Drone (TACAD)

A logistics delivery system aiming to address a historical tension in warfare between "tooth" (the combatants) and "tail" (those personnel who provide the logistical and other necessary support for those combatants). This relationship can have strategic consequences. Overland resupply routes, for example, greatly can increase the monetary and human costs of supply.

In the United States, Army and Marine Corps planners are supporting development by Logistic Gliders company of the LG-1000, a single-use glider made of plywood, incorporating a commercial off-the-shelf global positioning system (GPS) and autopilot capabilities and guided by a small number of servo motors. The glider likely would cost less than $3,000, and it could carry about 700 pounds. If deployed from cargo aircraft, the glider may represent a standoff logistics capability that would facilitate logistics operations in contested aerial environments.

Currently, the U.S. military relies on the Joint Precision Airdrop System (JPADS) for this capability. Although it can carry up to 10,000 pounds, it must be dropped very close to its intended landing area. It also can cost up to $30,000, and thus the military likes to recover it, requiring a soldier to carry a 30-pound unit until it can be returned. In many ways, the low-technology solution to a supply problem embodied in TACAD typifies the Marine Corps' institutional preference, in particular, for inexpensive technology when possible.

Heather Pace Venable

See also: Boston Dynamics BigDog; Boston Dynamics Handle Robot; Global Positioning System (GPS); Quadcopter; Unmanned Aerial Vehicle (UAV)

Further Reading

Ergene, Yigit. *Analysis of Unmanned Systems in Military Logistics.* Master's thesis. Monterey, CA: Naval Postgraduate School, 2016.

Kuckelhaus, Markus. *Unmanned Aerial Vehicle in Logistics: A DHL Perspective on Implications and Use Cases for the Logistics Industry.* Troisdorf, Germany: DHL Customer Solutions & Innovation, 2014.

McCoy, John V. *Unmanned Aerial Logistics Vehicles: A Concept Worth Pursuing?* Master's thesis. Ft. Leavenworth, KS: Command and General Staff College, 2003.

Tactical Language and Culture Training System (TLCTS)

A program designed to facilitate learning of selected languages by U.S. military personnel. The effort is part of a Defense Advanced Research Projects Agency (DARPA) program called DARWARS. Although the name plays on that of the DARPA sponsor and the popular Star Wars movie franchise, the objective purpose was to increase understanding of languages such as Iraqi, Pashto, and Dari, which grew in military relevance during the Global War on Terrorism following the terrorist attacks of September 11, 2001. The underlying concept, partly parallel to online and computerized language-learning tools in the civilian realm, is that the ubiquity and increasing capabilities of computers could be used to support language skills.

TLCTS involves self-paced courses on computers, built so that military members can learn aspects of strategically prioritized languages related to daily life and to military missions. These areas are of particular importance for easing tension between military personnel and civilian bystanders and for communicating in ways that will help avert avoidable violence. TLCTS courses reportedly use learning approaches in keeping with best practices in the field. Within the course, learners can speak and select gestures appropriate to the situation at hand.

Exercises are designed to show learners when multiple alternative expressions can reflect similar meanings, and skill builder exercises are written to help address common learning mistakes. Different settings exist whereby the learner can recalibrate the course level and gradually practice improved pronunciation. The "Tactical Iraqi" TCLTS course was ready in mid-2005, and "training with Tactical Iraqi . . . affected the trainees' confidence in their ability to understand and speak Arabic" (Johnson, 2010, 190). The Marine Corps complemented the use of TLCTS by also providing its active duty and reserve personnel access to RosettaStone commercial language software. Reportedly, the game rewards effective learners by setting the local characters to assist in tasks related to postwar reconstruction.

TLCTS is estimated to have cost $6 million to prepare, with the addition of follow-on capabilities costing between $300,000 to $600,000 per language. In addition to TLCTS, DARWARS also sponsored a computer-based multiplayer training simulator called *DARWARS Ambush!*, which aimed to reinforce understanding of tactical scenarios. Advocates of *DARWARS Ambush!* highlighted the opportunity for military users to flexibly shape training scenarios so that they could remain current and presumably retain a degree of fidelity and realism to the modeled scenarios.

Nicholas Michael Sambaluk

See also: Defense Advanced Research Projects Agency (DARPA)

Further Reading

Hou, Ming, et al. *Suitable Adaptation Mechanisms for Intelligent Tutoring Technologies.* Toronto: Defence R&D Canada, 2010.

Johnson, W. Lewis. "Serious Use of a Serious Game for Language Learning." *International Journal of Artificial Intelligence in Education* 20 (2010), 175–195.

Losh, Elizabeth. *In Country with* Tactical Iraqi*: Trust, Identity, and Language Learning in a Military Video Game.* Research paper. Irvine: University of California Irvine, 2005.

Polania, William G. *Leveraging Social Networking Technologies: An Analysis of the Knowledge Flows Facilitated by Social Media and the Potential Improvements in Situational Awareness, Readiness, and Productivity.* Monterey, CA: Naval Postgraduate School, 2010.

Terminal High Altitude Area Defense (THAAD)

A U.S. Army theater ballistic missile defense (BMD) system consisting of a cueing process, a ground-based missile track radar, and a series of interceptor missiles that work together to engage and intercept incoming ballistic missile warheads high in the atmosphere as they reenter en route to their target. The THAAD interceptor missiles are considered a "hit-to-kill" system, which means that the interceptor missile (or kill vehicle) flies to and destroys the incoming target missile warhead through high-speed kinetic impact and must impact the target to cause the kill. The THAAD system is intended to provide additional capability and protection by engaging the target missile at high altitude and extended ranges. As originally designed, the THAAD was intended to be integrated into a portfolio of theater ballistic missile defense capabilities and was not intended to be used as the sole defense mechanism for protected assets. Particularly, the THAAD was intended to target the high-altitude missiles at or above the atmosphere, leaving close-in target defense to other BMD systems. The THAAD system's radar for the ground acquisition, guidance, and tracking of target missiles is paired with an infrared seeker on the THAAD interceptor missiles. The two sensors working in concert allow improved tracking and target discrimination over that of a single sensor.

A THAAD engagement would proceed in the following manner:

1. An external sensor (airborne, satellite, naval, or other) detects adversary missile launch and alerts necessary theater authorities or commands.
2. The THAAD ground radar acquires and tracks the incoming target missile(s) and relays the information to interceptor guidance.

3. THAAD interceptor missile(s) are launched and guided to the incoming missile targets, with the ground radar monitoring to determine if the target missile is destroyed (killed).
4. Steps 2 and 3 repeat until the missile(s) are destroyed or until the incoming missiles are engaged by a local, lower-altitude ballistic missile or air defense system.

As a ballistic missile defense system, the THAAD may be deployed to regions of interest to protect U.S. or allied forces or interests assessed to be within range of medium- and long-range ballistic missile threats. However, presence of a BMD system may be interpreted as destabilizing an environment that is reliant on mutual deterrence by providing one side an advantage in survivability and thus reducing the risk to that party if they chose to pursue or initiate armed conflict. In particular, the deployment of the THAAD system to the Korean Peninsula in 2016 drew concerns and criticism from the international community. THAAD was deployed to locations in South Korea as a response to the increasingly bellicose actions and verbiage of North Korea, and especially in response to North Korean accelerated ballistic missile and nuclear testing in 2015 and 2016. The 2016 THAAD deployment sent a strong message to North Korea while at the same time providing U.S., South Korean, and regional allies with some protection from North Korea's substantial ballistic missile capability.

THAAD was developed due to the global proliferation threat posed by theater ballistic missiles, which were becoming more widespread throughout the late 1980s and 1990s. An increasing number of nations developed, purchased, or compiled ballistic missile systems to augment their military capabilities. For many nations, possessing ballistic missiles (especially mobile missiles), was a fairly simple way to greatly increase their military power projection range and provided either deterrence or threat to reinforce their regional political agenda. This created a need for a system that could be put in place to protect U.S. and allied interests, which could be held at risk by the ballistic missile threat. The THAAD system underwent design and was proposed to the U.S. Congress in 1994 with an expected development and testing cycle lasting into early 2000, and full production and system deliveries expected in the mid 2000s. The 1990s estimated cost for the total delivered THAAD capability of >1,400 interceptor missiles, 99 launchers, and 18 ground radars was $14 billion. However, missile defense is technically challenging, and after failed tests and missile interceptor redesign the THAAD program was $1 billion more expensive than anticipated and still undergoing testing in the 2009 time frame, albeit with higher success rates against representative adversary missile targets. Lockheed Martin was awarded the development contract in 2000 and the production contract in 2006; the program came under Missile Defense Agency oversight in 2001. The U.S. Army stood up its first operational THAAD battery in 2008.

Daniel J. Schempp

See also: Anti-Satellite Weapon (ASAT); Pukkuksong-1 Submarine-Launched Ballistic Missile (SLBM)

Further Reading

Government Accountability Office. *Ballistic Missile Defense: Issues Concerning Acquisition of the THAAD Prototype System*. Washington, D.C.: Government Printing Office, 1996.

Suh, J. J. "Missile Defence and the Security Dilemma: THAAD, Japan's 'Proactive Peace,' and the Arms Race in Northeast Asia." *The Asia-Pacific Journal* 15, no. 9 (April 2017).

United States Senate Committee on Foreign Relations. *Ballistic Missile Defense: Information on Theater High Altitude Area Defense (THAAD) System*. United States. General Accounting Office. NSIAD. Washington, DC, 1994.

Watts, Robert C. IV. "'Rockets' Red Glare'—Why Does China Oppose THAAD in South Korea, and What Does It Mean for US Policy?" *Naval War College Review* 71, no. 2 (2018).

Unmanned Aerial Vehicle (UAV)

Also variously known as remotely piloted aircraft (RPA) or as "drones," the vehicle portion of an overall unmanned aerial system (UAS), where the UAS consists of the UAV platform itself, as well as the operator and the communication link between the two. Whereas the term "drone" is commonly distained by UAV personnel, RPA is institutionally preferred by the Air Force, the term overtly connects UAV operators with the service's institutional culture celebrating pilots, and the emphasis of human involvement reinforces the point of the "man-in-the-loop" that distinguishes many types of vehicles from autonomous systems.

Since 2001, the UAV has come into prominence for its intelligence, surveillance, and reconnaissance (ISR) mission that leaders have come to rely on for the prosecution of conflicts like Operation Enduring Freedom (OEF) and Operation Iraqi Freedom (OIF). It has also acquired notoriety during the Global War on Terror (GWOT), particularly in regard to controversial uses in carrying out targeted strikes on alleged terrorists. The rise in artificial intelligence is expected to further elevate the prominence and importance of UAVs, while also catalyzing further ethical debates as UAVs become more autonomous, capable of using artificial intelligence to make decisions. Other trends in UAV research and development focus on developing swarms of microdrones, jet-powered versions, and UAVs with even more endurance.

Interest in the concept dates to the 1890s, and the U.S. Navy contracted for the first UAV as a gunnery training aid in 1917. The Army followed suit a year later with what it dubbed the "Bug," although these predecessors of modern cruise missiles came too late for combat use during World War I. Funding challenges shut down further research until the end of the 1930s. By 1942, the Navy forged ahead again with its pioneering efforts to place weapons on a UAV. The U.S. Army Air Forces (and Navy) also tried to use UAVs in combat again, but its Operation Aphrodite program struggled to hit targets. Combat-worn bombers were meant to be flown manually toward targets before crews would escape and leave the terminal guidance to radio-control systems. The program is perhaps most famous for the mishap that lead to the death of Navy Lieutenant Joseph P. Kennedy Jr. in 1944.

Following World War II, the U.S. military made significant technological improvements that enabled the UAV's primary purpose to shift away from assisting with target practice for ships and airplanes to reconnaissance. These types of reconnaissance UAVs first saw combat in the Vietnam War, flying thousands of missions. The increase in the number of dangerous surface-to-air missiles (SAMs) in North Vietnam targeting airplanes encouraged many pilots to welcome UAVs. Concurrently, the Air Force began researching how UAVs could perform high-altitude, long-endurance missions. The Army, by contrast, more naturally focused on enabling UAVs to serve the tactical level of war at lower echelons.

Concurrently, Israel also began making major strides in UAV development, particularly in pursuing a trend toward miniaturization that would bear greater fruit in the 21st century. Israeli employment of UAVs included use as decoys in 1982 during Operation Peace for Galilee so that friendly aircraft could identify and destroy Syrian SAM batteries when the Syrians turned on their antiaircraft radar to track and target the UAVs. Operation Desert Storm in 1991 saw U.S. UAVs providing near-real-time intelligence to ground forces. A Pioneer UAV discovered Iraqi Army forces moving to launch a surprise attack during the Battle of Khafji, and this intelligence enabled the United States to respond with decisive air strikes that brought the attack to a halt. Some Iraqi soldiers even sought to surrender to a drone during the war.

The Air Force's Predators, or what might be considered the first modern UAVs, were test flown in 1994 and deployed to Bosnia the following year. Representing a significant financial savings, a first-generation Predator cost less than one-fifth the price of the F-16. Predator UAVs provided immensely valuable reconnaissance information, and debate about whether to integrate a strike capability into the systems eventually resulted by 2001 in the weaponizing of Predators by mounting pylons for a pair of Hellfire air-to-ground missiles. Predators were also redesignated, from RQ-1 to MQ-1, reflecting the shift from the reconnaissance role to their employment for multiple mission types. These had synthetic aperture radar (SAR), which could fly above clouds and still "see" targets on the ground. An increasingly capable satellite link also allowed real-time video to be transmitted to various users on the ground. Although this ability greatly has transformed battlefield awareness, it has important drawbacks to note. Many compare the narrow perspective provided by UAVs as similar to that of a soda straw, although that view has expanded and improved over time. The availability of video feed has caused other problems, with some arguing that commanders have become too reliant on it, so much so that some describe it as "Predator crack." These unnecessary demands have costs in terms of the required bandwidth as well as what some operators contend is meddling and excessive interference at the tactical level of war.

In many ways, the UAV became the symbol of the Obama administration's approach to the use of military force. Within two days of taking office, President Barack Obama not only continued President George W. Bush's use of UAVs but dramatically increased them, ordering his first strike that reportedly killed one terrorist as well as ten civilians, including up to five children. Undeterred, the administration became more reliant on UAV strikes. Although Bush only authorized about 50 UAV strikes, Obama ordered more than 500, which some estimate killed more than 3,000 terrorists and almost 400 civilians. With the transition from the Obama to the Trump administration, U.S. "drone strikes" have increased in some nations, particularly Yemen and Somalia. It is estimated that strikes in Somalia in 2017 grew from 15 to 30, in Yemen from 44 to 120, and in Pakistan from 3 to 8. The United States has also built new bases for UAVs, reportedly including in Nigeria.

These strikes have been the subject of much debate, with some analysts questioning their strategic effectiveness in the Global War on Terror (GWOT). Some argue that, despite the great tactical and operational success that the United States has had in identifying and killing terrorists, these actions have only spawned additional terrorists due to unintended killings of civilians as well as the violation of sovereign territory. Others despairingly describe this type of decapitation strategy as akin to playing "whack-a-mole."

These debates have not affected the U.S. military's acquisition of UAVs. The U.S. Air Force has recently phased out its MQ-1 Predators in favor of the MQ-9 Reaper and the RQ-4 Global Hawk. The operational range and armament capacity of the MQ-9 significantly exceeds that of its predecessor, while the RQ-4 Global Hawk occupies a similar reconnaissance role to the Air Force's high-altitude U-2 manned aircraft. Army and the Marine Corps UAVs provide ISR at the tactical level and are smaller vehicles. The U.S. Army is pursuing the acquisition of smaller UAVs for individual soldiers through the easily carried Soldier Borne Sensor. Other battlefield missions are being developed for UAVs, to include the logistical task of resupplying troops in the field.

Future military trends are expected to include further miniaturization, the employment of UAVs through swarming tactics, as well as tactics integrating UAVs with manned assets. Some even suggest that the F-35 may prove to be the last manned fighter the United States will build, based on anticipated developments in UAV technology as air-to-air combat platforms. The U.S. Navy currently is developing the MQ-25 Stingray (previously the X-47B, or the Unmanned Carrier-Launched Airborne Surveillance and Strike (UCLASS)) to provide aerial refueling. Its stealth capabilities would increase its chances of surviving in highly contested areas where it is challenging to establish air superiority. Although some might consider stealth to be a requisite for expensive UAVs, it is of less concern for smaller systems, which will be more affordable and will be equipped with swarming technology. In 2017, for example, F/A-18 jets successfully released hundreds of microdrones to participate in a synchronized swarm. This trend coincides with how engineers are taking inspiration from nature for the design and tactics of these UAVs. Scientists, for example, recently experimented with using living dragonflies as maneuverable UAVs. Other developments push the concept of a loitering munition, or a "kamikaze drone." Combining aspects of cruise missiles and UAVs, these expendable UAVs are designed to remain over a battlefield before identifying and destroying their targets in a way that makes them the descendent of the Army's "Bug." Other capabilities being developed reach across the domains of air, land, and space. For example, a submarine might launch a UAV into the air domain.

There is also a continuing push to develop faster types as well as those that can stay in the air much longer. The United States is pursuing designs that fly at Mach 0.8, and it hopes to field hypersonic UAVs by 2040. Endurance also continues to increase, with a diesel-powered UAV staying aloft for a record-breaking 121 hours, which approximates something akin to satellites with the added benefit of being able to avoid orbit in order to remain largely stationary. Researchers are hoping to keep UAVs aloft for weeks at a time, and some are pursuing solar energy to achieve this goal.

The cost and potential cyber vulnerabilities of UAV sensors and data links raise questions about the perceived benefits of relying on UAVs. The video feeds from UAVs undergoing reconnaissance missions in Iraq were reportedly sometimes hacked by insurgents. The Pentagon's upcoming release of its twenty-five-year roadmap for UAVs may help illuminate some answers to the future direction of UAVs. Still, the Department of Defense significantly has increased its funding requests by 26 percent in fiscal year 2019. Much of that budget is paying for the MQ-9 as well as the development of the MQ-25. The Pentagon is also requesting about 3,070 smaller UAVs, making it the largest purchase request it has made in years.

The proliferation of UAVs and their continuingly increasing capabilities and applications poses threats as terrorists and other violent nonstate actors join the circle of UAV operators. In January 2018, for example, unidentified actors coordinated an attack of more than a dozen UAVs against two Russian bases in Syria. Traditional capabilities like radar are designed for much larger aircraft and thus are helpless against small UAVs. Furthermore, there are so many filling the skies of Syria that the United States cannot control airspace lower than 3,500 feet. Many of these are being produced by one of the leading commercial producers of UAVs: China's Da-Jiang Innovations company, which has responded by updating its software to prevent its products being used militarily in Syria.

Since for every weapon there is an effective countermeasure, major strides have been made in the last few years to defend against these developments. Currently, more than 155 companies in 33 nations have begun producing various counter-drone products. New technology relies on detection, interdiction, or a combination of the two to detect and/or destroy UAS. These systems can either be located on the ground, on a UAV, or handheld, with most currently being ground-based. This technology is in a very nascent stage of development given its recent emergence on the market.

Although improvements in sensors and other capabilities, such as the speed of UAVs, have not advanced dramatically, the autonomous capabilities of these vehicles are on the verge of probably resulting in revolutionary change. Moreover, the race toward artificial intelligence could advance the trend toward autonomous UAVs and raise new questions about what kind of decisions these machines will make if they lose communication with humans, particularly in terms of releasing weapons.

Heather Pace Venable

See also: Counter Unmanned Aerial Systems (C-UAS); IAI Heron; MQ-1 Predator; MQ-9 Reaper; Precision-Guided Munitions (PGMs); Quadcopter; RQ-4 Global Hawk; RQ-170 Sentinel; Swarm Robotics; Tactical Air Delivery Drone (TACAD); Unmanned Ground Vehicle (UGV)

Further Reading

Cortright, David, Rachel Fairhurst, and Kristen Wall. *Drones and the Future of Armed Conflict: Ethical, Legal, and Strategic Implications*. Chicago: University of Chicago Press, 2015.

Friedman, Norman. *Unmanned Combat Air Systems: A New Kind of Carrier Aviation*. Annapolis, MD: Naval Institute Press, 2010.

Kreps, Sarah. *Drones: What Everyone Needs to Know*. Oxford: Oxford University Press, 2016.

Peters, John E., et al. *Unmanned Aircraft Systems for Logistics Applications*. Arroyo, CA: RAND, 2012.

Walsh, James Igoe, and Marcus Schulzke. *The Ethics of Drone Strikes: Does Reducing the Cost of Conflict Encourage War?* Carlisle, PA: Strategic Studies Institute, 2016.

Williams, Brian Glyn. *Predator: The CIA's Drone War on Al Qaeda*. Lincoln: University of Nebraska Press, 2013.

Wills, Colin. *Unmanned Combat Air Systems in Future Warfare: Gaining Control of the Air*. New York: Palgrave McMillan, 2015.

Woods, Chris. *Sudden Justice: America's Secret Drone Wars*. Oxford: Oxford University Press, 2015.

Unmanned Ground Vehicle (UGV)

A vehicle operating on the ground without a human being on board. UGVs can be directed remotely by human operators, can work autonomously, or can be designed with a combination of both modes so that a human is "on the loop." UGVs operate in several contexts; in civilian settings its operations tend to be where work is highly repetitive or dangerous, with jobs in the latter category including the Opportunity and Spirit rovers operating on the surface of Mars. Deployment of military PackBot robots for work mitigating the 2011 Fukushima nuclear disaster in Japan reflects the crossover applications for UGV technologies between civilian and military realms.

Nicola Tesla's experimentations with radio control began in the early 1900s, working with small boats and hypothesizing that radio control could direct an explosives-laden vehicle toward a targeted ship. Isolated examples of UVG experiments during World War I yielded desultory results, although a French type reportedly saw combat use in mid-1916. Further investigation of UGVs during the interwar period helped set the stage for the examples that served during World War II. Soviet radio-controlled Teletanks could be controlled to a distance of 1,500 yards. They were equipped with weapons whose deployment would have been dangerous or harmful to the user, such as flamethrowers, explosive charges, and chemical weapons, or alternatively with machine guns. Although chemical weapons Teletanks were not used in combat, the Soviet Union did deploy the other variants in combat against Finland during the Winter War of 1939–1940.

A more famous and numerous early UGV was Nazi Germany's Goliath tracked mine. Reputedly vulnerable to small arms fire, these could be deployed in wooded or urban areas, crawling toward targets such as enemy tanks and pillboxes and then detonating themselves and the target. In all, more than 12,000 units were produced during the war, making the Goliath far more numerous than preceding or contemporary UGV types. Its first variant was powered by a pair of small electric motors and carried a 130-pound warhead, while the successor type was driven by a motorcycle engine and a 220-pound munition, and different units were directed by trailed wire or by radio signal. Germany's Borgward IV was much heavier, carried nearly 1,000 pounds of explosives, and was powered by a larger engine. Borgwards were combat tested in the mine clearance role, but with disappointing results. However, fewer than 1,200 Borgwards were ever constructed, making it less than a tenth as numerous as the Goliath.

Remote guidance has remained important in UGV development since World War II. During the 21st century, technological leaders have since preferred aerial approaches for precision-guided munitions (PGMs), while less technologically rich adversaries have recently favored emplaced improvised explosive devices (IEDs) and vehicle-borne improvised explosive devices (VBIEDs) directed by fanatical or coerced drivers accompanying the explosives. As a result, few direct successors to the Goliath exist. However, the mine clearance role has seen significant exploration exceeding the roller-absorber approach attempted by the Borgward in the 1940s. U.S. forces in Iraq faced continuing challenges from insurgent-emplaced IEDs, and a technological race to devise and deploy tools allowing the neutralization and the employment of IEDs ensued. Although protective technologies for humans included new vehicles like the mine-resistant ambush-protected (MRAP), large numbers of explosive ordnance disposal (EOD) UGVs appeared in theater as well. U.S. deployment of such vehicles expanded thirtyfold between 2004 and 2005 as insurgent activities increased, and as of 2013 the U.S. Army alone had acquired 7,000 UGVs and had lost more than 10 percent to combat.

A notable U.S. UGV technology is the Talon, first used in 2000. Battery operated and directed via radio signal or fiber optics, it was designed to function for over 8 hours. Its sophisticated track system allows it to move as quickly as a running human, and it has reportedly demonstrated the ability to climb stairs. First deployed in an OED role during Balkan peacekeeping missions, it was also successfully used for rescue and recovery at Ground Zero in New York City in the wake of the September 11, 2001, terrorist attacks.

The rugged durability and effectiveness of the Talon enabled the development of the Special Weapons Observation Reconnaissance Detection System (SWORDS). This was a Talon chassis with a weapons mount for using a squad automatic weapon, a large-caliber antimaterial rifle, or a multiple barrel grenade launcher. Directed by a human user working remotely from a console, SWORDS could drive and shoot according to the commands received from the user at the console. SWORDS's developer is said to be designing a more ergonomic controller system incorporating virtual reality goggles. Three machine-gun-mounted SWORDS units were sent to Iraq in 2007, but although this was the first deployment of a remote-controlled gun-mounted system in a combat zone since the Teletank, SWORDS were not used in combat.

Although systems such as SWORDS are human-controlled, the development of armed UGVs has triggered debate in parallel with the invention and use of armed unmanned aerial vehicles (UAVs) and their role in aerial attacks starting in 2001. Sharp controversy surrounds the notion of autonomous and armed systems. Systems like the ship-borne Phalanx air defense system, an automated fire-control system using a 20-millimeter Gatling cannon to destroy incoming anti-ship missiles, was prototyped in the 1970s and has been deployed since 1980. Thus autonomous lethal systems were pioneered nearly half a century ago, because addressing the requirement for close-in defense against fast projectiles precluded meaningful human interaction with the decision loop.

However, perhaps due to the incorporation or mobility as well as lethality, armed and autonomous UGVs (or UAVs) have inspired concern, both in terms of ethical issues and practicality. Conceivably, machine learning could help autonomous systems function more effectively over time. In keeping with debate in other high-tech areas, technology advocates sometimes suggest either that U.S. leadership in the field would provide an opportunity to set precedent and restraint or that U.S. neglect in the field would allow for hostile nations to develop an unstoppable weapon. Although the United States is a notable leader in UGV technology, several other countries are engaged in the field, some as developers and even more as users or potential purchasers of UGV technologies.

Developers of armed UGVs have sought to address this widespread discomfort with the notion of armed robots, not only by considering ways that humans can remain "in the loop" or "on the loop" before weapons are fired, but also by exploring reduced lethality weapons. The Modular Advanced Armed Robotic System (MAARS), made by the same Foster-Miller company that created Talon and SWORDS, released to the U.S. Marine Corps for testing in 2008. MAARS was designed so that it could be equipped with a range of weapons including nonlethal and less-lethal devices ranging from a loudspeaker to a laser dazzler, although MAARS can alternatively be equipped with a manipulator arm for EOD activities or with a lethal weapon such as a machine gun. Although reportedly lacking some mobility attributes desired by its Marine testers, MAARS suggests some of the directions of UGV technology.

The Defense Advanced Research Projects Agency (DARPA) has fostered robotics development since the height of the Cold War, and Boston Dynamics is among the companies that has advanced UGV technology through DARPA support. Boston Dynamics technologies include the quadruped Cheetah robot, which can move at speeds exceeding 25 miles per hour, the quadruped experimental logistics BigDog meant to carry supplies for deployed soldiers traversing rugged terrain, and the Handle research robot equipped with wheels and legs to enable versatility in its movements.

Nicholas W. Sambaluk

See also: Boston Dynamics BigDog; Boston Dynamics Handle Robot; Defense Advanced Research Projects Agency (DARPA); Improvised Explosive Device (IED); Vehicle-Borne Improvised Explosive Device (VBIED)

Further Reading

Everett, H. R. *Unmanned Systems of World Wars I and II.* Cambridge, MA: The MIT Press, 2015.

Hambling, David. *We: Robot: The Robots That Already Rule Our World.* London: Aurum, 2018.

Jasthi, Sainadh, et al. "Unmanned Ground Vehicle for Military Purpose." *International Journal of Pure and Applied Mathematics* 119, no. 12 (2018), 13189–13193.

Lin, Patrick, et al. *Autonomous Military Robotics: Risk, Ethics, and Design.* Arlington, VA: Office of Naval Research, 2008.

Nardi, Gregory J. *Autonomy, Unmanned Ground Vehicles, and the US Army: Preparing for the Future by Examining the Past.* Ft. Leavenworth, KS: School of Advanced Military Studies, 2009.

Ray, Jonathan, et al. *China's Industrial and Military Robotics Development.* Vienna, VA: Center for Intelligence Research and Analysis, 2016.

Scharre, Paul. *Army of None: Autonomous Weapons and the Future of War.* New York: Norton, 2018.

Springer, Paul J. *Military Robots and Drones.* Santa Barbara, CA: ABC-CLIO, 2013.

V-22 Osprey

Tilt-Rotor transport aircraft developed jointly by Bell Helicopters and Boeing Helicopters and operated by the U.S. Marine Corps, Navy, and Air Force. Because of its design, the V-22 Osprey is capable of both vertical and forward flight. The Marine Corps is the primary operator of the Osprey. The MV-22 offers greater carrying capacity and speed than the CH-46 Sea Knight or other conventional helicopters. Proponents have argued that the Osprey is vital to the Marines' expeditionary maneuver warfare concept, which emphasizes speed and rapid concentration of forces. The Marine Corps expects 240 Ospreys to be operational in 2020. Although the Osprey's speed and carrying capacity exceeds that of conventional helicopters, the program's cost and safety issues made it a target for elimination multiple times over the course of its development.

The origins of the V-22 Osprey date to the late 1970s. In 1977, Bell Helicopter developed a tilt-rotor prototype known as the XV-15 that provided the basis for the eventual V-22. The V-22 program began in 1981. The program was created in response to the failure of American helicopters in 1980's Desert One mission in Iran and the Marine Corps' need to replace the Vietnam-era CH-46 Sea Knight helicopter. Although initially an Army program, the V-22 program came under Navy and Marine Corps leadership in 1982. The Osprey's first flight was in 1989.

Cost concerns led the George H. W. Bush administration to seek cancellation of the program by removing funding from the federal budget. Secretary of Defense Richard Cheney argued the V-22 program was too costly and that conventional helicopters could perform the same role. Support in Congress and from the Marine Corps, however, ensured its continuation. A later estimate placed the program's lifetime cost, adjusted for inflation, at $54 billion or $13 billion more than projected in 1982.

Safety concerns also affected the V-22's development, and four prototypes crashed during the V-22's development. Three of these crashes resulted in fatalities. The first fatal crash, attributed to hydraulic failures and engine design issues, occurred in July 1992 in Quantico, Virginia, and killed seven Marines. Two other crashes occurred in 2000, when a steep descent during a training mission killed nineteen during a Marine training mission in Arizona, and in December when hydraulic and software failures killed four Marines including the program's most experienced pilot. These incidents, and a scandal regarding falsification of maintenance records, preempted the Osprey's intended entry into service in 2001.

The Osprey entered service in 2007. Three variants of the V-22 are planned or in service. The principal variant is the Marine Corps operated MV-22. A CV-22 variant is operated by the U.S. Special Operations Command (USSOCOM) through a joint-funding agreement with the Air Force. The Navy plans to operate a separate variant of the V-22 for search and rescue missions. Although similar in design, there are differences in capabilities between the variants. The MV-22 is capable of transporting twenty-four Marines at a cruising speed of around 250 knots. Due to concerns about the Osprey's relatively small firepower, the MV-22 can be fitted with a 7.62-mm belly-mounted machine gun. The CV-22 has a range of up to 500 miles with the use of auxiliary fuel tanks, carries up to eighteen personnel, and will serve as a platform for special forces.

Its first combat action occurred in September 2007 when the Marine Corps deployed the MV-22 variant of the Osprey to Iraq's Al Anbar province. In November 2009, the Marine Corps began operating the MV-22 in Afghanistan. CV-22s were deployed to Mali in December 2009. Commanders deemed the deployments successful. In total, 458 Ospreys are projected to enter service. The Marine Corps plans to operate 360. Fifty Ospreys are expected to be operated by the Air Force and forty-eight operated by the Navy.

Andrew Harrison Baker

See also: F-22 Raptor; F-35 Lightning II

Further Reading

U.S. Library of Congress. Congressional Research Service. *V-22 Osprey Tilt-Rotor Aircraft,* by Christopher Bolkcom. RL31384, 2007.

U.S. Library of Congress. Congressional Research Service. *V-22 Osprey Tilt Rotor Aircraft: Background and Issues for Congress,* by Jeremiah Gertler. RL31384, 2011.

Weinstein, Cliff J. *Sink or Swim: The Marine Corps Capacity to Conduct a Marine Expeditionary Brigade Amphibious Assault Using Expeditionary Maneuver Warfare.* Ft. Leavenworth, KS: United States Army Command and General Staff College, 2010.

Whittle, Richard. *The Dream Machine: The Untold Story of the Notorious V-22 Osprey.* New York: Simon & Schuster, 2011.

Vehicle-Borne Improvised Explosive Device (VBIED)

Homemade bombs deployed via a transportation system. These weapons have gained notoriety in modern warfare for their use in Iraq against American and coalition forces and installations, especially in automobiles and trucks. VBIEDs offer the ability to employ large amounts of ordnance against the designated target and increase damage and fatalities, while blending in with other traffic prior to an attack. VBIEDs are often constructed with some form of initiation device attached to a detonator and explosives. VBIEDs may be prepositioned in a chosen area or driven by humans into the objective, thereby ensuring accuracy of the delivery system.

The first realistic use of VBIEDs dates to 1585, when Frederigo Giambelli used ships as VBIEDs to aid the Senate of Antwerp against their Spanish occupiers. Giambelli's ships targeted Spanish fortifications and ships on the Scheldt Estuary that threatened to close off the vital port. Detonated by a slow match fuse, the exploding vessels caused nearly 1,000 casualties and demonstrated the psychological value of VBIEDs as tools of terror due to their seemingly innocuous nature. In 1920, anarchists deployed a horse and cart bomb, with 100 pounds of explosives and 500 pounds of iron slugs, detonating on Wall Street in New York City in 1920, killing more than thirty people and injuring more than one hundred others. In perhaps one of the most famous VBIED attacks in history, on October 23, 1983, the terrorist group Hezbollah detonated two truck bombs against the barracks of U.S. Marines and French paratroopers in Beirut, killing nearly 300 French and U.S. servicemen.

Using a transportation form common to the area helps make the delivery method inauspicious as well as maneuverable. Packing large amounts of ordnance in a confined space magnifies an explosion, maximizing casualties and damage in an attack. VBIED attacks often occur in two forms: prepositioned and driven. Prepositioned attacks rely on a vehicle packed with explosives to be triggered by a device sending a signal to the detonator similar to other improvised explosive device attacks. These attacks are either initiated by an individual triggering the device or by a timer set to initiate the device at a predetermined time, often during periods of high volumes of traffic. The other form of delivery involves a VBIED being driven into its target. Although often suicidal, this method increases the likelihood of the bomb causing the most amount of damage and achieving its desired result. Difficult to detect once operational, VBIEDs are more likely to be defeated prior to creation by targeting the individuals and cells building them; once built, the best case scenario is detection and neutralization prior to being employed.

Christian Garner

See also: Improvised Explosive Device (IED); Vehicle-Mounted Active Denial System (VMADS)

Further Reading

North Atlantic Treaty Organization Glossary of Terms and Definitions (AAP-06). Brussels, Belgium: NATO Standardization Office, 2018.

Revill, James. *Improvised Explosive Devices: The Paradigmatic Weapon of New Wars.* Cham, Switzerland: Palgrave Macmillan, 2016.

Smith, Andrew. *Improvised Explosive Devices in Iraq, 2003–09: A Case of Operational Surprise and Institutional Response.* Carlisle, PA: Strategic Studies Institute, 2011.

Vehicle-Mounted Active Denial System (VMADS)

A nonlethal/less-than-lethal weapon that directs radio frequency (RF) millimeter wave energy at targeted personnel, causing pain and discomfort through microwave heating. VMADS is primarily an active defense system and may be used for area denial, as a warning or deterrence to approaching adversaries, to disperse hostile crowds/mobs, or in combination with escalating nonlethal or lethal defensive measures. The system consists of an onboard high-power electrical generation capability that powers a millimeter wave-producing RF system; the resultant wave is fed into a dish antenna atop the vehicle, which focuses directed energy on designated targets.

The use of directed RF radiation and its effects on humans was studied by the U.S. Air Force in the late 1990s, along with other options for active denial technology (ADT) such as acoustic, electrical, and nonlethal kinetic weapons. Potential applications of such active denial systems (ADS) include by police and naval units for defense and deterrence.

VMADS operates on a higher frequency than conventional "cooking" microwaves, reducing energy penetration and lethality. The directed energy penetrates only layers of skin, but the effects are painful, and although they are nonlethal, directed microwave energy weapons may still produce significant long-term negative effects or injury such as permanent eye damage and potentially skin burns. The chance of significant or lasting injury increases the closer the targeted personnel are to the VMADS, as they receive a correspondingly higher energy dose of focused radiation. Development and testing of ADS raises the question of ethical testing on animals and humans, and the legality of such weapon systems in armed conflict.

Daniel J. Schempp

See also: Counter Unmanned Aerial Systems (C-UAS); Directed-Energy Weapon

Further Reading

The Active Denial System: Obstacles and Promise. Williamsburg, VA: The Project for International Peace and Security, 2013.

LeVine, Susan. *The Active Denial System: A Revolutionary, Non-Lethal Weapon for Today's Battlefield.* Washington D.C.: National Defense University Center for Technology and National Security Policy, 2009.

Lewer, Nick. "Non-Lethal Weapons: Operational and Policy Developments." *The Lancet* 362 (2003), S20–S21.

X-37B

The first iteration of the U.S. Air Force's (USAF) orbital test vehicle (OTV), greatly resembling the earlier National Aeronautics and Space Administration's (NASA) piloted shuttles (1981–2011) in appearance. Produced by Boeing, the X-37B is one-quarter the size of the shuttle, standing about 9.5 feet tall and about 29 feet long with an almost-15-foot wingspan. A key difference from the earlier manned shuttle is that it is an unmanned and thus entirely robotic vehicle that the Air Force claims is used primarily to test developing technologies.

Contemporary to the NASA shuttle had been an autonomous test of a similar craft (the Buran space shuttle) by the then–Soviet Union in 1988. However, like its American competitor, the Space Shuttle, the Soviet craft did not have the capability to stay in space for more than a few weeks. By contrast, the X-37B's solar array allows extended stays in space. As of April 2018, the USAF is in the midst of its fifth OTV testing mission. Mission durations have steadily increased, standing at a record of 717 days in space as of April 2018.

The Air Force has responsibility for the United States' two X-37Bs. Air Force interest in hypersonic flight dates to the 1950s and its Dyna-Soar program (also known as the X-20). The service explored the concept of extended missions through its Manned Orbital Laboratory program. Pressure from the Department of Defense led to the cancellation of both Dyna-Soar and the Manned Orbital Laboratory in the 1960s.

The X-37B is a development from the X-37A, which NASA began developing in 1999 in pursuit of low-cost access to low-earth orbit (LEO) and improvements in "thermal protection systems and re-entry systems" (Clark). The Defense Advanced Research Projects Agency (DARPA) took over the X-37A program in 2004, and its successor came under the Air Force's purview in 2006 following further budget and program challenges. A drop test was conducted in 2006, and its first space launch occurred in 2010.

The Air Force has released limited information about the purpose of the X-37B, fueling speculation about its potential offensive capabilities. For example, some claimed in 2012 that the X-37B was spying on China's Tiangong-1 experimental space station. Others have viewed the program within the larger context of the United States seeking to turn space into a warfighting domain. Given the USAF's interest in global projection in particular, some suggest that the X-37B might be used as a space "fighter" or "bomber" aircraft, perhaps even targeting Earth from space. These critics see the program as highly destabilizing, particularly because of the extent to which the USAF has cloaked the program in secrecy.

The Air Force, however, has explained that it provides the ability to return payloads from space while running experiments. The USAF has been far more reticent about the purpose of its rocket engine, although some have speculated this might be useful for "significant orbit-changing and de-orbit maneuvers" (Clark). Its relatively large size, however, allows for easy tracking, which makes it unlikely that it is designed for offensive activities to interfere with an adversary's satellite. Better alternatives for this purpose, some suggest, might include the ZSS-11 and MiTEx.

Those who argue for the X-37 B's limited offensive capability point to its very small payload bay (about the size of a truck bed) as well as range limitations. Indeed, the X-37B has been tested primarily in LEO, although it was launched at a higher inclination in its last launch in order to continue to push past its perceived orbital limitations.

The successor step, called X-37C, has been under way since 2011. The vehicle is expected to be about 175 percent the size of the X-37B. This spacecraft potentially would provide the United States with the capability of launching its own astronauts into space again—up to six astronauts as well as gear that could be transferred to and from the International Space Station.

Simultaneously, Boeing is also working on the Experimental Spaceplane program (XS-1), or the so-called Phantom Express for DARPA, which would be a hypersonic spacecraft also designed to launch satellites quickly, affordably, and with limited risk. Unlike the X-37B, it would launch vertically without the assistance of any additional rockets and, ideally, would be able to relaunch in just a few hours. The need for two such programs or the differences between the Phantom Express and the X-37C are a matter for more speculation.

Heather Pace Venable

See also: Unmanned Aerial Vehicle (UAV); X-51 Waverider

Further Reading

Clark, Stephen. "Air Force Spaceplane Is an Odd Bird with a Twisted Past." *Spaceflight Now.* https://spaceflightnow.com/atlas/av012/100402x37update.

Ghoshroy, Subrata. "The X-37B: Backdoor Weaponization of Space?" *Bulletin of the Atomic Scientists* 71, no. 3 (2015), 19–29.

Grantz, Arthur C. "X-37B Orbital Test Vehicle and Derivatives." *AIAA SPACE 2011 Conference & Exposition.* Long Beach, CA: AIAA, 2011.

Lambeth, Benjamin. *Mastering the Ultimate High Ground: Next Steps in the Military Uses of Space.* Santa Monica, CA: RAND, 2003.

X-51 Waverider

An experimental (X designator) U.S. hypersonic cruise missile (HCM) development effort to create a maneuverable air-launched cruise missile capable of sustained, guided flight at speeds exceeding Mach 5 within the atmosphere. The X-51 is considered an "air-breathing" hypersonic cruise missile, meaning that it has an air duct or scoop to intake oxygen and an internal scramjet engine that forces high-speed and high-pressure combustion of injected fuel with the oxygen to produce efficient high-power thrust.

A hypersonic cruise missile resembles a traditional cruise missile, is maneuverable, high speed, and may contain a warhead. The X-51 is air-launched and uses a combined series of engines to accelerate to supersonic (>Mach 1 or >761 mph) and then hypersonic (>Mach 5 or >3,806 mph) velocities. This presents a significant technical design challenge, as this single type of engine cannot accelerate the X-51 in both the supersonic and hypersonic realms, and the vehicle must be able to maintain aerodynamic control at both speeds. In the X-51, a rocket engine is used to accelerate the missile to and through supersonic flight, with a scramjet engine activating for hypersonic flight. A scramjet engine uses direct injection of fuel into the supersonic engine air stream, which combusts and sustains high-speed flight.

As an HCM, the X-51 should operate at the highest achievable hypersonic speed at which control is still possible, in order to increase range and flexibility while reducing target response time. The X-51 has achieved speeds greater than Mach 6 in test flights and is envisioned to sustain Mach 7 in operation. Flight at higher hypersonic speeds causes extreme environmental effects and technical difficulties for propulsion, control, guidance, and vehicle structure. The high speed through the atmosphere causes friction resulting in extreme heating, which must be absorbed or mitigated by the vehicle. The X-51 uses hydrocarbon fuels (ethylene and JP-7) for the scramjet, which also serve to cool the engine. The X-51 was successfully flight tested in May 2010 and sustained record-breaking sustained speed just short of Mach 6. After the 2010 test, the X-51 underwent modifications and continues development and flight test.

The X-51 was developed for the United States by Boeing, in cooperation with Pratt & Whitney and Rocketdyne, who developed the scramjet engine system used in the X-51. The X-51 is the culmination of joint U.S. Air Force, National Aeronautics and Space Administration (NASA), Defense Advanced Research Projects Agency (DARPA), and aerospace contractor hypersonic development efforts spanning fifty years (1960s to 2010s).

Daniel J. Schempp

See also: BrahMos-II; Hypersonic Missiles

Further Reading

Lewis, Mark. "X-51 Scrams into the Future." *Aerospace America* 48, no. 9 (2010), 27–31.

Norris, Guy, and Graham Warwick. "Hyper Hopes." *Aviation Week & Space Technology* 172, no. 12 (2010), 58.

Speier, Richard H., George Nacouzi, Carrie Lee, and Richard M. Moore. *Hypersonic Missile Nonproliferation: Hindering the Spread of a New Class of Weapons.* Santa Monica, CA: RAND, 2017.

Vick, Tyler. *Geometry Modeling and Adaptive Control of Air-Breathing Hypersonic Vehicles.* Master's thesis. Cincinnati: University of Cincinnati, 2014.

Technology Bibliography

Ballistic and Cruise Missile Threat. Wright-Patterson AFB, OH: Defense Intelligence Ballistic Missile Analysis Committee, 2017.

Black, Jeremy. *War and Technology*. Bloomington: Indiana University Press, 2013.

Blom, John David. *Unmanned Aerial Systems: A Historical Perspective*. Ft. Leavenworth, KS: Combat Studies Institute, 2010.

Brahmbhatt, Nihar S. *Design and Optimization of an Electromagnetic Railgun*. Thesis. Houghton: Michigan Technological University, 2018.

Choi, Hyeg Joo, et al. *Effects of Body Armor Fit on Marksmanship Performance*. Natick, MA: Army Soldier Research, Development and Engineering Center, 2016.

Chow, Brian G. "Stalkers in Space: Defeating the Threat." *Strategic Studies Quarterly* 11, no. 2 (Summer 2017), 82–116.

Christensen, Thomas J. "Posing Problems without Catching Up: China's Rise and Challenges for US Security Policy." *International Security* 25, no. 4 (Spring 2001), 5–40.

Coyle, David M. *Analysis of Additive Manufacturing for Sustainment of Naval Aviation Systems*. Thesis. Monterey, CA: Naval Postgraduate School, 2017.

Cummings, M. L. *Artificial Intelligence and the Future of Warfare*. London: Chatham House, 2017.

Dillon, Matthew John. *Implications of the Chinese Anti-Satellite Test for the United States Navy Surface Forces*. Monterey, CA: Naval Postgraduate School, 2008.

Edgerton, David. *The Shock of the Old: Technology and Global History since 1900*. Oxford: Oxford University Press, 2007.

Ernhard, Thomas P. *Strategy for the Long Haul: An Air Force Strategy for the Long Haul*. Washington, D.C.: Center for Strategic and Budgetary Assessments, 2009.

Fisher, Richard D. *China's Progress with Directed Energy Weapons*. Washington, D.C.: US-China Economic and Security Review Commission, 2017.

Geis, John P. *Directed Energy Weapons on the Battlefield: A New Vision for 2025*. Maxwell AFB, AL: Air University Press, 2003.

Gertler, Jeremiah. *Air Force F-22 Fighter Program*. Washington, D.C.: Congressional Research Service, 2013.

Gertler, Jeremiah. *F-35 Joint Strike Fighter (JSF) Program*. Washington, D.C.: Congressional Research Service, 2018.

Gillespie, Paul G. *Weapons of Choice: The Development of Precision Guided Munitions.* Tuscaloosa, AL: University of Alabama Press, 2006.

Gipson, Issac G. *The Effectiveness of the US Missile Defense Capabilities as a Deterrent to the North Korean Missile Threat.* Thesis. Monterey, CA: Naval Postgraduate School, 2007.

Good, Joshua. *Blackstarting the North American Power Grid after a Nuclear Electromagnetic Pulse (EMP) Event or Major Solar Storm.* Thesis. Harrisonburg, VA: James Madison University, 2012.

Ilachinski, Andrew. *AI, Robots, and Swarms: Issues, Questions, and Recommended Studies.* Arlington, VA: Center for Naval Analyses, 2017.

Johnson-Freese, Joan. *Space Warfare in the 21st Century.* New York: Routledge, 2016.

Kapur, Vivek. *Stealth Technology and Its Effect on Aerial Warfare.* New Delhi: Institute for Defence Studies and Analysis, 2014.

Knox, MacGregor, and Williamson Murray. *The Dynamics of Military Revolution, 1300–2050.* Cambridge: Cambridge University Press, 2001.

Lewis, Larry. *Insights for the Third Offset: Addressing Challenges of Autonomy and Artificial Intelligence in Military Operations.* Arlington, VA: Center for Naval Analysis, 2017.

Lewis, Larry, and Diane Vavrichek. *Rethinking the Drone War: National Security, Legitimacy, and Civilian Casualties in US Counterterrorism Operations.* Quantico, VA: Marine Corps University Press, 2016.

MacMillan, Ian. "Fighter Jets, Supercars, and Complex Technology." *Strategic Studies Quarterly* 11, no. 4 (Winter 2017), 112–133.

Mettler, S. A., and D. Reiter. "Ballistic Missiles and International Conflict." *Journal of Conflict Resolution* 57, no. 5 (2013), 854–880.

Michel, Arthur Holland. *Counter-Drone Systems.* Annandale-on-Hudson, NY: Center for the Study of the Drone, 2018.

Moltz, James Clay. *Crowded Orbits: Conflict and Cooperation in Space.* New York: Columbia University Press, 2014.

Morgan, Richard A. "Military Use of Commercial Communication Satellites: A New Look at the Outer Space Treaty and Peaceful Purposes." *Journal of Air Law and Commerce* 60, no. 1 (1994), 237–326.

Nielsen, Philip E. *Effects of Directed Energy Weapons.* Washington, D.C.: National Defense University Press, 1994.

O'Rourke, Ronald. *Navy Lasers, Railgun, and Hypervelocity Projectile: Background and Issues for Congress.* Washington, D.C.: Congressional Research Service, 2016.

Parker, Philip M. *The 2019–2024 World Outlook for Directed Energy Weapons.* Las Vegas, NV: Icon Group, 2018.

Pepi, Marc S. *Advances in Additive Manufacturing.* Aberdeen, MD: Army Research Laboratory, 2016.

Pilger, Michael. *China's New YJ-18 Antiship Cruise Missile: Capabilities and Implications for US Forces in the Western Pacific.* Washington, D.C.: US-China Economic and Security Review Commission, 2015.

Pinkston, Daniel A. *The North Korean Ballistic Missile Program.* Carlisle, PA: Strategic Studies Institute, 2008.

Ray, Jonathan, et al. *China's Industrial and Military Robotics Development*. Washington, D.C.: Center for Intelligence Research and Analysis, 2016.

Rowley, Gary D. *Armed Drones and Targeted Killing: Policy Implications for Their Use in Deterring Violent Extremism*. Thesis. Washington, D.C.: National Defense University, 2017.

Russell, Robert W. *Does the MRAP Meet the US Army's Needs as the Primary Method of Protecting Troops from the IED Threat?* Thesis. Ft. Leavenworth, KS: Command and General Staff College, 2009.

Saalman, Lora. *Prompt Global Strike: China and the Spear*. Independent Faculty Research, 2014.

Sambaluk, Nicholas Michael, ed. *Paths of Innovation: From the Twelfth Century to the Present*. Lanham, MD: Lexington, 2018.

Scharre, Paul. *Army of None: Autonomous Weapons and the Future of War*. New York: Norton, 2018.

Shaw, I.G. "Robot Wars: US Empire and Geopolitics in the Robotic Age." *Security Dialogue* 48, no. 5 (August 2017), 451–470.

Singer, P. W. *Wired for War: The Robotics Revolution and Conflict in the 21st Century*. New York: Penguin, 2009.

Speier, Richard H., et al. *Hypersonic Missile Nonproliferation: Hindering the Spread of a New Class of Weapons*. Santa Monica, CA: RAND, 2017.

Springer, Paul J. *Outsourcing War to Machines: The Military Robotics Revolution*. Santa Barbara, CA: Praeger, 2018.

Sunde, Yokohama H., S. T. Ellis-Steinborner, and Z. Ayubi. "Vehicle Borne Improvised Explosive Device (VBIED) Characterisation and Estimation of Its Effects in Terms of Human Injury." *International Journal of Protective Studies* 6, no. 4 (December 2015), 607–627.

Van Atta, Richard H., et al. *Assessment of Accelerated Acquisition of Defense Programs*. Alexandria, VA: Institute for Defense Analysis, 2016.

Veronneau, Simon, et al. *3D Printing: Downstream Productive Transforming the Supply Chain*. Santa Monica, CA: RAND, 2017.

Wallace, Ryan J., and Jon M. Loffi. "Examining Unmanned Aerial System Threats & Defenses: A Conceptual Analysis." *International Journal of Aviation, Aeronautics, and Aerospace* 2, no. 4 (2015).

Watts, Robert C., IV. "'Rockets' Red Glare'—Why Does China Oppose THAAD in South Korea, and What Does It Mean for US Policy?" *Naval War College Review* 71, no. 2 (2018), 79–107.

Whittle, Richard. *Predator: The Secret Origins of the Drone Revolution*. New York: Henry Holt, 2014.

Yang, Andrew S. "Using the Land to Control the Sea?—Chinese Analysts Consider the Antiship Ballistic Missile" *Naval War College Review* 62, no. 4 (2009), 53–86.

Zikidis, Konstantinos, Alexios Skondras, and Charisios Tokas. "Low Observable Principles, Stealth Aircraft and Anti-Stealth Technologies." *Journal of Computations & Modelling* 4, no. 1 (2014), 129–165.

Zimmerman, Brock A., and Allen E. Ellis III. *Analysis of the Potential Impact of Additive Manufacturing on Army Logistics*. Monterey, CA: Naval Postgraduate School, 2013.

Chronology

2000

A disgruntled computer scientist, using a computer and a radio transmitter, launches the world's first proven case of a successful computer attack against supervisory control and data acquisition (SCADA) systems, targeting the computers operating the Maroochy Shire sewage control system in Queensland, Australia, and causing a local public health threat and local environmental damage.

U.S. Congress approves a modernization effort, called GPS III, for its global positioning system. President Bill Clinton ends the policy of "selective availability," which had previously maintained superiority for military access and capability over that available for civilian GPS use.

A U.S. Air Force program office called Big Safari is authorized to examine ways to equip the MQ-1 Predator unmanned aerial vehicle (UAV) with armament.

An experimental aerial surveillance mission is flown over Afghanistan to test an ability to monitor the terrorist financier Osama bin Laden.

The U.S.-guided missile destroyer USS *Cole* is struck by an explosive suicide boat, killing seventeen U.S. sailors. It continues a pattern of isolated bombings by jihadi terrorists in the 1990s, which targeted U.S. civilians in New York, U.S. diplomats in Kenya, and U.S. personnel in Saudi Arabia.

The X-35, a concept demonstrator for the upcoming F-35 Joint Strike Fighter, makes its first flight.

2001

The U.S. Air Force's Big Safari team successfully test fires Hellfire missiles from a laser-designator-equipped MQ-1 Predator. The Predator's eventual replacement, the MQ-9 Reaper, is first flown.

An estimated 80,000 Chinese patriotic hackers launch low-sophistication attacks against U.S. government websites in the wake of a diplomatic flap between the United States and the People's Republic of China after a Chinese pilot dies colliding with a U.S. military aircraft in international airspace.

The first combat-capable F-22 Raptor is flown by the U.S. Air Force.

The Code Red Worm marks the first major wave of malware to spread globally during the 21st century. It is swiftly surpassed in terms of its spread and economic damage wrought by the Nimda worm, which targets a variety of contemporary Microsoft operating systems.

Al Qaeda terrorists launch coordinated attacks against targets in Washington, D.C., and New York City, killing nearly 3,000 people and triggering the start of the Global War on Terror.

The first armed Predator missions are flown over Afghanistan.

Punctuated by U.S. special forces activity and air support, a combination of U.S. and rebel Afghan forces begin a campaign that overthrows Taliban and Al Qaeda control of Afghanistan.

President George W. Bush announces that the United States will withdraw from the 1972 Anti-Ballistic Missile Treaty, citing the need to facilitate development of missile defense systems that can protect against small-scale ballistic missile attacks by nuclear-armed rogue states.

2002

The newly created U.S. Department of Homeland Security is tasked with physical and cyber defense of the nation.

The Onion Router is released as a tool for protecting online browsing privacy through successive encrypted protocols.

An Iraqi MiG-25 illicitly operating in Iraq's no-fly zone shoots down a Predator in the first combat encounter between a conventional warplane and a missile-carrying UAV.

2003

Surveillance and exfiltration, dubbed Titan Rain, of an estimated 20 terabytes of data from U.S. government and corporate systems occurs.

SQL Slammer worm is released, briefly catalyzing the global collapse of Internet traffic for a period of 12 hours.

U.S. and allied forces invade Iraq, tasked with deposing Saddam Hussein and eliminating his suspected weapons of mass destruction program and support of various terrorist activities.

An electrical power fluctuation obscured by a software glitch results in outages for fifty-five million people across the northeastern United States and east-central Canada. At the time it was the second largest electrical outage in history, and the episode implies the damage possible through a deliberate cyberattack targeting critical infrastructure.

Harvard University responds to complaints that a social media website "Facemash," created by then-student Mark Zuckerberg, has abused personal information of students by using pictures without authorization to create hot-or-not voting prompts.

Hackers, believed to be working for the PLA, initiate a three-year cyberespionage campaign against U.S. defense contractors and national laboratories. Upon subsequent discovery by U.S. authorities, the intrusion will become known as Operation Titan Rain.

2004

Mark Zuckerberg, alongside four college friends, establishes Facebook. At first limited to the student population at Harvard, the site grew rapidly and its aperture opened, and by 2006 anyone claiming to be at least thirteen years old could join.

Anonymous, the amorphous association of anti-secrecy agents acting online, is established.

2005

The F-22 achieves initial operational capability, although an expert panel within the Air Force recommends adjustments to address reported oxygen supply challenges.

2006

Operation Shady RAT, so-named for the remote access tool used by hackers in the PRC, begins. Cybersecurity experts at McAfee will announce their discovery of the action in 2011, noting that the hackers' targets included dozens of businesses around the world as well as international entities such as the United Nations and the International Olympic Committee.

USS *Freedom*, the first of the U.S. Navy's monohull Littoral Combat Ships, is launched.

The F-35 Joint Strike Fighter, an A-model for the U.S. Air Force, is first flown. The first Marine's B-model will fly in June 2008, and the first Navy C-model will fly in June 2010.

Tactical Iraqi, a language-acquisition project within the U.S. military's Tactical Language and Culture Training System, is developed for use by U.S. forces preparing for deployment to Iraq.

2007

The People's Republic of China (PRC) demonstrates an antisatellite (ASAT) weapon capability by destroying one of its own satellites, which was no longer functional.

The Idaho National Laboratory confirms the Aurora vulnerability of hardware infrastructure to physical damage via malware. Industry is notified within months of this discovery.

Estonian governmental, media, and business websites are targeted by a wave of cyberattacks ascribed to Russia. The attacks came after an Estonian decision to relocate a grave marker monument dating to Soviet occupation, which proved a flashpoint between Russian government-encouraged hacktivists and Estonia.

The MQ-9 Reaper enters service with the U.S. Air Force. It is soon joined by the RQ-170 Sentinel, a stealthy UAV.

U.S. Defense Secretary Robert Gates declares that acquisition of more Mine-Resistant Ambush-Protected (MRAP) vehicles is his department's highest priority for the remaining months of the fiscal year.

The V-22 Osprey tilt-rotor aircraft enters U.S. service.

Israeli Air Force jets conduct a kinetic strike against Syria's Al Kibar suspected nuclear facilities. The jets successfully infiltrate Syrian airspace through the launch of a cyberattack dubbed Operation Orchard, which uses a "man in the middle" ploy to confuse Syrian air defense radar.

Following a series of reductions in ordered numbers of F-22 in response to rising program costs, the U.S. government rolls back some of its earlier reductions by bringing the F-22 order back to 183 aircraft.

2008

The video sharing social media website YouTube suffers a 2-hour outage. Afterward, the outage is traced to Pakistani efforts to block domestic access to a posted video deemed inflammatory and anti-Islamic.

USS *Independence*, the U.S. Navy's first trimaran Littoral Combat Ship asserted to possess relative fuel efficiency advantages over the competing Freedom class, is launched.

The North Atlantic Treaty Organization establishes a Cyber Defence Centre of Excellence and establishes its headquarters in Tallinn, Estonia, in response to Russian cyberattacks targeting Estonia the previous year.

The Caucasian country of Georgia is targeted in cyberattacks including denial-of-service actions against government websites. The cyberattacks, spanning from July 20 through August 14, precede and overlap with Russian kinetic strikes and invasion of breakaway portions of Georgia.

A breach of U.S. military networks, achieved through malware introduced by USB flash drives, is discovered and a response effort dubbed Operation Buckshot Yankee is initiated.

The Conficker worm spreads rapidly across Microsoft operating devices, although the absence of an apparent dangerous payload reportedly baffles cybersecurity experts trying to discern a motivation for the malware that is believed to have originated in Ukraine.

The Gaza War, also called Operation Cast Lead, begins. Spanning three weeks from December 2008 to mid-January 2009, the combat is paralleled by mutual online perception management campaigns aimed at impacting international opinion.

Production begins of the Terminal High Altitude Area Defense (THAAD) terminal phase of the ballistic missile defense system in the United States.

2009

U.S. Cyber Command (USCYBERCOM) is established, with Admiral Keith Alexander serving in the dual-hat role to head it as well as the National Security Agency, whose responsibilities involve global data monitoring including in the cyber domain.

For the first time, a member of the hacking entity Anonymous pleads guilty to charges of causing "unauthorized impairment of a protected computer."

2010

Google announces that advanced persistent threats connected to the PLA had undertaken extensive surveillance against more than one hundred targets, including defense and aerospace companies, financial institutions, software companies, and the online activities of human rights advocates critical of PRC policies.

Then-Army specialist Bradley Manning exfiltrates data, which Manning subsequently delivers to WikiLeaks. Illicit publishing begins in April 2010, and Manning will spend from July 2010 to January 2017 in federal prisons for the deliberate release of sensitive information.

The U.S. Air Force's X-37, an unmanned space place, is launched into orbit to begin an eight-month mission.

Cybersecurity experts identify the Stuxnet virus. Believed to have been under development since 2005, the malware was specifically designed so that it would cause damage to the kind of gas centrifuges used by Iran to enrich uranium and so that the damage wrought could occur subtly enough to resemble random hardware failures rather than malicious software.

Facebook reaches 500 million users.

2011

PRC's fifth-generation Chengdu J20 makes its first flight.

Tunisia's government falls on January 14 in the face of popular demonstrations, marking the first success of antiauthoritarian activities quickly called the Arab Spring. The Egyptian and Yemeni governments fall weeks later, while civil war erupts in Libya that will result in Muammar Gaddafi's ouster in August and an ongoing civil war in Syria.

Hackers sympathetic to Bashar al-Assad's regime establish the Syrian Electronic Army and begin launching denial-of-service attacks and malware, mostly against organizations deemed supportive of rebels fighting a civil war against Assad's government.

Israel's missile defense system Iron Dome is declared operational, and estimates suggest that it intercepts 90 percent of incoming rockets.

Al Qaeda leader Osama bin Laden is killed in a U.S. special forces raid on his covert home complex in Pakistan.

U.S. forces officially end their mission in Iraq, which had begun with the ousting of Saddam Hussein's regime in 2003 and the search for suspected weapons of mass destruction.

2012

Social media giant Facebook purchases the social media photo sharing company Instagram.

The 195th and final F-22 is delivered by Lockheed Martin to the U.S. Air Force.

Computers belonging to two Middle Eastern energy companies are made inoperable by the Shamoon virus, attributed to Iran and believed to potentially constitute retaliation following the uncovering of the use of Stuxnet to hinder Iran's nuclear enrichment program.

A cyberattack group called the Qassam Cyber Fighters launches the first of three phases of attacks to interrupt online activity of U.S. financial institutions, ostensibly in retaliation for the online presence of the film *The Innocence of Muslims* for its antagonistic tone toward Muslims.

PRC's stealth J31 Gyrfalcon, outwardly resembling the United States' fifth-generation F-35 and believed to be intended as a competitor export for defense sales, makes its first flight.

The *Tallinn Manual,* an unofficial and nonbinding study of international law regarding cyber conflict, is completed after three years of development by a group of scholars led by the chairman of the U.S. Naval War College's international law department.

Production of MRAP vehicles ends, with more than 12,000 having been deployed to Iraq and Afghanistan for use by U.S. forces, and reportedly a total of 20,000 remaining in U.S. inventory, either with nondeployed brigades or with training units or in storage.

After repeated setbacks, North Korea successfully test launches an Unha "Galaxy" rocket and places a satellite into orbit, simultaneously implying a potential intercontinental ballistic missile capability.

2013

The cybersecurity firm Mandiant releases its APT1 report, which outlines extensive cyberespionage activities undertaken by a group in the PRC eventually identified as People's Army Liberation Unit 61398.

Facebook enters the Fortune 500 list, less than a year after its initial public offer the previous summer.

The PRC launches a suborbital rocket that it identifies as scientific but which the U.S. government concludes is part of a ground-based antisatellite system, complementing a separate ASAT technology that PRC demonstrated in 2007.

Edward Snowden, a contractor with access to National Security Agency data, exfiltrates sensitive data to the anti-secrecy website WikiLeaks. U.S. authorities respond by charging Snowden with violating the Espionage Act dating to World War I, although Snowden's initial escape to Hong Kong leads to subsequent arrival in Russia and an ongoing and legally ambiguous existence in that country evading extradition to the United States.

2014

U.S. Air Force spokespeople assert that backup oxygen systems, demanded in response to reported complications with the standard oxygen system, would be completed by spring of 2015.

Boko Haram jihadi fighters raid the Nigerian town of Chibok and kidnap more than 200 schoolgirls. An international campaign to raise awareness and action for their rescue includes use of Twitter through the hashtag #BringBackOurGirls.

The NSA identifies PLA Unit 61486, which actively engages in spear phishing and other attacks connected with military and economic espionage. The hacker group is dubbed "Putter Panda" by experts working for the cybersecurity company CrowdStrike.

The Islamic State of Iraq and Syria (ISIS) surges across northern Iraq and propagandizes its battlefield successes through a Twitter campaign known by its hashtag, #AllEyesOnISIS. Simultaneous with its battlefield advances, ISIS also unveiled an English-language online propaganda magazine called *Dabiq*, which it would release approximately every two months for the following two years.

U.S. military personnel are deployed to Iraq in support roles to assist Iraqi forces attempting to stem ISIS advances across northern Iraq toward Baghdad.

North Korean hackers calling themselves the "Guardians of Peace" leak embarrassing and sensitive data exfiltrated from Sony Pictures. The attack is a North Korean response to the scheduled release of *The Interview*, a comedic movie premised on a U.S.-sponsored assassination of North Korean dictator Kim Jong Un. These events prompt reconsideration about whether such an act constitutes an attack on critical national infrastructure.

The Laser Weapon System (LaWS) is installed on the USS *Ponce* and is successfully tested.

2015

The Internet-hosting service GitHub is targeted in the first attack of the "Great Cannon," a PRC cyberattack tool capable of redirecting global Internet traffic in order to contribute to a distributed denial-of-service attack.

The Joint Land Attack Cruise Missile Defense Elevated Netted Sensor System program is suspended.

Russia conducts its first test of a new direct ascent ASAT missile, called PL-1 Nudol.

The hacktivist entity Anonymous announces the start of operations against ISIS in response to ISIS-affiliated terrorist attacks in Paris, which occurred on November 13. Anonymous's activities include the doxing of suspected ISIS recruiters and the December 11 launch of "ISIS Trolling Day" in order to ridicule and combat the jihadi organization online.

A portion of Ukraine's electrical power grid is taken offline for several hours as a result of a cyberattack experts attribute to Russia. The attack exploited a known vulnerability in Microsoft software identified over a decade earlier. The incident illustrated the challenge of updating and protecting critical infrastructure from cyberattack.

2016

Federal investigators and Apple engage in legal battles when the latter refuses to assist the former in gaining access to a smartphone used by a dead jihadi gunman who had launched a mass shooting attack in San Bernardino in December 2015. In the wake of Apple's repeated refusals to cooperate, the U.S. government gained access through paying a hacker to break into the device.

DDoS attacks across much of the United States and parts of Europe interrupt the domain name provider Dyn in three waves during a single day, and two of the waves each bring outages lasting 2 hours each.

Donald J. Trump is elected 45th president of the United States, and the surprise outcome coincides with varied accusations involving Russian efforts to manipulate the election. Partisan claims allege that Russian activity occurred with the collusion of Trump or his campaign. Meanwhile, forensic evidence mounts in the following months of Russian work to use social media platforms to foment tension in the United States, which analysts note could lead to erosion of U.S. dedication to alliance commitments.

The People's Liberation Army Navy declares the Liaoning aircraft carrier combat ready.

2017

ISIS is reported to experiment with using commercially available quadcopters as bomb-equipped small-scale guided weapons.

The fourth mission of an X-37 series vehicle ends with the landing of the X-37B after a mission just thirteen days short of two years in orbit.

WannaCry ransomware spreads globally. Experts assert that the malware was developed through the use of weaponizable vulnerabilities that had earlier been leaked. By year's end, cybersecurity analysts in the United States and elsewhere traced WannaCry attacks to North Korean hackers.

Japan's acceptance of a U.S.-built V-22 Osprey marks the first export of the type to a foreign military.

The People's Liberation Army Air Force officially accepts the J20 into service. Analysts do not agree about whether the plane is meant primarily as a fifth-generation jet fighter for the air superiority role or as a multirole aircraft.

2018

At least two Su-57 fifth-generation Russian jets are deployed as part of a contingent of more than a dozen aircraft to assist Assad's regime in combating rebels in Syria's civil war.

Russian president Vladimir Putin announces an expansion of his country's nuclear arsenal, arguing that it is a counter to U.S. interest in ballistic missile defense technology.

The MQ-1 Predator is retired from the U.S. inventory, having been replaced by the more capable MQ-9 Reaper.

The heaviest distributed denial-of-service attack to date occurs on March 1, when GitHub is subjected to a peak of 1.35 terabits per second. Four days later, the record is broken by an attack against a customer of the Massachusetts-based Internet service provider Arbor Networks, when a similar attack peaks at 1.7 terabits per second.

The Israeli Air Force conducts the first-ever combat sortie with an F-35.

Buffeted by controversy about its policies having permitted the controversial and illicit use of user data, Facebook on July 26 becomes the first company in history to lose $100 billion in stock value in a single day. The following day, Facebook suspends the account of political pundit Alex Jones, citing his earlier statements as hate speech. In September, the company would announce its willingness to cooperate with organizations associated with the North Atlantic Treaty Organization to prevent manipulation of elections in Brazil.

A Taiwanese semiconductor manufacturer's production is interrupted by a new outbreak of a changed variant of the WannaCry ransomware, which had struck computers worldwide the previous summer.

Observers report a new ASAT being carried by a Russian MiG-31 jet.

A Marine F-35B becomes the first U.S.-operated plane of that type to be used in combat. Operating from the amphibious assault ship USS *Essex*, it strikes a Taliban target in Afghanistan. The first reported F-35 crash occurs the next day, during a mishap of a separate F-35B in South Carolina.

YouTube suffers an hour-long global outage on October 17. No official explanation is forthcoming, although speculation by media and others note apparent similarities with the YouTube outage in February 2008 triggered by censorship by the government of Pakistan.

Schedule for the first launch of GPS III vehicles connected to modernization approved in 2000.

Contributor List

VOLUME EDITOR

Dr. Nicholas Michael Sambaluk
Associate Professor
eSchool of Graduate Professional Military Education
Maxwell Air Force Base, Alabama

CONTRIBUTORS

Dr. Andrew Akin
Assistant Professor
eSchool of Graduate Professional Military Education
Maxwell Air Force Base, Alabama

Andrew Harrison Baker
Doctoral Student
Auburn University

Dr. Amy Baxter
Assistant Professor
eSchool of Graduate Professional Military Education
Maxwell Air Force Base, Alabama

Sean Nicholas Blas
Instructor
Air Command and Staff College
Maxwell Air Force Base, Alabama

Dr. John Blumentritt
Assistant Professor
eSchool of Graduate Professional Military Education
Maxwell Air Force Base, Alabama

Daniel Campbell
Doctoral Candidate
Auburn University

Lisa Jane de Gara
Board of Governors
Grant MacEwan University

Dr. Melvin G. Deaile
Associate Professor
Air Command and Staff College
Maxwell Air Force Base, Alabama

Christian Garner
Instructor
United States Military Academy

Dr. Nicole Marie Hands
Clinical Assistant Professor
Purdue University

Dr. Michael Hankins
Assistant Professor
eSchool of Graduate Professional Military Education
Maxwell Air Force Base, Alabama

Andrew O. Hunstock
United States Air Force
Maxwell Air Force Base, Alabama

Elia G. Lichtenstein
Independent Scholar

Dr. Deonna D. Neal
Department Chair, Leadership
Associate Professor
eSchool of Graduate Professional Military Education
Maxwell Air Force Base, Alabama

Kathy Nguyen
Doctoral Candidate
Texas Woman's University

Dr. Melia T. Pfannenstiel
Assistant Professor
Air Command and Staff College
Maxwell Air Force Base, Alabama

Karine Pontbriand
Doctoral Student
University of New South Wales

Diana J. Ringquist
Independent Scholar

John P. Ringquist
United States Army

Nicholas W. Sambaluk
Independent Scholar

Rose E. Sambaluk
Independent Scholar

Dr. Margaret Sankey
Department Chair, Research
Associate Professor
Air War College
Maxwell Air Force Base, Alabama

Daniel J. Schempp
Deputy Department Chair, Strategy
eSchool of Graduate Professional Military Education
Maxwell Air Force Base, Alabama

Dr. John A. Springer
Associate Professor
Purdue University

Dr. Paul J. Springer
Department Chair, Research
Air Command and Staff College
Maxwell Air Force Base, Alabama

Kate Tietzen
Doctoral Candidate
Kansas State University

James Tindle
Doctoral Candidate
Kansas State University

Luke Wayne Truxal
Doctoral Candidate
Kansas State University

Stephanie Marie Van Sant
Independent Scholar

Dr. Heather Pace Venable
Associate Professor
Air Command and Staff College
Maxwell Air Force Base, Alabama

Dr. Jonathan K. Zartman
Associate Professor
Air Command and Staff College
Maxwell Air Force Base, Alabama

Index